Aeolian Geomorphology:

An Introduction

Addison Wesley Longman Limited
Edinburgh Gate, Harlow
Essex CM20 2JE, England
and Associated Companies throughout the world

First published 1996

British Library Cataloguing in Publication Data
A catalogue entry for this title is available from the British Library.

ISBN 0-582-08704-X

Library of Congress Cataloging-in-Publication data
A catalogue entry for this title is available from the Library of Congress

Set by 22 in 9/11pt Times

Produced by Longman Singapore Publishers (Pte) Ltd.
Printed in Singapore

Erratum

In Fig. 2.1 (p. 9) the expression should read $u. \propto \frac{x}{y}$

Contents

Preface

We have aimed in this book to produce a concise, accessible introduction to aeolian geomorphology. The recent flourishing state of our subdiscipline is manifest in a flurry of papers in scientific journals. The rate of activity is revealed by the huge number of papers published in the last two years that we have referred to. It is true that there have been a number of specialized collections of papers, one or two monographs, and chapters and papers in other books (mainly about desert geomorphology), but most of this work is fragmented, and not highly accessible to the general geomorphologist, much of it being at a very technical level. In fluvial, glacial, periglacial, coastal and slope geomorphology, by contrast, there is a large choice of specific volumes which make the subjects accessible to non-specialists, particularly undergraduates. We hope to redress the balance.

At first sight it may seem strange that a book on the geomorphological work of the wind should be written by two geomorphologists from such a wet place as England, but there are two perfectly good reasons. First, the concentration of geomorphologists on glacial, fluvial and coastal landforms has blinded most people to the quite considerable extent of our coastal sand dunes, our almost as extensive ancient inland dunefields, and the less extensive but widespread deposits of loess, as well as to the huge contribution that the wind has made to soils everywhere in Britain. Our islands were much more wind-blown 20 000 years ago than they are today. Even today we suffer, albeit on a much smaller scale than places like Kazakhstan, from wind erosion on agricultural fields and from outbreaks of desert dust (in our case from the Sahara). In fact, the work of the wind in Britain may be almost as obtrusive in everyday life as that of rivers, and more conspicuous than that of slopes.

Second, we have a flourishing, though small community of aeolian geomorphologists in Britain, largely the intellectual inheritance of Dick Grove, who is still at work in Cambridge. One of us was taught by Dick Grove himself, and the other in turn by one of his students, Andrew Goudie. We owe them a considerable debt for kindling our interest in deserts and aeolian processes.

We have been at pains throughout the book to emphasize that aeolian geomorphology is a rapidly expanding topic, in which there are very many uncertainties. We hope that we have capitalized on these, for they do present advantages. They encourage interesting debates, stimulate further enquiry, and as a teaching medium allow us to emphasize the fragile nature of most scientific study, the need to keep an open and enquiring mind, and the need to develop the appropriate techniques. Aeolian geomorphology is a good medium for thought in three other respects. It requires a great breadth of related disciplines to be focused on its problems. It operates at a range of scales, from the microscopic to the global and from processes that last less than a second to those that continue for millennia. Finally, when it comes to application, successful aeolian geomorphology, like the application of other parts of the discipline, needs attention to social and political analysis. By providing a summary of the state-of-the-art, our intention is to stimulate further interest and enquiry.

Acknowledgements

We are grateful to the following people for permission to reproduce copyright material:
Fig. 1.3 *Philips Modern School Atlas*, Harold Fullard (ed.), Reed Consumer Books Ltd.; Table 2.1 *United States Geological Survey, Professional Paper* **1052**: 137–169 United States Geological Survey (Fryberger, 1979); Fig. 2.5 *Acta Mechanica* **63**: 267–278 Springer-Verlag (Mitha S., Tran, M.Q., Werner, B.T., & Haff, P.K., 1986); Fig. 2.6 *Journal of Geophysical Research* **76**: 2880–2885 American Geophysical Union (Hsu, 1971) also with *Dynamics of wind erosion*: II. Initiation of soil movement *Soil Science* **60**: 397–411 William and Wilkins (Chepil, 1945); Fig. 2.7 and 2.10 Clifford, N. J., French, J. R. and Hardisty, J. (eds.) *Turbulence: perspectives on sediment transport*, reprinted by permission of John Wiley and Sons Ltd. (Butterfield, 1993); Fig. 2.8 Abrahams, A. D. and Parsons, A. J. (eds.) *Geomorphology of desert environments*, Chapman and Hall (Lancaster & Nickling, 1994); Fig. 2.12 *Bulletin of the Geological Society of America* **97**: 1270–1278 Geological Society of America (Anderson, 1986); Fig. 2.13 *United States Geological Survey Professional Paper* **1052**: 137–169 United States Geological Survey (Fryberger, 1979); Fig. 3.4 *Geographical Journal* **126**: 18–31 Royal Geographical Society (Grove, 1960); Fig. 3.6 *Proceedings of the Geologists' Association* **100**: 83–92 Geologists' Association (Goudie, 1989); Fig. 3.7 *Carolina Bays and their origins, Bulletin of the Geological Society of America* **63**: 167–224 Geological Society of America (Prouty, 1952); Fig. 3.9 Relic drainages, conical hills . . . , *Journal of Geophysical Research* **87**: 9929–9950 American Geophysical Union (Breed, C.S., McCauley, J.F. & Grolier, M.J., 1982); Table 4.1 *Science Geologiques, Memoires* **88**: 23–32 Science Geologiques, Memoires (Chester, 1990); Fig. 4.2 Dust emissions . . . , *Sedimentology* **40**: 859–863 Blackwell Science Ltd. (Nickling & Gillies, 1993); Fig. 4.3 Dust storms and related phenomena . . . , *Earth Surface Processes and Landforms* **12**: 415–424 reprinted by permission of John Wiley and Sons Ltd. (McTainsh & Pitblado, 1987); Fig. 4.4 Accumulation of Asian long-range . . . , *Catena Supplement* **20**: 25–42 Catena Verlag (Inoue & Naruse, 1991); Fig. 4.5 and 4.12 Frostick, L. E. and Reid, I. (eds) *Desert sediments, ancient and modern, Special Publication* **35** Blackwell Science Ltd (Pye & Tsoar, 1987); Fig. 4.7 Assessment of the African . . . , *Journal of Geophysical Research* **97** (D2): 2489–2506 American Geophysical Union (Dulac, F., Tanre, D., Bergametti, G., Buat-Menard, P., Debois, M. & Sutton, D, 1992); Fig. 4.8 Dust intrusion . . . , *Journal of Applied Meteorology* **30**: 1185–1199 American Meteorological Society (Dayan, U., Hefter, J., Miller, J. & Gutman, G, 1991); Fig. 4.9 *Proceedings of the Soil Science Society of America* **34**: 296–301 Soil Science Society of America (Frazee *et al.* 1970); Fig. 4.11 Nickling, W. G. (ed.) *Aeolian geomorphology*, Chapman and Hall (Middleton, N.J., Goudie, A.S. &

Wells, G.L, 1986); Fig. 4.16 *Catena Supplement* **20**: 1–14 Catena Verlag (Zhang Linyuan, Dan Xuerong & Shi Zhentao, 1991); Fig. 4.17 *Palaeogeography, Palaeoclimatology, Palaeoecology* **78**: 217–227 Elsevier Science (Rea, 1990); Table 5.1 Pye, K. (ed.) *The dynamics and environmental context of aeolian sedimentary systems*, Geological Society of London (Livingstone & Thomas, 1993); Table 5.2 and 5.3 *Progress in Physical Geography* 17: 413–447 Edward Arnold (Sherman & Bauer, 1993); Fig. 5.8 *Earth Surface Processes and Landforms*, in press reprinted by permission of John Wiley and Sons Ltd. (Wiggs, G.F.S., Livingstone, I. & Warren, A, 1995); Fig. 5.9 *Boundary-Layer Meteorology* **36**: 319–334 Kluwer Academic Publishers (Wippermann & Gross, 1986); Fig. 5.15 *Desert Geomorphology* UCL Press (Cooke, R.U., Warren, A & Goudie, A.S., 1993); Fig. 5.22 Abrahams, A. D. and Parsons, A. J. (eds) *Geomorphology of desert environments* Chapman and Hall (Lancaster, 1994), reprinted with permission from *Nature* **304**: 337–339 Copyright 1983, Macmillan Magazines (Wasson & Hyde, 1983); Fig. 5.23 *Sedimentology* **37**: 673–684 Blackwell Science Ltd (Rubin & Ikeda, 1990); Fig. 5.24 *Geography* **73**: 105–115 Geographical Association (Livingstone, 1988) and also with *Sedimentology* **30**: 567–578. Blackwell Science Ltd (Tsoar, 1983a); Fig. 5.25 *Earth Surface Processes and Landforms* **14:** 317–332 reprinted by permission of John Wiley and Sons Ltd (Livingstone, 1989) and also with *Earth Surface Processes and Landforms* **18**: 661–664 reprinted by permission of John Wiley and Sons Ltd (Livingstone, 1993); Fig. 5.26 Nickling, W. G. (ed.) *Aeolian geomorphology* Chapman and Hall (Livingstone, 1986); Fig. 5.28 *Zeitschrift für Geomorphologie NF* **34**: 19–36 Gebrueder Borntraeger (Tseo, 1990); Fig. 5.29 *Journal of Oman Studies Special Report* **3**: 169–181 Office of the Adviser for Conservation of the Environment, Diwan of Royal Court, Sultanate of Oman (Warren, 1988); Fig. 5.30 *Sedimentology* **36**: 273–289 Blackwell Science Ltd (Lancaster, 1989); Fig. 5.34 Plant succession . . . , *Ecology* **46**: 765–780 Ecological Society of America (Chadwick & Dalke, 1965); Fig. 5.35 Dune forming factors . . . , *Catena Supplement* **18**: 1–14 Catena Verlag (Klijn, 1990); Fig. 5.38 *Aeolian processes . . .* , *Journal of Coastal Research* **10**: 189–202 Coastal Education and Research Foundation (Arens & Wiersma, 1994); Fig. 6.1 and 6.2 *Sedimentary Geology* **10**: 77–106 Elsevier Science (Wilson, 1973); Table 6.1 *Progress in Physical Geography* **5**: 420–428 Edward Arnold (Fryberger & Goudie, 1981); Table 6.2 *Sedimentary Geology* **10**: 77–106 Elsevier Science (Wilson, 1973); Fig. 6.3 Desert sandflow . . . , *Geographical Journal* **137**: 180–199 Royal Geographical Society (Wilson, 1971); Fig. 6.5 *Bulletin of the Geological Society of America* 78: 1039–1044 Geological Society of America (McCoy, F.W., Nokleberg, W.J. & Norris, R. M, 1967); Fig. 6.6 *Journal of Geophysical Research* 95: 15463–15482 American Geophysical Union (Blount, H.G., Smith, M.O., Adams, J.B., Greeley, R. & Christensen, P.R., 1990); Fig. 6.7 *Australian Geographer* **19**: 89–104 Geographical Society of NSW (Wasson, R.J., Fitchett, K., Mackey, B. & Hyde, R., 1988); Fig. 6.8 Brookfield, M. E. and Ahlbrandt, T. S. (eds) *Eolian sediments and processes* Professor M E Brookfield (Lancaster, 1983); Fig. 6.9 *Earth Surface Processes and Landforms* **10**: 607–619 reprinted by permission of John Wiley and Sons Ltd (Lancaster, 1985); Fig. 7.4 Frostick, L. E. and Reid, I. (eds) *Desert sediments, ancient and modern* Blackwell Science Ltd. (Livingstone, 1987); Fig. 7.5 reprinted from: *Palaeoecology of Africa – and the surrounding islands* (E. M. van

Zinderen Bakker and J. A. Coetzee eds). Volume 15 Southern African Society for Quaternary Research, proceedings of the 6th Biennial conference, Pretoria, 26–29 May 1981, (J. C. Vogel, E. A. Voight & T. C. Partridge, eds) 1982. 25 papers, 235 pp., Hfl.150, A. A. Balkema, PO Box 1675, Rotterdam, Netherlands (Lancaster, 1982); Fig. 7.7 *Sedimentology* **7**: 1–69 Blackwell Science Ltd. (McKee, 1966); Fig. 7.8 reproduced with permission from the *Annual Review of Earth and Planetary Sciences* **19**: 43–75 1991, by Annual Reviews Inc. (Kocurek, 1991); Fig. 7.9 *Sedimentology* **32**: 147–157 Blackwell Science Ltd. (Rubin & Hunter,1985); Fig. 7.9 *Journal of Sedimentary Petrology* **52**: 823–832 Society for Sedimentary Geology (Tsoar, 1982); Fig. 7.10 *United States Geological Survey Professional* Paper 1120A: 1–24 United States Geological Survey (Ahlbrandt & Fryberger, 1980); Fig. 8.1 reprinted with permission from *Nature* **361**: 432–436 Copyright 1993 Macmillan Magazines (Taylor, K.C., Lamorey, G.W., Doyle, G.A., Alley, R.B., Grootes, P.M., Mayewski, P.A., White, J.W.C. & Barllow, L.K., 1993); Fig. 8.3 *An analysis of the morphological variation of linear sand dunes and of their relationship with environmental parameters in the Southwest Kalahari*, unpublished PhD Thesis, University of Sheffield (Bullard, 1994); Fig. 8.4 *Sedimentary Geology* **65**: 139–151 Elsevier Science (Clemmensen, 1989); Fig. 8.5 Abrahams, A. D. and Parsons, A. J. (eds) *Geomorphology of desert environments* Chapman and Hall (Tchakerian, 1994); Fig. 8.6 reprinted with permission from *Nature* **272**: 43–46 Copyright 1978 Macmillan Magazines (Sarnthein, 1978); Fig. 8.7 *Zeitschrift für Geomorphologie Supplements band* **10**: 154–179 Gebrueder Borntraeger (Warren, 1970); Fig. 8.8 Brookfield, M. E. and Ahlbrandt, T. S. (eds) *Eolian sediments and processes* Professor M.E. Brookfield (Ahlbrandt, T.S., Swinehart, J.B. & Maroney, D.G., 1983); Fig. 8.9 *Geografiska Annaler* **5**: 113–143 Scandinavian University Press (Högbom, 1923); Fig. 8.11 Pye, K. (ed.) *The dynamics and environmental context of aeolian sedimentary systems* Geological Society of London (Carter & Wilson, 1993); Fig. 9.3 *Journal of Agriculture, Ecosystems and Environment* **22/23**: 41–69 Elsevier Science (Heisler & Dewalle, 1988); Fig. 9.4 *Dust Bowl: The Southern High Plains in the 1930s* Copyright © 1979 by Oxford University Press Inc. reprinted by permission (Worster, 1979); Fig. 9.5 reprinted with permission from *Nature* **316**: 431–434 Copyright 1985 Macmillan Magazines (Middleton, 1985); Fig. 9.7 van der Meulen, F., Jungerius, P. D. and Visser, J. (eds) *Perspectives in coastal dune management* SPB Academic Publishers bv (Doody, 1989); Fig. 9.16 *World atlas of desertification* Edward Arnold (United Nations Environment Programme UNEP, 1992).

Whilst every effort has been made to trace the owners of copyright material, in a few cases this has proved impossible and we take this opportunity to offer our apologies to any copyright holders whose rights we may have unwittingly infringed.

CHAPTER 1 Wind environments

The work performed by the winds in the atmosphere appears hardly to have received its due share of attention.

(Udden 1894: 318)

The work of the wind

A century after Udden wrote these words, aeolian geomorphologists may still feel neglected, but they must also acknowledge that their branch of the discipline has experienced a recent surge of interest, if measured by the geometric increase in the volume of literature over recent decades.

The renaissance has a number of elements. The space programme, which was extraordinarily well funded in the 1960s and 1970s, sought terrestrial analogues for aeolian features on other planets, most notably Mars (Greeley and Iversen 1985), and also supplied surveys of Earth from satellite images that revealed the vast extent and astonishing patterns of aeolian landforms (McKee 1979a). Microprocessor technology has provided better monitoring equipment, which can now be relied upon in hot, cold, sandy, salty and dusty environments (Chapter 2). Better vehicles, better roads and better navigational equipment have made it much easier and safer to travel in deserts. Since about 1980, new discoveries have made the dating of aeolian deposits very much easier (Chapter 8). Finally, the management of wind erosion on agricultural land has continued its stimulus to research into aeolian geomorphological processes (Chapter 9).

Shortly after Udden's seminal writing, there was a weak attempt to kindle interest in aeolian geomorphology when, early in the twentieth century, some authorities tried to promote the idea that wind is as important an agent of erosion as water, in deserts at least. Keyes (1912) spoke of 'eolation' as a process that had planed off vast surfaces (Chapter 3). The patent exaggeration of these claims and the clear evidence of the activity of water meant that almost all subsequent workers looked only to fluvial and slope wash processes to explain even desert landscapes. With the glaring exception of Bagnold's mould-breaking *The Physics of Blown Sand and Desert Dunes* in 1941, the domination of geomorphology by fluvialists was sustained in the 'quantitative revolution' in the geomorphology of the 1950s and 1960s when measurement of processes, particularly in small fluvial catchments, became the vogue. Chorley *et al.* in 1984 could suggest with impunity that the drainage basin was the fundamental unit for geomorphology, yet in doing so they were clearly ignoring the work of the wind.

Neglect was possible because, in the absence of the recent advances in earth observation and methods of measurement, few geomorphologists, if any, guessed at the enormous extent of aeolian landforms or at the magnitude of their role in earth surface sediment movement. It will be shown later in this book that aeolian landforms (with little fluvial influence) can conservatively be estimated to cover 20–25 per cent of the terrestrial land surface; to dominate deep-ocean sedimentation; to compete with (though not surpass) the amounts of sediment carried by rivers; to have played an even greater role in geomorphology during some recent phases of the Pleistocene; and to play what is probably a dominant role in the geomorphology of some other planets. The aim of this book is to report these recent discoveries and to change the balance within geomorphology in favour of aeolian processes.

Aeolian geomorphologists no longer make extravagant claims for the power of the wind, but, dealing as they do with these intrusive and potentially damaging processes and this great variety and extent of features, they can justly claim that the effect of the wind on landscapes, within and without the deserts, is a major force in earth surface sculpture.

Fig. 1.1 The scope of aeolian geomorphology: (a) wind-eroded features known as yardangs (Chapter 3) (photo: Carol Breed); (b) a dust storm (Chapter 4); (c) deposited dust, which is known as loess, in the valleys of the Matmata Plateau, Tunisia (Chapter 4); (d) a desert sand dune, Namib Desert (Chapter 5).

The scope of aeolian geomorphology

Aeolian geomorphology (sometimes spelt 'eolian' in North America) is the study of the ways in which the wind has formed the land by erosion, transport and deposition (Fig. 1.1). Because the wind moves sandy material more readily than coarser or finer particles, it operates most effectively where there is material of this size at the surface and where it is not held down by vegetation or moisture. Deserts, being dry and relatively bare, are therefore the prime area of aeolian activity, but it is by no means restricted to these areas. Places where the supply of the right kind of sediment is abundant, such as coasts and the edges of glaciers, are also susceptible, and places where the protective vegetation layer has been removed, especially by agricultural practices, may also become very vulnerable.

The scale of aeolian processes ranges from the entrainment of an individual grain of dust to the movement of dunes composed of many hundreds of thousands of tonnes of sand, and the scale of aeolian features extends from the minute pits created by the impact of saltating sand grains to the world's major sand seas, many as large as medium-sized nation states. Aeolian processes include the insidious stripping of topsoil from fields and the transport of great quantities of sediment in dust storms; the deposition of this material far from its source; the transport of great volumes of sand by saltation just above the desert surface; the abrasion of the surfaces that these sand streams encounter; the accumulation of this material into dunes; and the seemingly inexorable movement of these dunes. The aeolian features that these processes produce include eroded ridges, or yardangs, some of them tens of metres high; extensive deposits of wind-blown dust, or loess, found over perhaps 10 per cent of the Earth's land surface; equally important dust deposits in the oceans; much less obvious, though ecologically very significant, additions of dust to most of the world's soils; and, of course, the sand dunes of deserts and coasts.

The layout of the book

This book is an introduction to the work of the wind in the landscape. This first chapter briefly introduces the environments in which wind is important. Chapter 2 looks at the structure of the wind close to the surface and its effect on the fundamental processes of entrainment, transport and deposition of sand and dust as well as the development of ripples. Chapter 3 considers the erosive effect of the wind, and describes the landforms such as yardangs, ventifacts, pans and desert pavements that this erosion produces. Because dust and sand react differently to transport by the wind, each is treated separately. Dust is investigated in Chapter 4, while Chapter 5 covers the formation of sand into dunes. Chapter 6 describes the amalgamation of dunes into sand seas, dunefields and sand sheets. Chapter 7 then looks at the sedimentary characteristics of sandy aeolian deposits. Chapter 8 describes the operation of aeolian processes in the past and their legacy of landforms. Chapter 9 reports some of the effects of aeolian activity on human activity, and vice versa, and the attempts to manage these processes.

Global circulation and wind patterns

Wind is the consequence of differences in air pressure from place to place. As a general rule, air moves from areas of high pressure to areas of low pressure, although other factors act to complicate this simple rule.

At the broadest level, the global circulation of air masses is thermally driven. Air warmed at the equator rises and moves poleward, although it does not actually make it to the poles. It sinks in the subtropics, creating two thermally driven cells on either side of the equator, which extend to latitudes at about 30° north and south and are known as the Hadley Cells. Beyond these, in the mid-latitudes and within the polar circles, are two further cells which are less directly driven by radiation differences (Fig. 1.2).

Atmospheric high pressure exists where air subsides and surface low pressure exists where air diverges. Thus there are high-pressure cells, or anticyclones, formed in the subtropics and around the poles; and low-pressure areas, also called depressions or cyclones, at the equator and in the mid-latitudes.

Because of the need to conserve angular momentum on the spinning surface of the globe, air flow is not directly from high pressure to low pressure, but around these cells. Angular momentum is greatest at the equator which is the latitude furthest from the Earth's axis of rotation. Because the Earth spins towards the east, near-surface air moving towards the equator is deflected to the right in the northern hemisphere and to the left in the southern hemisphere, creating a pattern of easterlies near the equator in both hemispheres. In the northern hemisphere, air moves clockwise around high pressure and anticlockwise around low pressure; in the southern hemisphere, the reverse is true. The surface wind pattern which develops is of easterly trade winds between the tropics; of mid-latitude westerlies between the subtropical highs and the polar front; and of another zone of easterlies close to the poles. This general pattern is further disrupted by the disposition of land and sea on the earth's surface and by the topographic variation of the land to give the pattern in Fig. 1.3.

The global distribution of wind energy does not demonstrate overwhelming latitudinal patterns, but indicates that the windiest places are along coastlines and the calmest in continental interiors (Eldridge 1980).

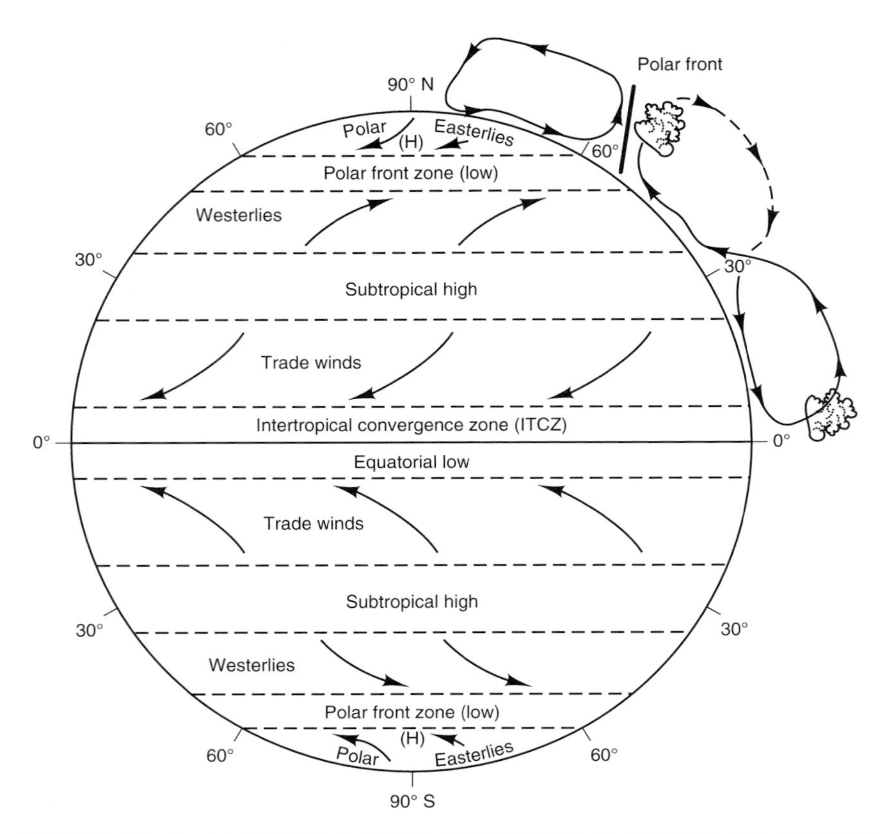

Fig. 1.2 Idealized representation of patterns of atmospheric air circulation from equator to pole.

Desert wind systems

Much aeolian geomorphological activity, though by no means all, takes place in deserts. The largest of the tropical and subtropical deserts are dominated by the subtropical high-pressure systems and may therefore be subject to the relatively low-energy wind systems associated with the centre of the anticyclones, at least for part of the year (e.g. Kalahari, southern Africa). Others, however, are subject to the relatively high-energy environments of the tropical easterly trade wind belts which blow around the anticyclones and which blow over nearly half the globe. The intertropical convergence zone (ITCZ), which separates the wet air of the equatorial belt from the dry air of the subtropical and mid-latitude deserts, moves seasonally from tropic to tropic, altering the strength and position of the subtropical high-pressure wind systems (Fig. 1.3). Areas that are relatively calm under the influence of subtropical highs for part of the year may be subject to incursion by stronger wind regimes, including hurricanes and the monsoons, for the other part.

The trade winds are strongest on land in winter when the anticyclones are best developed. In the Sahara the trades blow in winter as the *Harmattan* which is responsible for considerable aeolian activity, moving both dust (Chapter 4) and sand (Chapter 6). A weaker system, associated with the anticyclones is the *Etesian* wind system, which operates on the northern parts of the Sahara in summer.

The other major weather systems to affect the tropical and subtropical deserts are the monsoons, drawn in towards the continents in summer. The Asian monsoon is the strongest of these, drawing south-westerly winds across the eastern tip of Arabia, into southern Pakistan and across western India. This system also draws a north-westerly wind, known as the *Shamal*, across parts of Arabia and Iran, and is the major cause of dust storms in this area. There are weaker monsoon systems in northern Australia, southern Africa and the West African Sahel, and California is

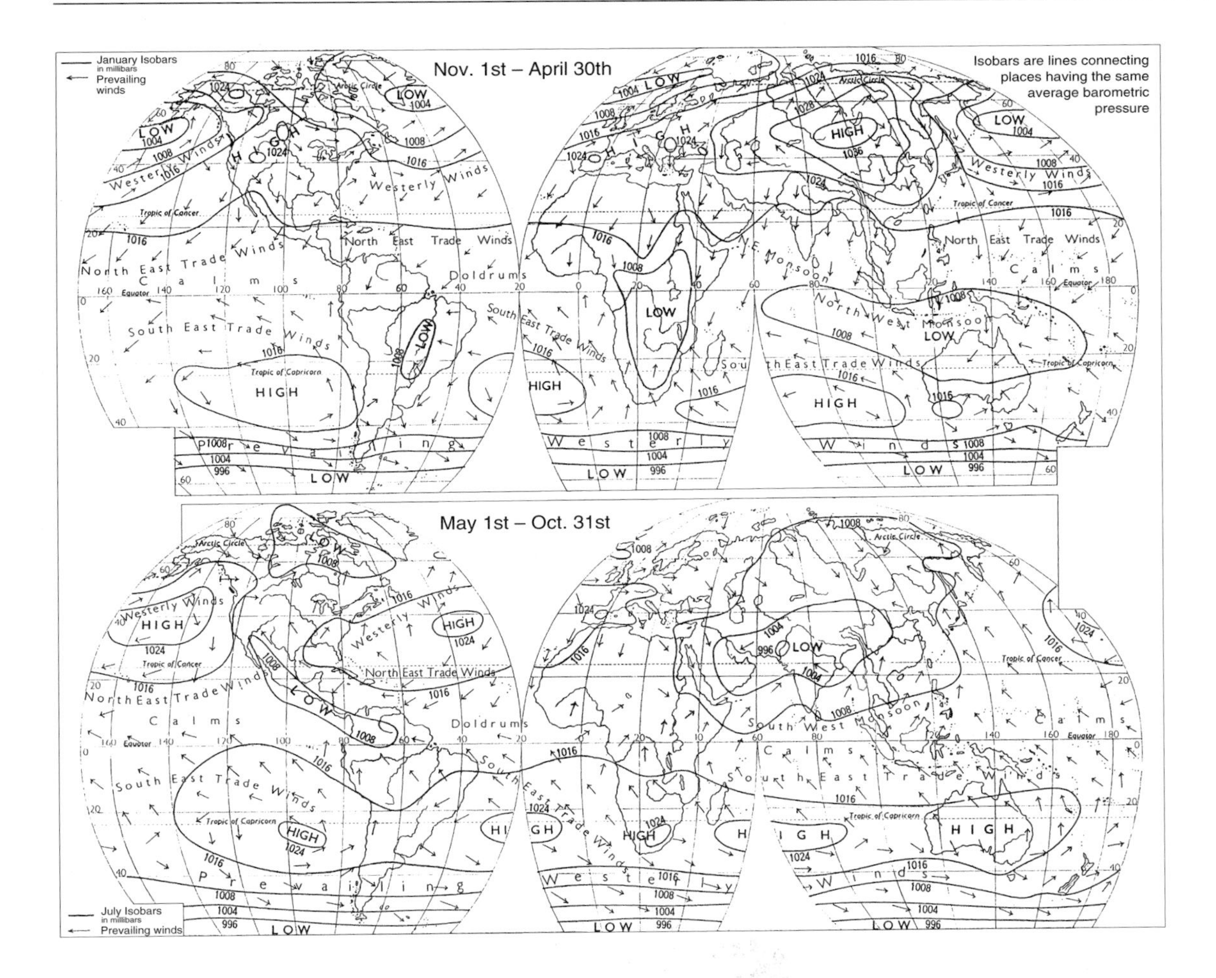

Fig. 1.3 Map of the world's major wind systems.

the subject of a yet weaker summer monsoon. Though the winds in these monsoons are generally gentle, there are occasional bursts of high energy associated with thunder-clouds, which can give rise to quite violent dust storms known as *haboobs* (Chapter 4).

The mid-latitude deserts in each hemisphere are affected by 'Ferrel' westerlies. These winds affect sand and dust movement on the poleward margins of the tropical and subtropical deserts, as in west coast deserts such as the Namib in southern Africa and the Atacama in South America, in the northern Sahara and the northern Arabian deserts, and in the southern Australian deserts. They are also the main systems in the mid-latitude deserts, as in the USA, central Asia and Patagonia. These winds are most active in winter and spring, which is the main dust-storm season in these areas.

These large-scale wind patterns are responsible for large-scale patterns of aeolian activity. The Harmattan, for instance, is responsible for entraining a huge volume of dust from the Sahara and depositing it in the Atlantic Ocean. It has also been credited with the pattern of sand flow throughout the Saharan sand seas. The pattern of dunes in the southern hemisphere deserts of southern Africa and of Australia has been attributed to the anticlockwise wind flows around the anticyclones centred on those continents, although not without contention.

Aeolian activity is important in cold environments as well as hot. Eldridge's (1980) work showed that wind energies are often greatest in the poleward parts of the continents, and Cailleux (1967) reported mean velocities in Antarctica of 22 m s^{-1}, with a maximum at Mirny (66°S 96°E) of 62 m s^{-1} and at Cape Denisan

(67°S 143°E) of 87 m s^{-1} among the highest recorded anywhere on Earth. Ten per cent of sediment in the Arctic Ocean has an aeolian origin. Although aeolian geomorphology has been dominated by work in the low- and mid-latitude deserts and along coastlines, there have also been studies in cold environments, most notably in the dry valleys of the Antarctic.

Local winds

Within the global pattern of wind systems there are local winds which are the consequence of two main effects: the effect of topography, both because of the intrusion of a hill or mountain into the flow and because of the effect of wind flow up or down a slope; and the effect of land/sea interfaces (Simpson 1994). The consequences of thermal effects, such as thunderstorms and dust devils, are covered in Chapter 4.

Topography

A mountain massif causes an oncoming wind to diverge around it and accelerate over it. One of the largest of the effects of topography is the splitting of north-easterly winds by the Tibetan Plateau so that the wind blows both west into the Tarim Basin and on into the (former) Soviet deserts. In the lee of a massif, complex eddy patterns can be created which can include zones of very low wind velocity. Mountain passes, in contrast, can funnel winds and cause considerable acceleration. One of the consequences for aeolian geomorphology of large topographic obstacles is that sand piles up against the windward faces of mountains as 'climbing dunes' or 'sand ramps' and is deposited on the lee slopes of uplands as 'falling dunes' (Chapters 5 and 6). Another consequence is erosion, for in zones of accelerated flow, such as passes and summits, sand is not deposited, but may be transported with enough force to form spectacular wind erosion features. Examples include the mega-yardangs of Tibesti and the ventifacts of California (Chapter 3). There are other consequences of topography for the deposition of loess, though these are in some dispute (Chapter 4).

The effects of topography may also be felt in upslope (anabatic) and downslope (katabatic) winds, which often intensify other wind patterns. In general, anabatic winds occur as a result of greater heating of upper slopes than of valley floors, while katabatic winds occur as colder air drains into depressions.

Land and sea breezes

Apart from deserts, the other major location of aeolian activity is coastlines. This high level of activity is in part a result of a plentiful supply of material exposed in the littoral zone. It is also because winds blowing onshore, having travelled over the very low friction surface of the sea, have higher velocities than are experienced in nearby locations inland. In addition, however, the different thermal properties of land and water cause land and sea breezes.

Because the land surface responds more rapidly to daytime heating than the sea, a pressure difference is created and air is drawn inland, or onshore. At night the more rapid cooling of the land reverses the effect and winds are from land to sea, or offshore. The onshore air movement, or sea breeze, is generally the stronger and is most commonly felt from mid-morning into the evening. Like anabatic and katabatic winds, these effects often intensify larger-scale wind patterns. They may be very important in the formation of coastal dune systems (Chapter 5).

Wind regimes in the past

Global wind regimes have not always been as they are today. Climate changes have included major and minor shifts of wind belts, and changes in wind velocity. For example, the extension of ice sheets towards the equator also brought the high-energy environments associated with periglacial regimes. The effect of shifting regimes and climate change are discussed in later chapters, and especially in Chapter 8. The evidence shows some much windier periods in the past, most notably at the time of the last glacial maximum, when even more dust and sand was being transported than today.

Measuring and describing wind patterns

Most meteorological stations record wind speed and direction, and these data are potentially of very great value to aeolian geomorphologists. The records of speed are made with anemometers, the details of which are discussed in Chapter 2.

Some care needs to be exercised when dealing with these data, particularly when they are located in remote locations. First, the measured velocity is dependent on the height at which the reading is taken, especially in the few metres closest to the ground (because there is a gradient of increasing wind speed away from the Earth's surface). The WMO recommends that anemometers are mounted 10 m from the ground and that they are sited in an open situation, but in reality this is often not the case. Where the height of the anemometer is known, some adjustment may be possible by using estimates of the nature of the velocity profile.

Second, the method of recording varies greatly. The most sophisticated data logging is by microprocessor which can then be downloaded at the site or even relayed back via a satellite link. This is a particular advantage when meteorological stations are in remote locations. In older equipment the logging is mechanical, whereby a trace representing speed and direction is recorded on a paper roll which is changed weekly or monthly. Wind speed information may be collected manually from a totaliser anemometer which records the total number of revolutions since the last record, often every 6, 8 or 12 hours, but data like these, averaged over a number of hours, hide major peaks and calms of wind speed. Occasionally wind speed and direction are recorded by an operator from an anemometer and vane which give instantaneous readings of the current situation. Each of these three methods of recording – continuous logging (either electronic or mechanical), totaliser or instantaneous reading – gives different results, and consequently some care must be exercised when comparing data sets.

The great variety of wind climates created by global and local wind patterns can be represented by diagrams of the frequency of winds blowing from different compass directions over a year. These frequency diagrams are called *wind roses*. Aeolian geomorphologists usually convert the data that are used to create a wind rose to show the potential of the wind to move sand or dust, and the resulting diagrams are called sand or dust roses. The conversion involves equations that relate wind speed to potential sediment transport rate, which are discussed in Chapter 2.

Conclusion

We have moved a considerable way in recent years towards an understanding of the effect of the wind on landforms, but few aeolian geomorphologists would now disagree that they should know yet more about winds. There are two major constraints. First, winds are nowhere as well measured and recorded as streamflow is for fluvial geomorphologists. Second, the dynamics of the wind are the domain of meteorologists, and few aeolian geomorphologists are adequately aware of the state of thinking among modern meteorologists. It is to be hoped that both these constraints may be overcome, the one by new forms of data capture, as from satellites, the other as a new generation of aeolian geomorphologists realises the importance of meteorology to their subject.

Further reading

The recent renaissance of aeolian geomorphology is reflected in the publication of other textbooks on more restricted topics (e.g. Pye and Tsoar (1990) on sand and dunes; Pye (1987) on dust) as well as books on deserts which include sections on aeolian geomorphology (Thomas 1989b; Cooke *et al.* 1993; Abrahams and Parsons 1994). There have also been proceedings and reports from a number of relevant conferences (Brookfield and Ahlbrandt 1983; Barndorff-Nielsen *et al.* 1985; Nickling 1986; Gimmingham *et al.* 1989; Bakker *et al.* 1990; Nordstrom *et al.* 1990; Barndorff-Nielsen and Willetts 1991; Pye 1993a; Pye and Lancaster 1993). Bibliographies have recently been provided by Busche *et al.* (1984), Horikawa *et al.* (1986) and Lancaster (1988a).

Greater depth is given to the discussion of atmospheric circulation and wind patterns by a number of books including Barry and Chorley (1992) and Oke (1990). The structure of winds in the boundary layer is given fuller coverage by Greeley and Iversen (1985).

CHAPTER 2

Grains in motion

Introduction

The physics of erosion, transportation and deposition by the wind, which this chapter examines, are at the heart of aeolian geomorphology. In essence, the chapter is an elaboration of a very simple statement about these processes: strong winds can entrain and carry more particles than gentle ones. The chapter concludes with an introduction to ripples, which is included here because of the close connection between ripples and sand transport mechanisms at the scale of individual grains.

Fluid flows

Wind is the movement of air. As a fluid, air behaves in many ways like water, another fluid, but air is 1000 times less dense, and this restricts the size of material it can move. The definition of a fluid is that it cannot resist stress, or an applied force. Pushing against a solid such as a piece of rock at first causes little deformation, although as the force is increased, the solid may eventually break or deform catastrophically. Fluids, in contrast, always deform or flow when a force, however small, is applied. For there to be a wind, a force must have been applied by the difference in air pressure from one place to another.

Viscosity

Viscosity is the capacity of a fluid to resist stress. All fluids display viscous behaviour when they deform under any stress, but a given force deforms different fluids to different extents. Fluids that are more viscous are more able to resist stress. Viscosity is temperature-dependent, so that the same fluid becomes less viscous (more readily moved by stress) when it is heated.

Flow types

Fluid flows can be described as being *laminar* or *turbulent*. Laminar flow is rare either in water or in air. Turbulent flow is the usual condition, and comprises eddies, which, while making overall downstream or downwind progress, have internal patterns of motion and a certain independence of behaviour. Turbulence is induced by friction between a boundary and the fluid (discussed further below), by obstructions in the flow or by thermal effects.

Individual molecules or pockets of air in a turbulent flow move in complex swirling patterns at a range of spatial and temporal scales, up, down, from side to side, and often in the opposite direction to the net flow. Velocity measurements in these conditions can only be statistical averages over time, and are usually only expressed in the downwind direction. Notwithstanding this convention, the other components of flow can be very important in aeolian geomorphology. The upward component, for example, is crucial to the suspension or long-distance transport of dust (Chapter 4).

The degree of turbulence is expressed by the *Reynolds number*, devised by a nineteenth-century physicist, Osborne Reynolds. The Reynolds number is defined thus:

$$Re = \frac{\rho h U}{\eta}$$

where ρ is fluid density, h is flow depth or the thickness of the boundary layer (the zone in which velocity is constrained by the effect of the boundary), U is mean flow velocity, and η is fluid viscosity.

Re is a dimensionless ratio in which higher values indicate greater turbulence. Laminar flow occurs when the effect of viscous forces (those resisting deformation) dominate the inertial forces (those resisting changes of speed); turbulent flow occurs when the opposite is true. Because the flow depth in air in the atmosphere is always relatively large and viscosity is low, *Re* values in air are always high even in fairly gentle winds. In air, flow is turbulent when *Re* exceeds about 6000.

Interaction of the wind and the bed

Wind shear

The energy to lift and carry grains of sand and dust comes from the wind as it 'shears' the Earth's surface. Shear occurs when one body (in this case the air) slides over another (in this case the ground). The shear force per unit area (in $\mathrm{N\,m^{-2}}$) is termed τ_0 ('tau-zero').

The shear force of the wind on the ground surface has been virtually impossible to measure directly until very recently because instrumentation had not been developed that could record what was happening at the surface, even in wind tunnels. A probe that is flush with the surface has now been developed, and this does give an idea of the surface shear (Castro and Wiggs 1994), but it is not yet able to measure all the kinds of problem associated with aeolian geomorphology, especially in field conditions. Before the probe can be adapted (which may take several years), geomorphologists and engineers must make do with other ways of estimating shear. In the absence of direct measurements, they have been accustomed to using an elegant, but not necessarily exact, method developed by the great early twentieth-century aerodynamicists Kármán and Prandtl.

The Kármán/Prandtl reasoning relies on two assertions about the nature of the velocity profile in the atmospheric boundary layer (Fig. 2.1). The first is that, because of the friction that occurs between a fluid and an 'aerodynamically rough' surface (which almost all natural surfaces are), there is a minutely thin layer of air immediately above the surface (or 'bed') in which flow is stationary or very slow. The depth of this 'viscous sub-layer' is related to the roughness of the bed, and is so small that it is only about 1/30th of the height of protruding irregularities such as sand grains. The depth can be thought of as equivalent to the 'roughness length', denoted by z_0.

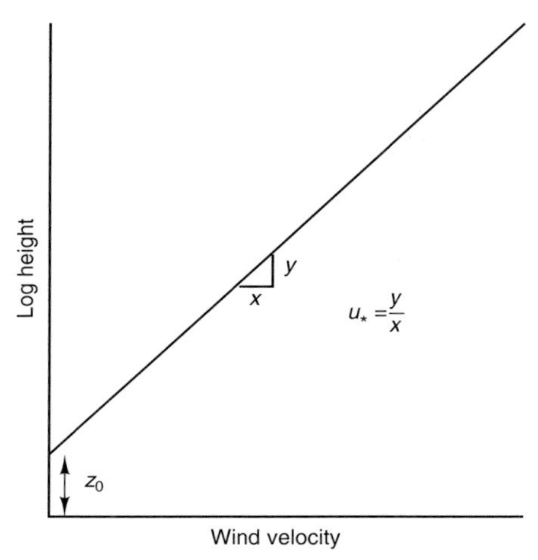

Fig. 2.1 The idealized height pattern of velocity (or velocity profile) of the wind over a smooth, flat surface. The notations are explained in the text.

The roughness length, being very small on most bare surfaces, is almost impossible to measure, and is usually derived by extrapolation from the wind profile, which is assumed to be logarithmic and must therefore reach the ground at a height above zero (Fig. 2.1). The logarithmic assumption, however, is itself dubious in many natural situations, and, in consequence of this and other problems, estimates of z_0 vary wildly. One set of five measurements of wind velocity above 0.4 m from the rough surface of a beach, yielding an acceptably logarithmic profile, could have been taken to predict a z_0 of 10^{-11} m (the wavelength of gamma rays), which is clearly absurd (Sherman and Bauer 1993). Notwithstanding, errors can be minimized and estimated, if careful procedures are followed (Bauer *et al.* 1992). Only now are more precise observations being made on the effect of the size and spacing of roughness elements on the position of z_0 and some are finding that many of the older assumptions about z_0 are very tenuous (Gillies and Nickling 1994). Moreover, these wind-tunnel examinations do not account for natural roughness elements in the field. Ripples are the most obvious (and ubiquitous) of these, and may increase the effective z_0 tenfold (Sherman and Bauer 1993). For all these problems, the reality of the roughness length concept has to be accepted for the moment.

The second Kármán/Prandtl assumption, sometimes termed the 'law of the wall', is the semi-logarithmic pattern of increase in velocity with height above z_0 (Fig. 2.1). This is another gross approximation of what actually occurs, but suffices in simple situations.

With these two principal assumptions, the Kármán/Prandtl approach allows the calculation of another quantity, the *shear velocity* (also sometimes termed *friction velocity*), which is commonly denoted as u_* (which can be said 'u star'). Shear velocity is related to shear stress thus:

$$u_* = (\tau_0/\rho_a)^{1/2}$$

where ρ_a ('rho-a') is the density of the air (in $\mathrm{kg\,m^{-3}}$). u_* therefore has the dimensions of velocity ($\mathrm{m\,s^{-1}}$), which is why it is termed 'shear velocity'.

The reason that u_* is useful for estimating τ_0 is that u_* is said in the Kármán/Prandtl approach to be related to the velocity profile, and velocity profiles are relatively easy to measure in the field and laboratory. The relationship of u_* to the slope of the velocity/log height curve is shown on Fig. 2.1 and can be written in the following way:

$$u_z/u_* = 1/\kappa \ln(z/z_0)$$

where u_z is wind velocity at height z; and κ ('kappa') is Kármán's constant, usually taken as 0.4.

If this reasoning is accepted, all that is needed to discover the surface shear force is to place a few anemometers at different heights above the surface and to derive the velocity profile from their output (Fig. 2.2). If a value for z_0 and the logarithmic nature of the profile are assumed, only one anemometer is needed. Both of these approaches have been common in recent aeolian geomorphology (in the absence of anything better), and although the discussions above and below show that both are based on many dubious assumptions, both methods have given useful results (Mulligan 1988; Burkinshaw *et al.* 1993; Wiggs 1993).

Micrometeorological anemometry

The continuing improvement in the design of anemometers has allowed increasingly better measurements of shear on the Kármán/Prandtl assumptions. The most popular design in the recent past has been the cup anemometer (Fig. 2.2). The cups rotate on arms, and operate a reed switch whose periodic signals are recorded on a logger. Cup anemometers have improved and now jam with sand less regularly, but they still have many disadvantages. They respond slowly to an increase in wind speed, and worse, keep on spinning when it decreases. Thus they can give only averaged wind speeds over times greater than a few minutes. Moreover, their bulk means that they cannot be placed close to the ground where wind-speed measurement is critical.

Hot-wire or film anemometers are a newer development. They work on the principle that higher wind

Fig. 2.2 Cup anemometers arranged so as to record a velocity profile. Three hot-wire anemometers are mounted on the centre post and the right-hand post is supporting a 'Sensit' probe, used to measure the number and energy of saltating particles. All the equipment shown records data on loggers which are downloaded to a computer.

speeds reduce temperature proportionately in a heated wire. They can be made much more sensitive to fluctuating wind speeds than can cup anemometers, thus recording much more of the turbulence. They are now the principle instrument used in wind-tunnel studies, but only recently have they been made sufficiently robust to withstand sand blast (for use in the field or a wind tunnel into which sand can be introduced); these are 'armoured hot-film probes' (for example, Kocurek *et al.* 1994). The state-of-the-art anemometer is an armoured cross-wire probe that records wind-speed variations in two orthogonal directions, allowing it to record stresses on the surface (Castro and Wiggs 1994).

None of these systems of anemometry can in general be used to record wind direction. This is usually done with a rotating vane.

Problems with the Kármán/Prandtl approach

The major assumption in the Kármán/Prandtl approach, the logarithmic profile, is very dubiously valid in field conditions. There are two particular complications. These occur when there is strong surface heating or cooling (as in deserts), or when the air flow is over topographic obstacles (like yardangs and dunes), or when there are changes in surface roughness (Chapter 5). In all these cases, velocity profiles diverge seriously from the logarithmic model. Natural environmental conditions are rarely stable enough or homogeneous enough for semi-logarithmic profiles to develop, though they may pertain in wind tunnels.

The second dubious assumption in the Kármán/Prandtl approach is that turbulence is assumed to be fairly uniform. It is easy to see why this issue was avoided, for turbulence is very hard to measure, being three-dimensional and at a number of different scales. But turbulence has a number of effects on the entrainment and movement of sediment. Perhaps most important, it causes fluctuations of shear on the bed; for example, Rasmussen *et al.* (1985) found that turbulence over periods of 240 s caused u_* to vary by 10 per cent. Turbulence shakes and loosens particles, though the exact importance of this process is hard to gauge. Turbulence appears to be much more significant to the entrainment of dust than to that of sand (Chapter 4; Nickling 1978).

Turbulence also produces the almost universal 'flurries' of grain movement that can be seen in any cloud of sand or dust, even in the tightly controlled conditions of a wind tunnel. These are almost certainly very important in initiating movement, for they apparently raise a flush of grains on their leading edge, and these grains are probably the ones that initiate more general saltation (Williams *et al.* 1990; Butterfield 1993). The flurries probably follow the 'burst-sweep' pattern first identified under water by Grass (1971), and it is also probable they that leave shallow flow-parallel traces on the surface. On a beach or in the desert, moving sand organises itself into long streamers (Fig. 2.3), which probably also

Fig. 2.3 Sand streamers on the low-tide beach at Bridlington, UK. The lighter-coloured saltating sand is blowing in well-defined trails across the darker, wet beach. (Photo: S. Tribe.)

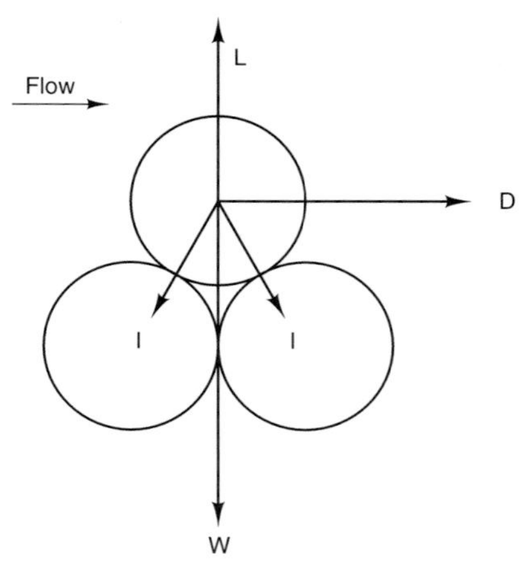

Fig. 2.4 The forces acting on a stationary particle resting in a fluid flow. The fluid exerts lift (L) and drag (D), and these are resisted by the particle's weight (W) and interparticle forces (I).

follow these burst-sweep patterns. These may manifest themselves as flow-parallel vortices, which break up and re-form in an irregular fashion. They create conditions in which sand flux is extremely variable at any one site, falling to zero during some 5 s periods, even in very strong winds, and this makes the detailed prediction of sand flux, on small temporal or spatial scales, very difficult, if not impossible (Arens 1995). In spite of their importance to the understanding of sand movement and sand flux, very little is known about these types of movement.

Furthermore, and at a larger scale, turbulence is associated with some weather patterns more than others. Thus the passage of fronts or thunderstorms can bring very turbulent winds which rarely allow the surface to stabilise, and thus render it much more susceptible to movement (Helm and Breed 1994). Helm and Breed found that sand movement in Arizona was related to much the same range of weather systems as was dust movement (Chapter 4), each with its distinct turbulence characteristics.

When considering the force delivered by the wind to the bed, aeolian geomorphologists are left in a rather unsatisfactory position. There is now enough information to confirm that Kármán/Prandtl assumptions about the character of boundary layers are too simplistic, but as yet there is insufficient theory or technology to be able to offer a fully developed alternative, although some of the technological problems may soon be overcome. This will therefore be an area of considerable study for the immediate future.

Entrainment

The beginning of sand movement is achieved in a number of ways. These are very difficult to observe or to model, and, despite some theorizing and observation, little is known about them. Recent work shows that the processes are very complex.

In theory, loose particles on a surface over which a wind is blowing experience a vertical lift force, which, if sufficient, can overcome two types of resistance (Fig. 2.4). The first form of resistance is the gravitational force

$$(g(\rho_p - \rho_a)d^3)$$

where g is the acceleration due to gravity ρ_p is the particle density, ρ_a is air density, and d is the particle diameter. The second type of resistance is a group of other forces that include friction and cohesion. These arise because the grains are tightly packed together and they are proportional to $\rho_a d^2 u_*^2$ (Iversen *et al.* 1987).

Lift

Higher velocities are accompanied by lower pressures (Bernoulli's equation), so that where the velocity profile is steepest, bringing high and low velocities close together in the airstream, as it does near the surface, there are also great pressure differences. Thus grains protruding into this zone of the flow experience pressure changes which may induce some lift. Protrusion compounds the pressure effect by forcing the flow to accelerate over the particle (yet further lowering pressure). However, lift can only be important very close to the bed, and even then can raise grains only slightly. It may well be more effective on rough beds in the field than on smooth beds in wind tunnels, on particles already set in motion by drag than on static ones, and in conditions in which turbulence can provide sudden very low velocities.

Drag

Most experiments have shown that drag is more powerful than lift. The two components of drag are 'surface drag', which is skin friction between the particle and the air; and 'form drag', produced by the difference in pressure between the windward and lee sides of the particle (especially if there is flow separation, as over angular particles, and/or at high Reynolds numbers, as is general in fast flow). Because surface drag is greatest on top of the particle, it tends to roll it, while form drag causes the particle to both

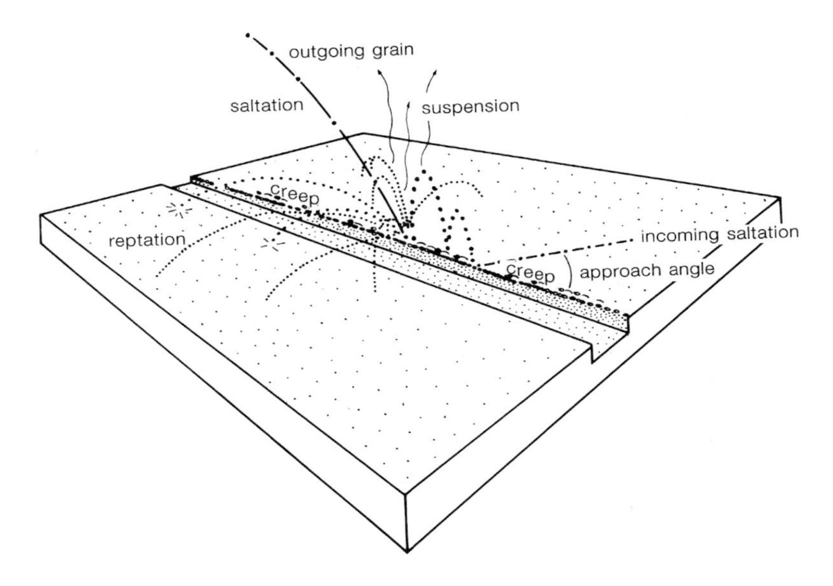

Fig. 2.5 Modes of grain transport (based partly on data from high-speed filming by Mitha *et al.* (1986)). The terms are explained in the text.

roll and slide. Drag may even induce ejection when moving particles collide or are dragged over small projections (Nalpanis 1985).

Bombardment

Most authorities maintain that once a few grains have been lifted and begin bombarding the bed on their return, shear on the bed is lowered to a value at which entrainment by lift or drag is not possible, and further entrainment is almost wholly by bombardment (Owen 1964; Anderson and Haff 1988, 1991), though this has yet to be proved (Butterfield 1993). Bombardment is the process by which saltating grains (see page 15) bring the momentum that they have acquired from higher levels in the air stream back to the surface, and the initiation of further movements by the collisions between these and other grains.

When the wind meets a sand patch, bombardment begins almost immediately (see below), so that lift and drag probably operate alone only in a very narrow zone. High-speed films (Werner 1990) show that a descending grain usually produces one high-energy ejection (as pictured in Fig. 2.5), which takes some 50 per cent of the impact energy (the ejectum is often the impacting grain itself) and about ten lower-energy, short-distance movements from a zone around the original impact (the 'splash' effect, known as reptation, discussed on page 17), which take much of what energy is left.

The threshold of movement

One of the most important concepts in aeolian geomorphology is the minimum force required to move particles of a given size. This is known as the 'threshold (or critical) shear force' (τ_t) for that size of particle. Because of the practical problems encountered in measuring surface shear forces (outlined on pages 9–10), it is more usual for the 'threshold velocity' (u_t), or better still the 'threshold shear velocity' (u_{*t}), for any size of particle to be calculated or measured.

If a succession of trays of particles, each with grains of different size, is exposed to a range of wind speeds in a wind tunnel, and the value of u_* at which the particles begin to move is plotted against the size of particle on the trays, a minimum value of u_{*t} is found, with threshold values increasing away from the minimum towards both finer and coarser grains (Fig. 2.6). For quartz grains the minimum threshold velocity is in the range 70–125 μm (micrometres or microns: 1 μm = 10^{-6} m). The minimum decreases with particle density (Iversen and White 1982). Many authorities have reviewed the theory behind this kind of behaviour (for example, Pye and Tsoar 1990).

This kind of experiment also reveals a difference between the *static* (or *fluid*) threshold and the *dynamic* (or *impact*) threshold. The static threshold marks the point at which grains first start to move (by drag and

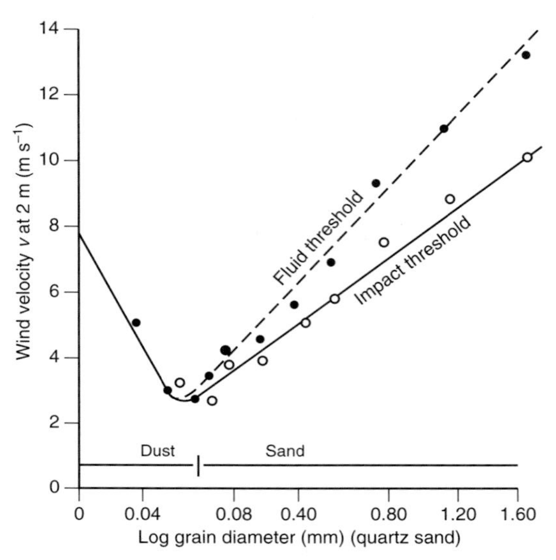

Fig. 2.6 Threshold curves relating the start of motion of grains of different size to wind velocity (after Chepil 1945; Hsu 1971).

lift). Once entrainment has started, it is bombardment that keeps the system moving (as explained above), and, because the descending grains bring down energy from above, less overall energy is required to maintain movement, and it can continue at a lower overall wind velocity than before. Consequently, the dynamic threshold, at which grain movements can just be maintained, is about 80 per cent of the static threshold (Anderson and Haff 1988). The two thresholds are compared in Fig. 2.6.

As predictors of the field behaviour of sediments, the threshold curves in Fig. 2.6 must be treated with caution. There are a number of significant complications. First, there is the size-sorting of the sediment, for mixtures of different sizes very much complicate the threshold effect, and the curves in Fig. 2.6 are derived from experiments with series of trays of particles each in a narrow size range. The curves may apply to most dune sands, which are generally well sorted, but when there are mixes of particle sizes (which is not uncommon on other surfaces) a whole range of complications appear, such as the protection of small particles by big ones, different packing densities, the bombardment of fine ones by moving larger particles and so on.

The most systematic approach to the relationship between sediment sorting and the threshold velocity was Nickling's (1988). Nickling studied threshold behaviour in a wind tunnel for a number sediments of different grain sizes and sorting coefficients, and found, in summary, that the more poorly sorted the sediments, the greater the distinction between the start of movement (by a few grains) and general overall movement.

Other factors that complicate field thresholds include surface crusts, roughened surfaces, and surfaces with different slopes (although on most dune surfaces this is a very minor complication) and particles of different shape and packing density. More important complications are soil moisture and the cover of vegetation (including almost invisible algae), which are discussed later in this chapter. Despite these reservations, the curves do give valuable indications about the behaviour of real sediments, and these require some explanation.

The threshold of movement for sand

The rising curve for the thresholds on the coarse-particle side of Fig. 2.6 is easily explained. Increasing size, and therefore mass (if density is constant), increases the gravitational force that must be overcome before movement can occur: heavier grains move less readily than lighter ones. Lettau and Lettau (1978) found that the threshold velocity was rather well defined in the field for sands predominantly between 125 and 177 μm. Below 4 m s^{-1} (measured at 0.63 m above-ground) there was rarely movement; between 4 and 5 m s^{-1} there was some creep; above 5 m s^{-1} there was some leaping movement; and above 5.5 m s^{-1} movement was 'uniformly strong'. This last velocity value was equivalent to a u_* of 0.25 m s^{-1}. These results agree with most wind-tunnel experiments.

The threshold of movement for silt and clay

At first sight, one might expect that the smaller (and therefore generally lighter) a particle, the more easily it would be entrained, yet Fig. 2.6 shows that this is not so for particles smaller than about 100 μm. There are a number of reasons for this behaviour.

One explanation, though probably not the most important one, is that fine particles lie in the zero-velocity layer near the bed or at least in a zone where wind velocity is very low. This might explain the experimental results in a wind tunnel, which are what the threshold curves are derived from. But on rough natural surfaces this is less likely to be a major factor. A second and somewhat more important reason why fine grains are difficult to move in the wind tunnel is the absence of bombardment by coarse moving particles. In the field, dust is seen to be most easily lifted into the wind if the surface is being bombarded by sand particles carried by the wind (this is further explored in Chapter 4). Indeed, when there is bombardment, the inverse relation of particle size and u_{*t} disappears (Gillette and Walker 1977).

A third and probably the most important reason for the inverse relation of size and threshold velocity in fine sediments (both in the wind tunnel and the field) is cohesion. It is more than coincidence that cohesive forces begin to equal gravitational ones when particles are less than about 100 μm. Fine sediments cohere better than coarse ones because they have greater packing density; higher numbers of other particles with which they are in contact; and greater inter-particle bond strengths. Thus cohesion is a function of $(1/d)^3$ (Smalley 1970). These properties, in turn, derive from: the platy shape of many fine particles, which creates more opportunities for contact, if packing is parallel; their greater specific surface (surface area per unit volume), again giving many contacts; greater surface activity per unit volume (encouraging weak chemical bonds, known as van der Waals forces); and electrostatic charges, in some circumstances (Iversen and White 1982; Nickling 1988). Furthermore, moisture is held more tightly between fine particles, and adds a considerable element of cohesion in many circumstances.

There are further complications with fine particles. The finest are mostly clay minerals, and these usually cohere in large clods which prevent erosion not only because of their size, but because they create a rough, pebble-like protection (Gillette 1986). In some circumstances, however, aggregation can decrease the threshold, for if the aggregates are the size of fine sand, they require a lower u_{*t} than their individual clay particles, and can then be entrained and transported as if they were sand (and are known then as 'pellets'). These may bombard the surface and further increase entrainment, though they break up much more quickly than single-grain sands.

The transport of fine particles as dusts is discussed further in Chapter 4, and the effects of moisture on thresholds and transport rates are discussed below.

Modes of transport

Once the threshold of movement is passed and grains are entrained, they travel in four distinct ways which, in increasing order of velocity, are creep (and related near-surface activity), reptation, saltation and suspension (Fig. 2.5). The key process is saltation, for once movement has begun, it is saltation that powers all the others, especially creep and reptation. Even the entrainment of the fine particles that later enter suspension is largely caused by the impact of saltating grains, and it is only when they have escaped from the saltation layer, as the great majority of them quickly do, that they follow independent trajectories. It is because of these critical roles that saltation is discussed first in what follows.

Saltation

The word 'saltation' comes from the Latin *saltare* ('to leap'). 'Leaping' best describes the paths of sand grains as they are ejected into the air, gather momentum from the higher wind speeds above the bed, and then descend back to the ground. Mean launch angles are between 30° and 50° downwind (Anderson 1989), but some grains even bounce back against the flow. Ejection velocities are related to u_* and are 50–60 per cent of impact speeds. Saltation velocities have a wide Gaussian distribution (Anderson 1987a; Anderson and Haff 1988). Velocities are greater from slopes inclined into the wind, such as the windward sides of ripples (Willetts and Rice 1989). This should mean that saltation is more vigorous on rippled than on plane beds.

Some large particles rise high into the saltation layer, perhaps because they have lower specific surface areas and therefore suffer proportionally less vertical drag (Jensen and Sørensen 1986). In a sand storm, fine particles of the order of 100 μm rise only a few centimetres, while a few with diameters of 1000 μm may reach 1.5–2 m (Sharp 1964). In a severe windstorm in the San Joaquin Valley of California, with wind speeds at 10 m above-ground up to 53 m s^{-1}, particles of 23 mm (23 000 μm) diameter were imbedded into a wooden telegraph pole at 0.8 m above ground (Sakamoto-Arnold 1981).

But these are apparently only a few maverick grains, for most authorities have found, both in the field and in the wind tunnel, that there is a rapid fall-off in the mean size of particles and also a rapid improvement in sorting with height above the ground. A new model developed by Anderson and Bunas (1993) suggests that fine grains are always ejected with greater velocity than coarse ones, the difference being greater for greater speeds of the incoming (impacting) grain. There is an upper limit to ejection velocities, which is higher for fine grains than coarse grains, giving much higher trajectories for the finer particles at this limiting velocity. Further increases in impact energy serve to increase the rate at which particles are ejected, rather than the speed of individual particles.

Saltation hop lengths are about 12 to 15 times the height of bounce, but have a very wide distribution. Although there is some disagreement about the relationship between grain size and distance of a leap, Anderson and Bunas's (1993) model shows that small grains are generally ejected with greater velocity, achieve higher altitudes, and thus have longer jump lengths. Trajectory lengths may be directly related to u_*, but perhaps only for these high-travelling grains (Werner 1990). The path of descent is determined by gravity and the drag imparted by the wind (Fig. 2.5). The momentum in grains larger than about 100 μm ensures that this path is smooth and undisturbed by turbulence. Flight paths are not much affected by collisions with other saltating grains, even though there may be a number of these. Lift induced by spin (the 'magnus effect') is generally thought only to be a second-order effect.

Most models and experiments show that the great majority of grains descend at angles of between 10° and 15° (Anderson 1989), and most experiments show that the mean angle decreases with grain size and u_*, as one might expect from what has been said above (Jensen and Sørensen 1986). The horizontal velocity at impact of large grains is close to that at the top of their trajectory, but small ones move at a velocity nearer the wind speed close to the bed (Nalpanis 1985). Most impact velocities are below 4 m s^{-1} (Anderson and Haff 1988, 1991). These velocities allow saltating grains, unlike reptating ones (see below), to have enough momentum to 'splash' up others when they return to the surface (this being the way in which reptation and creep are maintained and in which new grains enter saltation) (Fig. 2.5). This momentum is directly related to the prevailing value of u_* and to the size of the particle. Higher momentum particles eject more grains (Anderson and Bunas 1993).

The cloud of saltating grains has an effect on the velocity profile of the wind. Because it is abstracting energy from the wind, the velocity profile in the saltation curtain adopts a form that is quite different from that of a wind without sand. Modelling has now confirmed Bagnold's (1941) empirical finding that, when there is saltation, wind-velocity profiles have an inflexion at a different roughness height which he termed, z_0', and which is higher than z_0 over a fixed surface. The focus at z_0' has been called 'Bagnold's kink' by McEwan (1993). However, Bagnold's finding that z_0' was invariable with wind velocity is now

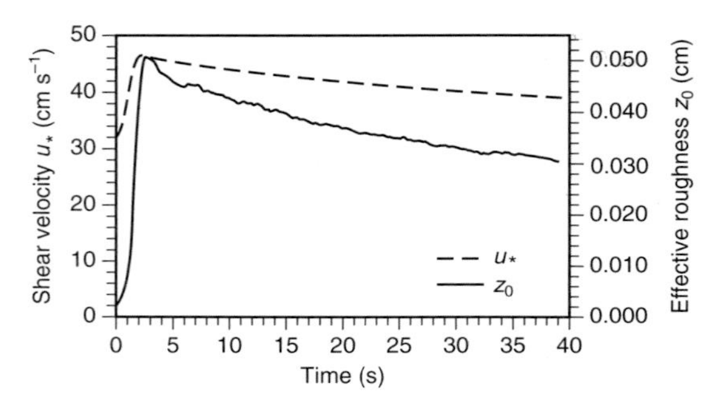

Fig. 2.7 Modelled changes in the shear velocity following the initiation of saltation (after Butterfield (1993) from a model by McEwan (1991)).

generally rejected. Owen (1964) made a widely quoted reformulation, but newer ones are now appearing, and give better results when compared with observations (Anderson and Haff 1988; Wilburg and Rubin 1989; Werner 1990; Sherman 1992; Raupach *et al.* 1993). Some models and observations now also suggest that the wind-velocity profile in the saltation layer is convex upward in shape (McEwan and Willetts 1991; McEwan 1993).

The changes in the wind-velocity profile in response to saltation must be manifestations of a feedback mechanism, for they approach a steady state quite quickly. Modelling and wind-tunnel experiments show that when wind velocity is suddenly raised through the threshold velocity over a sandy bed, the shear velocity rapidly responds and even 'overshoots' to a peak value, before declining to a steady level (Fig. 2.7; McEwan and Willetts 1991; Butterfield 1993). The rate of 'cut-in' is apparently related to u_* (Anderson and Haff 1991). In Butterfield's experiments this kind of 'steady' rate of transport was achieved in a matter of about 9 s after movement had begun.

However, Fig. 2.7 shows that a truly steady state is not achieved until after a further period, of the order of a minute or more. Between the initial rapid cut-in and this second kind of steady state, there appears to be a gentle decline in the transport rate. The explanation may be that an equilibrium has to be established between the saltating curtain and the flow above it, and that the change takes time to propagate through the curtain (McEwan and Willetts 1993). Although Butterfield's (1993) experiments did detect a form of longer-term adjustment (albeit with slightly different characteristics and over a longer period than theory predicted), he noted that in real winds, in which velocity fluctuates wildly, this second steady-state condition would be reached very seldom, and therefore that flow and transport were probably in permanent disequilibrium.

Creep

All grains travelling close to the surface were, until recently, categorized as 'creep'. Now that they can be better observed, other processes have been distinguished, as explained below, but two residual modes of grain movement can still be described as 'creep'. One is the rolling of coarse particles driven by the impact of finer grains in saltation (which was the original conception). The other is the rolling of these grains into craters created by the saltation impacts (induced by gravity). Creep, therefore, mostly involves the coarser particles, but the exact distinction in any one case between the size of the grains that take part in creep and those in other kinds of motion is probably a function of the prevailing mix of grain sizes and of u_*.

Willetts and Rice (1985a, 1986) filmed what they regarded as creep at 3000 frames s^{-1}, and then observed it in slow motion. With u_* at 0.48 m s^{-1}, they found that 355–600 μm diameter grains moved at about 0.005 m s^{-1}. Grains of the same size began their journey as a group, but rapidly dispersed as some moved more quickly than others, and the grouping disappeared completely within 3 min. Some creeping grains (like some in reptation and saltation) are buried for long periods of time, often in ripples (Barndorff-Nielsen *et al.* 1982).

The proportion of the total load travelling in these kinds of ways must vary with grain-size mix, though it may be independent of u_*. The creep:saltation ratio has been found to be between 1 : 1 and 1 : 3 (Willetts and Rice 1985a), but many of the results need to be treated with caution because of the difficulty of isolating creep from the other near-surface processes, particularly reptation (see next column).

Other near-surface activity

When a mobile surface is examined closely it is seen that there are modes of activity other than creep. First is the differential loss or accumulation of different size fractions, depending on whether the bed is eroding or accumulating. On an equilibrium surface, by definition, grain size does not change, but there are few, if any such surfaces in nature, for dunes are dynamic bodies composed of erosional and depositional portions. On an eroding dune surface, the probability that a grain will be removed is a logarithmic function of the logarithm of its size; thus small grains are much more likely to be removed. On an accumulating surface, by the same token, the probability of deposition is likely to be a logarithmic function of the logarithm of size; here fine grains are much more likely to be deposited (Bagnold and Barndorff-Nielsen 1980). Thus erosional surfaces are overloaded with coarse grains and depositional surfaces with fine grains. This kind of process on eroding beds has been modelled by Anderson and Bunas (1993), as will be explained on page 26. When these two probability distributions are combined, a log-hyperbolic function is produced (as explained in Chapter 7).

There are yet other processes at work on a mobile sand surface. One of these is the preferential movement of coarse particles up to the surface when a sand composed of a mixture of sizes is shaken (in this case by bombardment) (Sarre and Chancey 1990). The process probably extends no more than five grain diameters beneath the surface, and is probably very fast. It seems to be the result of a compressional–dilational wave which radiates from the point of impact of a saltating grain. Models of the shaking of size-mixtures show that the amplitude of shaking is directly related to the speed of sorting and that large particles rise by rotating and ratcheting themselves against the smaller ones, their roughness therefore being an important control on the rate of rise (Haff and Werner 1986). Yet another process occurs when some of the saltating grains hit the surface and tunnel along just beneath it, jettisoning other particles, sometimes many grain-diameters downwind (Fig. 2.5; Willetts and Rice 1985b).

There are still further processes at work in the surface. Bombardment both consolidates the surface, rendering it harder to mobilise, and elevates some grains to positions where they are more vulnerable to later dislodgement (Iversen *et al.* 1987). There is clearly much more to learn about what happens in the few layers of grains on moving beds of sand.

Reptation

Recent research has distinguished another distinct and important type of near-surface motion in aeolian transport. This is *reptation*, (from the Latin *reptare*: 'to crawl'), which is the 'splashing' or low hopping of grains dislodged by the descending high-energy particles (Anderson and Haff 1988). When a saltating grain hits the surface, it dislodges about ten reptating ones (Fig. 2.5; Werner and Haff 1988). Reptating grains differ from those in creep, because they continually pass between the reptation and saltation modes (Anderson 1987b). They differ from those in saltation mainly in their velocity distribution, which is strongly exponential, heavily weighted towards small velocities (Anderson 1987a). Nevertheless, the majority of all the grains in motion at any one time are in reptation (Anderson *et al.* 1991). As in saltation, the number of ejecta in reptation is strongly related to the impact speed of the incoming grain, and so to u_*.

Suspension

Suspension can be defined as the condition in which grains follow turbulent motion; in saltation they do not. This is not an entirely satisfactory definition, because suspension can occur in laminar flow, but it suffices for the present purpose. A simple measure of the distinction is if the ratio of u_* to u_f is greater than one (u_f being the fall velocity of a particle in air). u_f is a function of the balance between the weight of the particle and the drag of the air upon it. The vertical velocity in turbulence near the ground is approximately equal to u_*, so that, with $u_f < u_*$, particles stay aloft. Intermediate sizes of grain take part in 'modified saltation', where saltation paths are made irregular by turbulence (Anderson 1987a).

Observations over level surfaces show a sharp transition to suspension or modified saltation when grain size falls below about 100 μm, at quite a range of values of u_* (Nalpanis 1985). There is a fairly clear distinction between silts and clays, which can be held aloft in many winds, and sands, which rarely go into suspension (Pye 1987). There may, however, be complications near the bed. Saltation may dampen turbulence, and this may help to retain dust-sized particles within the saltation layer (Jensen *et al.* 1984). There are also undoubtedly complications when there is flow over undulating topography, as over dunes, for the turbulence induced by the topography may help to raise quite coarse particles. Thus sand undoubtedly enters suspension in the lee of transverse dunes, and has been found to be a major process in the formation of coastal foredunes, where suspension may carry quite coarse grains many metres inland in high winds (Arens 1994).

Distinctions can be made between particles in suspension. Particles larger than 70 μm cut through some of the turbulence. They may descend near the bed occasionally, but when they do, they experience enhanced lift forces and may again be swept aloft. Particles coarser than 50 μm in diameter (sometimes called 'coarse dust'), travel only a few tens of kilometres before settling (if and when the wind subsides). Loess is a deposit (Chapter 4), usually between 20 and 30 μm, most of which can be shown to have travelled no more than about 300 km from first suspension to deposition (in windstorms with common levels of turbulence). Fine dust, less than about 15 μm in diameter, can travel much further, and forms the dust haze that persists even in calm conditions (Pye 1987).

Transport rates

One of the central concerns of aeolian geomorphology is the *sediment transport rate*. The transport rate, usually denoted q (or Q), is defined as the mass of sediment passing through a plane perpendicular to the wind of unit width and of infinite height above the ground per unit time. In SI units it is measured in kg (m-width)$^{-1}$ s^{-1} (sometimes in m^3 (m-width)$^{-1}$ s^{-1}).

Because of the difficulty of directly measuring the transport rate (see below), both those interested in landform development and those trying to manage aeolian hazards place great value on being able to predict the transport rate from wind data collected at meteorological stations. Unfortunately, difficulties in understanding the saltation process, difficulties in measuring u_*, the highly turbulent nature of flow near the surface, and the difficulty of observing what happens close to the ground, along with complications such as moisture, vegetation and surface crusting, mean that there is still no single, simple relationship linking sediment discharge to wind velocity.

This lack of an established relationship is not a consequence of lack of study. It has been investigated in theory, in the wind-tunnel and in the field by many workers over the past century, with the result that a plethora of formulae now exist. These formulae have been reviewed by Greeley and Iversen, who provide a lengthy list (Greeley and Iversen 1985: 100, Table 3.5), and by Sarre (1987). All have noted how difficult it is to derive and confirm the formulae.

All the formulae agree that the relationship between the transport rate (Q) and shear velocity (u_*) is best expressed in the general form:

$$Q \propto u_*^{\,a}(u_* - u_{*t})^b$$

where the sum of the exponents, a and b, is 3.

One of the most widely used formulations, that of Lettau and Lettau (1978), is expressed in the form:

$$q = C(\rho_a/g)u_*^{\,3}(1 - u_t/u_z) \qquad (2.1)$$

where q is the discharge rate of sand in g (m-width)$^{-1}$ t^{-1} (t is a specified time period); C is an

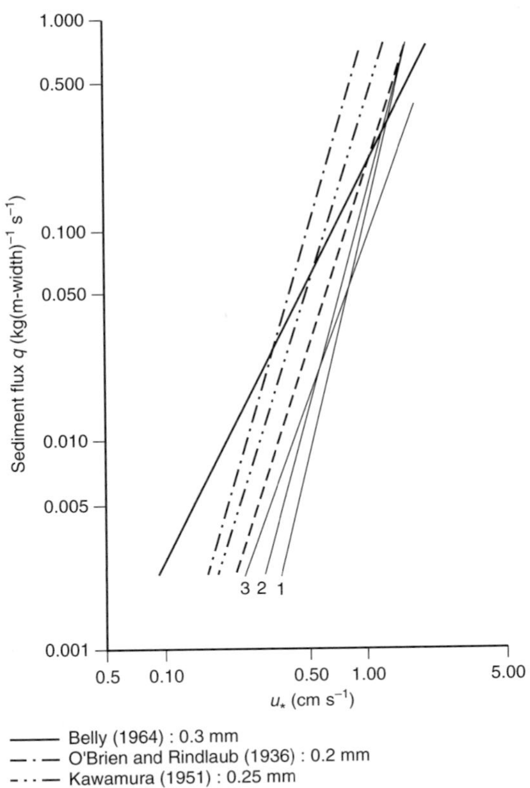

Fig. 2.8 The predictions of various sand transport formulae (after Lancaster and Nickling 1994).

empirical constant related to grain size, commonly of the order of 6.5; ρ_a is the density of the air; u_t is the threshold velocity; and u_z is wind velocity at height z. Here, u_t should be taken as the impact rather than the fluid threshold (Werner 1990).

Comparisons of the predictions of the various formulae available show that they diverge quite widely (Fig. 2.8). Furthermore, they were all derived for what Sherman and Hotta (1989) have termed 'ideal surfaces': essentially those that are horizontal, covered with well-sorted loose sand, unaffected by moisture, vegetation and crusting, over which a steady wind with a semi-logarithmic velocity profile is blowing. These situations rarely occur in nature, so that the formulae provide little more that an estimate of the maximum transport rate that could occur, although even this has sometimes been challenged by results which suggest that the formulae are under-predicting (Sarre 1987).

Slope has an effect on sediment transport, although there is some disagreement about the exact nature of the effect. Reviewing various formulae, Sherman and Bauer (1993) noted that correction factors would be between 0.400 and 0.960 for a 5° slope. Yet another correction factor was used by Howard *et al.* (1978). With some exceptions, however, the formulae predict that the slope angles found on the windward slopes of most desert dunes would have very little effect on transport rates. Only on steeper slopes, as on the eroded faces of coastal foredunes, or when a desert wind reverses over a former slip face, does slope appear to become an important factor in sand transport. The correction factors do not, however, take account of speed-up and jet-type flow over hills (Chapter 5) and may, therefore, be of rather limited value (Sherman and Bauer 1993).

The effects of moisture on the threshold of movement and transport rates

Except in extremely arid conditions, moisture is a common and very important control on the threshold of movement and the rate of transport by the wind. Complete saturation halts sediment movement even in very windy conditions, but as the soil dries out a whole battery of processes interact to control entrainment and movement. A crude estimate is that a soil must dry to 4 per cent water content before movement starts (the pores then being about 15 per cent full) (Azizov *et al.* 1979), but the issue is clearly more complex, as field measurements show. A priori, it is easy to see that the effect of moisture must depend on particle size and on factors like organic matter content and the influence of wind speed on evaporation.

Grain size affects the drying process through its influence on the behaviour of water in the pores. The meniscuses on the water trapped between the grains are more gently curving between sands than between silts or clays, so that the same amount of moisture holds sands less firmly (McKenna Neuman and Nickling 1989). Sands are also drier because they drain more quickly (Agnew 1988). There may be quite complex effects if the pores in a sediment have a narrow size-distribution, for then a high proportion of them may suddenly empty of water as the soil dries, and sediment may abruptly become available for transport by wind (Nickling 1988). The positive effect of wind speed on evaporation means that the threshold on a wet surface may be very little different from that on a dry one in a high wind (Hotta *et al.*

1984). Sandy surfaces dry out very quickly so that a shower has little long-term effect on the rate of movement, but moisture, if sufficient, can have a much longer-term effect on sand movement if it encourages the germination and growth of vegetation (see below) (Helm and Breed 1994).

There are yet more complicating processes. If strongly driven, rain may increase the transport rate by splashing particles into the path of the wind (Sarre 1988; de Lima *et al.* 1993; Chapter 5). Rain may also slake and pelletize fine soils and render them more susceptible to deflation when they dry (although this effect is not as immediate as splashing). Moreover, if saltation has begun, as on a drier patch of beach, it may continue over wetter patches, being maintained by bombardment (Sarre 1990). Sarre suggested that saltation itself was not inhibited until moisture content of the surface sediment reached above 14 per cent. The effects of atmospheric humidity are more demanding to measure than those of soil moisture, and its effects are therefore even more uncertain (Knottnerus 1980).

The effect of vegetation on sand transport

The effect of vegetation is to restrict the movement of sand, either by trapping sand and dust already being transported (see the sections on 'loess' in Chapter 4 and 'anchored dunes' in Chapter 5), or by preventing or restricting entrainment. Marshall (1973) reported that particle transport increased greatly where vegetation cover dropped below 15 per cent, although this analysis assumed a flat rather than undulating surface. The presence of vegetation, like any other roughness element, raises the roughness length so that, except in the sparsest vegetation covers, all sand- and dust-sized material rests in the zero-velocity zone at the ground surface. The roughness introduced by the vegetation means that z_0 is greater than a grain diameter.

Paradoxically, however, vegetated surfaces cause steeper velocity gradients and thus greater shear stresses than unvegetated surfaces. This is because there is a greater friction effect of the vegetation which causes greater drag on the flow. But this increased stress is not usually transferred to the ground surface, and is therefore ineffective in entraining sand or dust (Thom 1971; Jackson 1981). However, the effect of vegetation is further complicated by the fact that this rougher but also highly variable surface means that considerable gustiness is created. Raupach (1991) showed that 50 per cent of momentum transfer in plant canopies occurred during less than 5 per cent of the time. Thus full vegetation covers preclude aeolian entrainment, but geomorphologists are beginning to make tentative efforts to understand the effect of a partially vegetated canopy on particle entrainment by the wind (for example, Musick and Gillette 1990; Stockton and Gillette 1990; Wolfe and Nickling 1993; Wiggs *et al.* 1994), and some recent debate has been concerned with the level of dune activity that can occur when vegetation is present (Chapters 5 and 8). It cannot automatically be assumed that when vegetation is present aeolian activity ceases.

Measuring sand flow

For all the problems of calculating sand transport from wind data, it is still a more accurate method of estimation than the direct, physical measurement of transport. Measuring sand transport has three major inherent problems (and many minor ones). First, sand transport by wind is not unidirectional (as it is in a stream), so that a single sand trap must rotate with the wind and rotational mechanisms tend to become clogged with sand (let alone the problems of the response times to highly variable directions). Second, the surface on which the trap sits is liable to be eroded, in places quite naturally, as on a dune, but also because the trap disturbs the flow and increases the erosion rate locally (especially at the mouth of the trap). Thus traps tend to be undermined, or, if the immediate surroundings are protected, say with a film of oil, to become isolated on small plinths, surrounded by scoured hollows. In either case they cannot record the ambient drift accurately. The third and last problem is that traps deflect the flow that carries the saltating sand (they are said then not to be isokinetic), though this problem can be reduced by careful design. Minor problems include selectivity as to particle size, and the need for interference with the processes as the trap is emptied.

Attempts to balance these problems have produced a series of traps which vary vastly in design. Among the mechanical traps, the over-engineered extreme is the massive and very intrusive United States Geological Survey model in which a rotating catcher feeds sand from various heights into sectors of a large drum, about 1 m in diameter, buried beneath the surface. This model is only suitable for very stable surfaces and where is maintenance capability is on constant call. It can very easily become isolated on a plinth, deflecting the sand stream away from its apertures. The more popular 'Leatherman' trap (Leatherman 1978) also rotates, but cannot segregate the sand flow by height

(a)

(b)

Fig. 2.9 Sand trap designs: (a) 'Leatherman' sand trap (photo: R. Sarre); (b) Aarhus design (photo: Giles Wiggs).

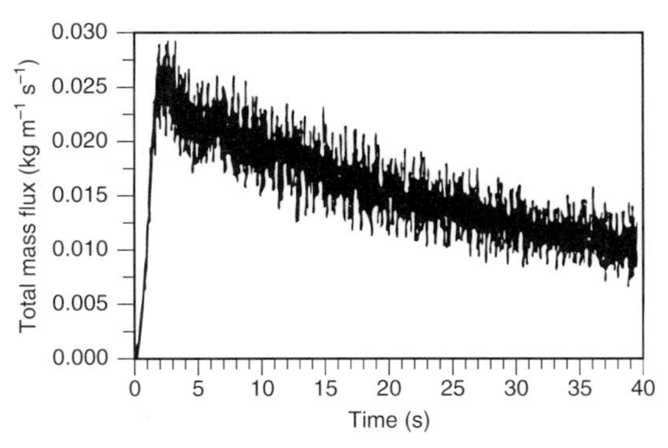

Fig. 2.10 Sand transport measured with a sand trap fitted with a load cell in a wind tunnel (after Butterfield (1993)).

and has a much smaller subterranean store (and therefore fills much more quickly) (Fig. 2.9a). The very simple 'Aarhus' trap (Fig. 2.9b) does segregate by height, and is cheap and easy to use, but because it fills within minutes and does not rotate, is useful only for short-duration studies in small areas. An omnidirectional trap for use in estimating total transport in coastal dunes was developed by Arens and van der Lee (1995), but was found to have low efficiency, presumably because of deflection of flow.

A promising mechanical trap has been used by Butterfield (1993). It collects sand in a shallow tray which is designed to have very little effect on the flow, but it cannot rotate, does not segregate by height and fills quickly. It is thus only suitable in wind tunnels or for short periods in the field. If the electronics are available (as in Butterfield's wind-tunnel study), a load cell can be attached to the trap to permit sand flux to be recorded continuously (Fig. 2.10). Another wedge-shaped trap has been developed by Nickling and McKenna Neuman (1995). At the low-tech, yet apparently adequate, end of the spectrum of mechanical devices is a pit, cut across the path of the wind, in which collecting trays can be inserted. A 3 m wide and 0.5 m deep pit was used by Greeley *et al.* (1994) in a study of aeolian sand movement on a beach.

Electronic devices are as yet in the early stages of development. Some of these use the principal that light is extinguished by saltating sand. Others record the sound made by the impact of saltating grains (an example being the 'saltiphone') (Spaan and van den Abeele 1991) or electronic pulses as they hit a piezoelectric device, as in the 'Sensit' (Fig. 2.2) (Stockton and Gillette 1990).

The problems in calibrating these devices and their expense have precluded any comprehensive empirical

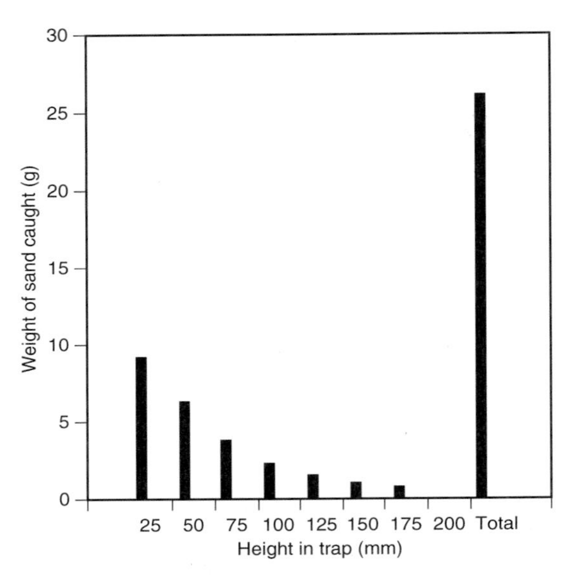

Fig. 2.11 The vertical variation in the volume of sand flux.

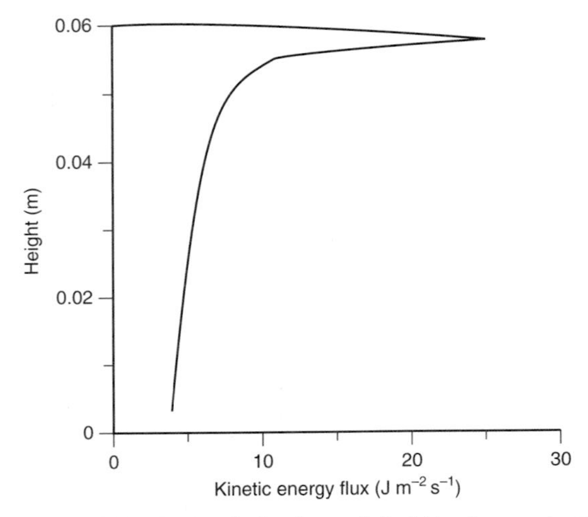

Fig. 2.12 Vertical variation in modelled kinetic energy of saltating grains (after Anderson 1986).

study of sand transport. Results are available only for small areas and over a short time periods, and few are strictly comparable one with another.

The vertical pattern of sand flux

The pattern of sand flux above the bed has geomorphological significance, for it can influence the pattern of abrasion on objects in the path of the grains. The exponential decrease in the mass flux of particles with height is well established. Butterfield's (1991) wind-tunnel study found 79 per cent of sand to be travelling below 0.018 m. Some results of a study of the vertical pattern of sand flux collected by using the Aarhus trap (discussed on page 21) are given in Fig. 2.11. The height of travel varies with the quality of the rebound surface. The pattern has now also been adequately modelled (Anderson and Haff 1988; Werner 1990).

An exponential pattern, however, does not apply to the vertical distribution of kinetic energy (which is the real control of abrasion), as the curved profiles of surfaces eroded by wind strongly hint (Chapter 3). Modelling confirms that the kinetic energy flux for saltating sand peaks above-ground (Fig. 2.12). If the model is set using $u_* = 1.0\,\mathrm{m\,s^{-1}}$ and particles of $250\,\mu\mathrm{m}$ diameter, the maximum energy flux is at about 0.8 m; with a mixture of grain sizes the peak is lower. The height varies partly because of differing rebound qualities of the surface. The modelling is confirmed by field measurements. The main reason for this pattern is that particles in saltation spend most of their time near the top of their trajectories (Anderson 1986). Anderson found that the maximum kinetic energy flux for suspended grains came at a much greater height, and was not as peaked as for the saltating particles. For both types of movement, kinetic energy was related to u_*^5.

The directionality of sand flux

The directional pattern of sand flux is a problem almost unique to aeolian geomorphology, for no other system of sediment transport has such variability in direction, few indeed having more than very slight variations about one dominant direction. The great variability of directionality in aeolian systems is, of course, due to the great variety of wind climates discussed in Chapter 1. There have been several attempts to categorize this variability, most of which are based on diagrams of the frequency of winds blowing from different compass directions over a year. These frequency diagrams are called *wind roses*.

Aeolian geomorphologists, however, must convert the data that are used to create a wind rose to show the potential of the wind to move sand or dust from different directions, thus converting the wind roses to *sand* or *dust roses*. The conversion involves equations, such as Lettau and Lettau's (equation (2.1), above), which relate wind speed to potential sediment transport rate.

The method most frequently used is that described by Fryberger (1979). Here, the potential maximum amount of sand which could be moved by the wind in a year is called the *drift potential* (DP) and is expressed in *vector units* (VU). Drift potentials for all compass directions can be used to calculate the overall net movement, expressed as an amount, the *resultant drift potential* (RDP), and a direction, the *resultant drift direction* (RDD). Fryberger calculated the potential sand drift by using an adaptation of the Lettau and Lettau equation (2.1) in the form:

$$Q \propto V^2(V - V_t) . t$$

where Q is directly proportional to the potential sand drift, V is the measured wind velocity, V_t is the threshold velocity for sand movement, and t is the duration of the wind of speed V, expressed as a percentage of the total time. It was because the expression is proportionate, that Fryberger used the term 'vector units' (VU) for the magnitude of the drift potential (DP). It is important to note that Fryberger's calculations used knots as the unit for wind velocity.

The Fryberger method, though useful at a general level, does have a number of pitfalls, recently highlighted by Bullard (1994). For example, where data were not complete, Fryberger (1979) used a linear regression to relate rate of sand drift to drift potential. Bullard's work on data from Kalahari meteorological stations showed that, because the pattern of magnitude and frequency of wind speeds varies from station to station, each station has its own relationship between sand drift and drift potential. It may not therefore be valid to use a single, averaged regression equation to transform the data.

With these reservations, wind data can be used to distinguish two main characteristics of wind climates. The first is the *wind energy*. Fryberger (1979) distinguished low-, intermediate- and high-energy environments according to their drift potential in vector units (Table 2.1). High-energy environments existed where drift potential was greater than 400 VU; intermediate energy wind environments were those where drift potential was between 200 and 399 VU; low-energy environments were where drift potential was less than 199 VU. In his 13 data sets for desert regions, Fryberger found only two that were high-energy environments, six that were intermediate-energy and five that were low-energy wind environments.

The second useful characteristic that can be taken from wind data is the pattern of *wind directions* experienced throughout the year. Fryberger described three main wind direction regimes which, with subdivisions of two classes, gave a five-class scheme of commonly occurring regimes (Fig. 2.13). In 'unimodal' regimes winds blow more or less from one direction throughout the year. If the distribution is restricted to a 45° arc of the compass the distribution is termed 'narrow unimodal' (Fig. 2.13a): a greater spread of a unimodal distribution is termed 'wide unimodal' (Fig. 2.13b). Distributions with two modes are termed 'bimodal'. These are subdivided between those where the angle between the two modes is less than 90°, which are termed 'acute bimodal' (Fig. 2.13c), and those in which the angle between modes is greater than 90° which are 'obtuse bimodal' (Fig. 2.13d). 'Complex' describes any regime with more than two modes or where no modes are easily distinguished (Fig. 2.13e). An index of directionality can be calculated by dividing the resultant drift potential by the drift potential. Values of RDP/DP are always between 0 and 1: a value of 1 would indicate a perfectly unimodal regime; values close to 0 indicate very complex regimes.

These distinctions between wind regimes, both by energy level and by directional regime, are important in the discussion of processes in later chapters; in particular, the pattern of wind directions is a major control of desert sand dune type (Chapter 5). Fryberger warned against using a single data set to represent the conditions in a large area, pointing to high DPs in northern Mauritania but intermediate and low values further south near the Senegal River. Work by Lancaster *et al.* (1984) and Lancaster (1985a) in the Namib and Bullard (1994) in the Kalahari illustrated the great variability of wind regime within quite short distances. Bullard also showed that wind energy levels varied at a station from year to year. Clearly, considerable care must be exercised when representing the wind regime of an area, for there is rarely an adequate network of meteorological stations, a long enough run of data, or data that have been collected in a reliable manner.

Ripples

Ripples (Fig. 2.14) are the smallest of aeolian bedforms, and as such have much in common with dunes. However, there are two good reasons for discussing them separately from dunes and in close conjunction with sand movement: they are a distinct class of bedform, whose sizes rarely if ever overlap with those of dunes; and their formation is closely

Table 2.1 Fryberger's (1979) examples of wind energy environments from a number of desert locations, expressed as drift potentials in vector units (see page 23).

Desert region	Number of stations	Jan.	Feb.	Mar.	Apr.	May	June	July	Aug.	Sept.	Oct.	Nov.	Dec.	Annual drift potential
High-energy wind environments														
Saudi Arabia and Kuwait (An Nafūd, north)	10	35	39	52	54	51	66	49	33	20	18	16	25	489
Libya (central, west)	7	40	42	48	64	51	41	20	18	24	24	22	37	431
Intermediate-energy wind environments														
Australia (Simpson, south)	1	43	40	27	17	13	10	18	26	52	56	46	43	391
Mauritania	10	45	49	45	38	33	40	26	19	20	20	19	30	384
USSR (Peski Karakumy, Peski Kyzylkum)	15	39	41	43	43	33	25	22	21	23	23	24	29	366
Algeria	21	21	27	37	48	32	27	18	13	15	16	16	23	293
South-West Africa (Namib)	5	8	2	6	17	13	50	19	22	27	44	17	12	237
Saudi Arabia (Rub' al Khali, north)	1	23	28	53	32	20	30	—	—	—	1	7	7	201
Low-energy wind environments														
South-West Africa (Kalahari)	7	14	11	8	10	9	11	18	24	26	26	17	18	191
Mali (Sahel, Niger River)	8	9	12	14	12	19	22	15	9	10	5	5	7	139
China (Gobi)	5	9	11	16	23	20	11	7	5	5	5	7	8	127
India (Rājasthān, Thar)	7	2	2	5	5	10	21	19	9	5	2	1	1	82
China (Takla Makan)	11	3	2	9	16	16	9	9	5	4	5	2	1	81

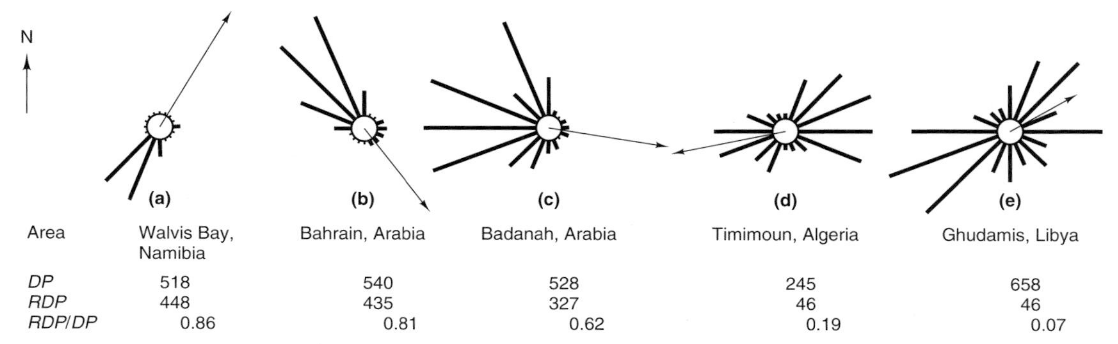

Area	Walvis Bay, Namibia	Bahrain, Arabia	Badanah, Arabia	Timimoun, Algeria	Ghudamis, Libya
DP	518	540	528	245	658
RDP	448	435	327	46	46
RDP/DP	0.86	0.81	0.62	0.19	0.07

Fig. 2.13 Sand roses showing five patterns of directional variability: (a) narrow unimodal, (b) wide unimodal, (c) acute bimodal, (d) obtuse bimodal, (e) complex (after Fryberger 1979).

linked to the modes of sand movement described in this section (thus requiring repeated reference to the same literature, as will be seen).

Ripples cover almost all dry, bare, sandy surfaces. They are absent only in four situations, which cover very small areas, individually and in aggregate: where there is very coarse sand; at high u_* (so high that this effect is rare in nature); where there is grain-fall into local areas of low wind velocity (as on gentle lee slopes); and on actively avalanching slip-faces. Most ripples are short-lived and travel much more rapidly than dunes. In a wind tunnel, ripples in well-sorted sand take less than 10 min to reach equilibrium with a new wind condition (Seppälä and Lindé 1978), though in the field (as will be seen) some mega-ripples may take years to develop and last for centuries.

Where ripples do occur, their wavelengths are between a few centimetres and tens of metres, and their heights vary from less than 0.01 m to about 0.30 m. Mean wavelength is strongly related to u_*, but as u_* increases, so does the range of wavelengths, and at high velocities many small, secondary ripples cover the large ones (Seppälä and Lindé 1978). Like dunes, ripples have gentle windward slopes (in general between 8° and 13°); lee slopes are up to 30° (Werner *et al.* 1986). Some ripples have sharp brinks; some are smooth. Most ripples are aligned roughly at right angles to the direction of the wind that formed them, although on sloping surfaces where the downwind component of grain movement is supplemented by gravity, they are slightly flow-oblique (Howard 1977).

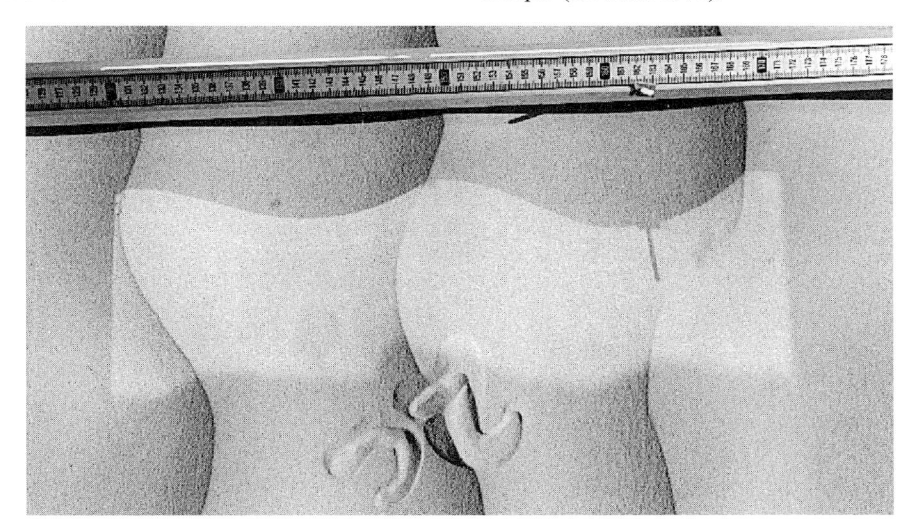

Fig. 2.14 Ripples on bare, dry sand. The photograph shows the technique of Werner *et al.* (1986) in which the shadows indicate the ripple profile. The darker, coarser grains on the crest of the ripple are evident. (Photo: J. Lenthall.)

Fig. 2.15 Mega-ripples (or granule ripples).

The controls on ripple form are well established. Wavelength and height are very obviously related to grain size: ripples in coarse sand are more widely spaced than those in fine sand (Seppälä and Lindé 1978). Sinuosity increases with grain size and wind speed. In winds of a constant speed, ripples of coarse sand are more curved than those of fine sand; in the same sized sand, ripples in strong winds are more sinuous than those in gentle winds. The slope of the underlying surface also has an effect: if all else is held constant, ripples have longer wavelengths on steeper wind-facing slopes. Wavelength is smallest on the steep downwind slopes (Werner *et al.* 1986).

The most common and striking sub-type is the *mega-ripple* (Fig. 2.15) among which wavelengths may reach 25 m. Mega-ripples are invariably composed of coarse sand, and are usually more symmetrical in cross-section than smaller ripples, perhaps because their great age means that they can be fashioned by winds from different quarters (Greeley and Iversen 1985). They may take anything between hours and centuries to form (Bagnold 1941; Sakamoto-Arnold 1981), depending on local wind conditions.

Wet conditions, as on exposed beaches exposed by the tide, or recently rain-soaked surfaces, develop 'adhesion' and 'rain-impact' ripples (Clifton 1977; Hunter 1980). '*Chiflones*' or 'sand streams' are wind-parallel features, which though widely observed, are the least researched of all ripple-sized phenomena (Simons and Eriksen 1953).

Ripple-forming processes

Saltation drives creep and reptating sand up the windward slope of a ripple, the rate increasing from the trough towards the crest; it then declines somewhat as the slope levels out. The increasing discharge rate stimulates the departure of finer grains, so that the surface becomes covered with coarser sand (Sharp 1963; Tsoar 1990a; Anderson and Bunas 1993). The almost universal coarseness of the sand on the crests (compared with sand in the troughs) is probably due to a decline in the creep rate at the crest and in the lee of the crest, where the bombardment by saltating grains is less intense (Bagnold 1941; Willetts and Rice 1989).

Coarse sand eventually rolls over and accumulates on the windward face. This process, though never involving avalanching, creates foreset bedding, as on dunes (Chapters 5 and 7). The ripple moves forward with the wind, much as do dunes, by erosion on the windward slope and accumulation in the lee. This movement incorporates the foreset beds into the body of the ripple, where they form its bulk, though these foresets are rarely preserved in ripple strata, which are composed of thin surface-parallel laminae. Forward movement shakes down fine sand, which accumulates in the core (Sharp 1963; Hunter 1977b). Ripples in uniform sand adjust quickly to new conditions; those of coarse sand adjust to stronger winds by a flattening of the crest, perhaps as the coarse grains on the crest go into saltation (Jensen and Sørensen 1986).

Like dunes, large ripples move more slowly than small ones (Seppälä and Lindé 1978). Sharp (1963) found a good fit of data to the formula:

$$U_{rm} = (U_4 - 15.5)/7$$

where U_{rm} is ripple celerity, and U_4 is wind velocity 4 ft (1.22 m) above ground. Hunter and Richmond (1988) found that ripples migrated about 5 m during a day in a moderate sea-breeze.

Hypotheses

It is surprising that a phenomenon as widely known and observed and as apparently simple as aeolian ripples should resist understanding for so long. The earliest hypotheses drew analogies between ripples on the surface of dry, loose sand and those on water, both of which occur when the wind shears across an interface between materials with different densities. But these hypotheses required the sandy material of the body of the ripple to act as a fluid, and because this is a barely tenable position, the model now has very few adherents (Kennedy 1969).

A somewhat more credible, but still unpopular model sees the curtain of saltating sand as the fluid whose upper boundary (with clear air) is deformed into waves, according to the Helmholtz equations. These waves are then transmitted to the surface beneath through the saltating curtain where they create rhythmic ripples (Brugmans 1983). The saltation curtain certainly has some of the properties of a dense fluid (Raupach 1991), but whether there is any true deformation of the upper boundary and whether this has any effect on the ground surface has yet to be proved. Yet other wave hypotheses invoke wave-like instability in the boundary layer above the bed, which is then transmitted to the surface. This notion has a long history (de Félice 1955; Wilson 1972a, b; Folk 1976), though little empirical support.

The most popular model over the last half-century, at least in the English-speaking world, has been the ballistic model (Bagnold 1941). It was based on a belief that there was a correspondence between ripple wavelength and the prevailing length of jumps in the saltating sand. The model postulated a chance distortion of the bed (of which there are many, even in a wind tunnel). More sand would be ejected from the windward slope of this bump than from the lee, and most of this would land one saltation path downwind to form an initial ripple. Saltation from this second site would land downwind to form the next ripple and so on. Saltation paths being believed to be much the same length where sand size and wind speed remained the same, the ripples formed in this way would have a regular wavelength.

The ballistic hypothesis had many attractions. It apparently explained the relation between wind speed and ripple wavelength, following the relationship between wind speed and saltation length. It also apparently explained the greater wavelength achieved by ripples with coarse sand at the crest: coarse sand, it suggested, provided a better rebound surface for saltation; this increased the length of saltation paths, and the wavelength of the ripples was adjusted accordingly. This last process led eventually, in this model, to the formation of mega-ripples (Ellwood *et al.* 1975).

There were always doubts about the ballistic hypothesis, even with its originator. Bagnold (1941) himself and later Greeley and Iversen (1985) described a class of ripple ('aerodynamic ripples') in fine sand under high winds that had wavelengths that were much longer than the saltation path-length. Further early doubts were raised by Sharp (1963), who noticed that, in a constant wind and with the same sand, ripples started small and increased in size, a sequence he believed to be incompatible with Bagnold's version of the hypothesis. Sharp believed that the parallel growth of height and wavelength showed that spacing was controlled more by the height of the ripple and the angle of the descending grains. Folk (1977) also pointed to the common habit of transverse ripples of meeting in Y junctions, a phenomenon he believed to be hard to explain with the ballistic hypothesis.

Much more fundamentally damaging to the ballistic hypothesis is the lack of evidence that ripple wavelength corresponds to saltation length (Walker and Southard 1982; Willetts and Rice 1989). Folk (1977) pointed out that a remarkably narrow peak in the distribution of saltation lengths would be needed if the ballistic process were to operate, and Anderson and Hallet's (1986) model of saltation produced no such peaks. Moreover, Anderson's (1987b) model showed that saltation trajectories of the length of ripple wavelengths did not have sufficient energy to drive reptation or creep, and could not, therefore, play a significant role in ripple formation.

It appears that Bagnold's picture of a 'rhythmic barrage of grains travelling trajectories equivalent to the ripple wavelength is not a correct image of the process' (Anderson 1987b: 954). Though it did have some attractive explanatory powers, the surprising thing, in view of the tenuousness of its assumptions, is that it became so widely accepted.

The best-developed model for ripple formation at present is that of Anderson (1987b). The model begins, like Bagnold's, with an initial irregularity in the bed. This generates perturbations in the population of reptating grains. Given a number of reasonable assumptions and an exponential (or gamma) distribution of reptation lengths, the model gives repeated ripples after about 5000 saltation impacts. The modelled ripples have a strong peak in wavelength in the order of six mean reptation lengths, which is realistic and has a reasonable relationship to u_*. The ripples grow in wavelength, as Sharp observed, and increase in size as small ones collide and merge. The reptation hypothesis is supported by the experiments of Willetts and Rice (1989), who found that as ripples grew, the length of hops of grains in reptation grew on upwind slopes and decreased on the downwind ones, producing rhythmic fluctuations and asymmetric ripple shapes. They found that the vertical component of velocity was greater in grains being ejected from the windward slopes of ripples, and the creep rate increased on these slopes. This suggested to them that ripples could even increase the total transport rate above that on a flat sand bed. Thus, ripples, like dunes, may be an 'equilibrium' response to transport processes (Chapter 5).

Anderson's early model involved the unrealistic assumption that the sand was all of one grain size. Anderson and Bunas (1993) have now developed a model which seems to show how coarse grains accumulate on the summits of ripples. The model incorporates two grain sizes (with a coarse : fine ratio of 1 : 2) which are fired at the bed one by one in a wind field that is compressed over the ripple crests (this, significantly, being necessary for ripples to develop in the model). The grains are ejected at velocities and travel on pathways that are determined according to some of the findings about saltation discussed above. The model shows, like the earlier one, that small initial undulations on the bed grow under this regime and become well-developed, regularly-spaced ripples after 20 million impacts or 'several minutes' of real time; in this case the model also shows that the ripples become coarser at their crests, as in reality.

Anderson and Bunas believed that their model confirmed earlier observations, mentioned above, by showing that impacts on the windward side of a ripple ejected more fine than coarse grains, so coarsening the surface. This sorting effect is greater for fine (low energy) than for coarse (high energy) impacting grains. The coarse grains in creep and reptation could not apparently escape the windward slope because of the lack of impacts in the lee of the ripple (also confirming the old hypothesis discussed above). It would be a tall order to expect that these models should explain Folk's 'Y' junctions or flow-parallel features, like chiflones, which will need much more study. But there is no doubt that the new models are a huge advance in the understanding of ripples.

Conclusion

This chapter has shown that research on sand movement, at the core of aeolian geomorphology, has made spectacular advances in the recent past, and that this gives us much greater confidence in extrapolating to larger-scale phenomena. But the discussion also shows that there are still vast uncertainties. The greatest advances have been in theory (especially in modelling) of sand movement and ripples, and there have also been major advances in studies in wind tunnels in these areas. But even here and even in such fundamental areas as the measurement of the roughness length, shear and the relation between the transport rate and wind speed, there remain enormous gaps. It is field observations, however, that are most in need of attention, for even in something as basic as the measurement of sand flux, there are very great inadequacies. Ripples have a long history of study, with many false starts, but with a clutch of new promising models. Though these bring us closer to an understanding than ever before, some aspects of ripple behaviour are still elusive.

Further reading

Some of the general issues concerning particle entrainment and transport in fluids are covered by Allen (1985: particularly Chapters 1, 3 and 4) and by Statham (1977: Chapter 6). Movement of sediment by wind is discussed by Greeley and Iversen (1985: Chapters 2.4 and 3), Pye and Tsoar (1990: Chapter 4) and Cooke *et al.* (1993: Chapters 17, 18 and 19), and in recent papers by Anderson (1986, 1987a, b; 1989), Anderson and Bunas (1993), Anderson and Haff (1988, 1991), Nickling (1988), Sarre (1987, 1988, 1990), and Sherman and Hotta (1989). The description of wind environments for aeolian geomorphology is covered by Fryberger (1979) and its application described by Breed *et al.* (1979a).

CHAPTER 3 Wind erosion

Introduction

Although the best-known and most widespread geomorphological imprints of the wind are depositional, the wind is also responsible for considerable amounts of erosion. The material deposited as loess or used to build dunes was, after all, originally eroded from somewhere, although not always initially by the wind. This chapter is concerned with the features left behind after material has been removed by the wind, and with the features created by bombardment with wind blown sand.

Figure 3.1 is a scheme of wind erosion processes and landforms. *Deflation* involves the removal of loose material from an area, while *abrasion* is the wearing down of more cohesive material by bombardment with wind-transported particles, generally sand. A third process is *attrition*, which is the comminution of grains in transport by their impact one on another: this third process is geomorphologically relatively less important. Between them, and sometimes in concert, deflation and abrasion create a number of landforms which include some very distinctive features, such as yardangs and ventifacts, but which also include pans and stone pavements where wind erosion is only one of a number of suggested origins and where its role is not always clear.

The controls on erosion

There are two sets of control on erosion by any process: the potency of the process (known as its *erosivity*); and the susceptibility of the surface (known as its *erodibility*). In aeolian geomorphology, erosivity is a function of wind energy and the time over which it is manifest.

Erodibility – the degree to which the surface is susceptible to erosion – is a function of several different characteristics. First, as explained in Chapter 2, is the grain size of the surface material. Figure 2.6 presents the classic threshold curve which shows that material of about 100 μm is the most liable to erosion. But erodibility has many other components, one of the most widespread of which is the degree to which the surface is free of vegetation because the presence of vegetation increases surface roughness and acts to slow the wind speed at the ground surface (Chapter 2). There is little aeolian activity in the humid tropical or temperate lands, largely because most soils are covered by vegetation. It is only where it has been cleared, as on agricultural fields, or where it is kept clear by waves and tides on the coast that there is a problem of wind erosion in these environments. Conversely, the absence of vegetation in deserts is the primary reason that the wind is so active in these environments. Other components of erodibility are roughness (Gillette and Stockton 1989), topography (which can provide wind shadows), soil moisture and surface crusts, all of which are discussed at greater length in Chapter 2. Attempts to assess erodibility, especially of agricultural soils, are discussed in Chapter 9.

The processes of wind erosion

Deflation

Deflation (from the Latin *deflare*: 'to blow away') is the net removal of material by the wind. Deflation operates effectively only in areas of unconsolidated

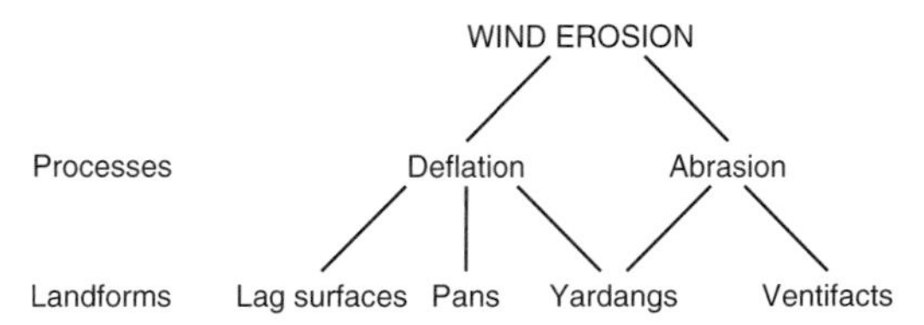

Fig. 3.1 Processes and landforms associated with wind erosion.

dust- and sand-sized particles. As one can deduce from the discussion in Chapter 2, the most favourable dust-producing surfaces are areas of bare, loose sediment with substantial amounts of sand and silt, but little clay, for clay provides too much cohesion and the absence of sand reduces the effects of bombardment. Gillette *et al.* (1980, 1982) ranked common desert soil surfaces in terms of their erodibility by deflation in order from the most erodible to the least erodible thus: disturbed soils, sand dunes, alluvial and aeolian sand deposits, disturbed playa soils, skirts of playas, playa centres, desert pavements.

Deflation can seldom proceed very far, for there are usually inhomogeneities in the eroding material or in the surface topography, and these sites form the nuclei of residual features. The most important deflational features are pans and stone pavements, but deflation may also be important in the formation of some yardangs. Most often where deflation is involved it works alongside other processes, either aeolian abrasion or non-aeolian processes.

Abrasion

Sand-sized material in saltation can impart considerable energy upon impact with other grains or rock surfaces. When the object of this bombardment is an individual clast it is called a *ventifact*, while features of the scale of landforms carved from bedrock by the impact of travelling grains are called *yardangs*. For both ventifacts and yardangs there is some debate about the nature of the abradant, and about the relative importance of abrasion and other processes, including deflation, in their formation.

Landforms of wind erosion

Although there are many landforms that can be attributed to wind erosion, their origin is rarely unambiguous or simple. Some features are the result of several different processes acting together, and for others the same landform can be created by several different combinations of processes, some of which are non-aeolian. The landforms of wind erosion are here dealt with in ascending size order.

Ventifacts

The bombardment of individual clasts of rock by wind-borne particles produces ventifacts. They are commonly pebble-sized (only a few centimetres across), but occasionally boulder-sized (more than 2 m across) (Fig. 3.2). Ventifacts were first described by Blake (1855) and the term 'ventifact' was first used by Evans (1911).

Smooth facetting of the surface is usually considered characteristic of ventifacts, but many display pits, flutes or grooves rather than polished surfaces (Whitney 1983; Laity 1994). Pits are closed depressions in the ventifact surface, often created by exploitation of pre-existing pitting such as the vesicles in basalt. Flutes are open at one end but closed at the other, mostly opening downwind, while grooves are open at both ends. Those ventifacts that do have smooth, wind-faceted surfaces may have a number of faces (up to 20 according to Higgins (1956)), but commonly have only two or three. The number of edges, or keels, is sometimes indicated by the German terms, *einkanter, zweikanter* and *dreikanter* meaning one-, two- and three-edged, respectively.

Some of the most impressive ventifacts occur in environments where large particles can be carried by the wind. One of these is the polar, periglacial environment, where the air is denser, because of its low temperature, and where, in combination with very high wind speeds it can carry larger particles (Selby *et al.* 1973; Whitney and Splettstoesser 1982; Hall 1989). Miotke's (1982) wind-tunnel experiments suggested that the potency of wind erosion was such that ventifacts might form in these environments in decades or, at most, centuries. He reported simulated rates of 5–20 mm yr^{-1}, while the summary by Greeley *et al.* (1984) showed abrasion rates between 10^{-4} and 10 mm yr^{-1}.

The form of ventifacts, such as the height distribution of the abrasion and the multiplicity of facets, has attracted attention for over a century, but the problem is still open to debate. The height-distribution problem appears to be nearer to solution than some of the others (Chapter 2). Observations show that the average height of maximum abrasion occurs

(a)

(b)

Fig. 3.2 Ventifacts: (a) Faceted dolerite ventifacts near Swartbank, central Namib Desert; (b) a 'dreikanter' ventifact from Oman.

somewhere above the ground; for example, at 0.24 m above-ground on telegraph poles after a severe blow in California (Sakamoto-Arnold 1981), or more generally between 0.07 and 0.50 m in the Mojave (Clements *et al.* 1963). These observations accord with laboratory experiments (Suzuki and Takahashi 1981), and with theory (Chapter 2).

Multifaceting is much more open to debate. Research over the years has thrown up at least five different hypotheses. The first, though now generally dismissed, is that it relates to the original, un-abraded form of the pebble (Higgins 1956; Kuenen 1960). It is now thought that the original shape only has an effect in the early stages of abrasion (Greeley and Iversen 1985). Another obvious recourse for theory is that the ventifacts have originated in multi-modal wind regimes or have been subject to successively different wind regimes (Kuenen 1960). The passage of mobile dunes, or the appearance of ephemeral features like bushes, could produce changes in the ground level wind regime near a ventifact even in an ambient wind regime that was unidirectional. Although variations in wind direction, for whatever reason, may explain some faceting, many multifaceted ventifacts are said to occur in unidirectional wind regimes, as in mountain passes (Sharp 1949); moreover multifaceting has been created in laboratory experiments in unidirectional winds (Schoewe 1932).

A more acceptable explanation is that faceted ventifacts have been repeatedly swivelled or overturned. Overturning is strongly suggested by the fact that where large and small ventifacts occur together, large, less-mobile ones are single-faceted, while most smaller, movable ones are multifaceted (Lindsay 1973). Sharp (1949) and Whitney and Splettstoesser (1982) found overlapping sets of flutes and grooves of different orientation on the same ventifact, which are yet another indication of rotation. Yet further evidence for overturning comes from dated sequences of ventifaction. In an Antarctic sequence, small ventifacts on the youngest exposures have many facets, presumably because abrasion has not yet eliminated the shapes of the original pebbles. Where exposure has been slightly longer, ventifacts have fewer facets, as the original shapes are eliminated. On the oldest exposures, the number of facets on pebbles increases again, and there is a decreasing relationship between facet orientation and wind direction, presumably as the probability of overturning has increased (Lindsay 1973). Overturning can also be achieved by wind erosion itself, if the pebble is resting on loose sand or soil (Sharp 1964; Mattsson 1976), or where winds are strong, as in Antarctica (Selby *et al.* 1973) and in ancient glacier-margin environments. Solifluction, frost heaving, swelling and shrinking clays, floods, or kicking and tunnelling by animals could also be responsible for rotation. None the less, rotation cannot explain consistent multifaceting (with systems of faceting facing similar directions on many pebbles), nor instances where multiply faceted ventifacts are imbedded in cemented rock where overturning is impossible (Higgins 1956).

Most controversially, some workers have suggested that facets, as well as pits, flutes and grooves, have been abraded by dust swirling round the clast, and abrading faces not at right angles to the wind. One argument has been that pitting and grooving is too intricate to have been produced by sandblasting. Abrasion by dust was Higgins' (1956) explanation for the faceting of supposed ventifacts embedded in a cliff face, although he later conceded that these might not have been ventifacts at all (Laity 1994). Others have produced further field evidence that dust and snow have eroded ventifacts (Whitney and Splettstoesser 1982). Whitney and Brewer (1968) even claimed that airborne ions could abrade. The strongest support for dust as the abradant of ventifacts has come from wind tunnel experiments by Whitney and co-workers. Whitney and Splettstoesser (1982) quoted experiments that appeared to show that erosion occurred on down wind facets that could not be reached by saltating sand. Dust was carried to these facets by complicated secondary flows. Others disagree, maintaining that saltating snow and suspended dust have little abrasive power compared to saltating sand (Greeley and Iversen 1985; Anderson 1986; McKenna-Neuman and Gilbert 1986; Laity 1994). The conclusion of Breed *et al.* (1989) was that most ventifact shapes and textures could be explained by impact-face sandblasting supplemented by fine-particle abrasion of all surfaces by subsidiary wind currents.

Yardangs

Bombardment is also responsible, at least in part, for the formation of yardangs, although in some yardangs there is also a considerable element of deflation, and there is some debate about the relative importance of these two processes. Whereas ventifacts are individual clasts, yardangs are cut from bedrock, and are therefore larger features (Figs 1.1a and 3.3). Perhaps because of the backlash against the views of eolianists such as Keyes (see page 37), yardangs have received remarkably little attention from geomorphologists.

The term 'yardang' was introduced by Hedin (1903) as the name given locally to narrow ridges in Chinese Turkestan. They appear in all the major desert regions except Australia, and although they are not a widespread landform, they are one of the very few which are unique to deserts. Major yardang locations include Rogers Lake in the Mojave Desert (Blackwelder 1934; Ward and Greeley 1984), Turkestan, central Asia (Hedin 1903), Kalut in Iran (Gabriel 1938), the Namib Desert (Stapff 1887; Kaiser 1926; Lancaster 1984; Corbett 1993), Peru (Bosworth 1922; McCauley *et al.* 1977a,b) and Egypt (Beadnell 1909; Walther 1924; Grolier *et al.* 1980). They also occur on Mars (Greeley and Iversen 1985).

There are two classes of these wind-eroded landforms which can be distinguished by size. The smaller ones (generally less than 100 m long) are the better understood, and some workers would confine the

Fig. 3.3 Small yardangs eroded from gypsiferous sands at Fatnassa, southern Tunisia. See also Fig. 1.1(a).

term yardang to these. The larger features are sometimes termed ridges or mega-yardangs.

The shape of the smaller yardangs is often likened to inverted ship hulls, oriented into the direction of the prevailing wind, and they often occur in closely packed arrays (or 'fleets'). They may have concave downward-tapering or convex and bulbous 'bows'. The top of the yardang (the keel of the boat) is sometimes flat. The highest and widest part of the structure is generally about one-third of the way between the bow and the stern in a well-streamlined yardang, and the downwind ends of yardangs are characterized by gently tapering bedrock surfaces or elongate sand tails. Commonly quoted length : width ratios range from 3 : 1 to 10 : 1 (McCauley *et al.* 1977b), and wind tunnel experiments suggested an ideal value around 4 : 1 (Ward and Greeley 1984). However, Halimov and Fezer (1989) expressed some scepticism about any general, idealized description of streamlined yardang forms. They described eight yardang types from a study site in the Qaidam depression in China – mesas, saw-tooth crests, cones, pyramids, very long ridges, hogbacks, whalebacks, and low, streamlined whalebacks – which underlined the variety of forms developed, and which they believed represented an evolutionary sequence.

The larger yardangs – *mega-yardangs* (Cooke *et al.* 1993) or *ridges* (Laity 1994) – are altogether more substantial features, but are reported only from the central Sahara and Egypt (Breed *et al.* 1979b; Mainguet 1968), the Lut Desert of Iran (Gabriel 1938), and parts of Peru (McCauley *et al.* 1977b). The best-described of the Saharan ridges are those in the Borkou area to the east, south and west of the Tibesti Mountains in Chad (Fig. 3.4; Capot-Rey 1957; Grove 1960; Hagedorn 1968; Mainguet 1968; Peel 1968; Mainguet *et al.* 1980). Satellite imagery has revealed that these ridge systems are extensive, covering hundreds of square kilometres, although individual ridges may only extend for a few kilometres. Ridges can therefore be distinguished from yardangs either because of their size – (up to 1 km in width), or because of the extensive occurrence of a regularly repeated feature, or both.

As with other wind-eroded landforms, notably pans, an obvious prerequisite for yardang or ridge development is an erodible substrate. Consequently, Goudie's (1989) list of lithologies in which yardangs are developed was dominated by soft sediments which are moderately cohesive but erode readily, such as Cenozoic aeolian, fluvial and lacustrine sands, silts and clays, although a few harder rocks were included.

Fig. 3.4 Wind-eroded 'mega-yardangs' cut into hard sandstones in the foothills of the Tibesti Mountains in the central Sahara (after Grove 1960).

However, many of the ridges of the Sahara are developed in highly lithified sandstones and limestones, some as old as the Cambrian (Mainguet 1968).

Beyond the necessity of easily eroded material, the process of yardang or ridge development is the subject of some debate. Most yardangs occur in unidirectional wind regimes, and are usually parallel to these winds. Much of the debate centres on the relative importance of deflation and abrasion. Although an abrasion hypothesis is sometimes attributed to Blackwelder (McCauley *et al.* 1977a), he was in fact of the opinion that abrasion alone could not form such major features, and that deflation might be very important in desert areas. Blackwelder (1934) was also aware of the important role that running water might play in the surface sculpture of yardangs. Whitney (1985) even argued that where abrasion was significant these features would be removed. The conclusion of McCauley *et al.* (1977a) was that abrasion contributed to undercutting at bows and flanks, but that the aerodynamic shape, where it existed, was a consequence of deflation, and that particle dislodgement was by weathering rather than abrasion. Ward and Greeley (1984) argued that abrasion dominated the initiation of the Rogers Lake yardangs, while deflation might have combined with abrasion to maintain the aerodynamic shape. Clearly, it is unlikely that abrasion is important above the level of saltation across the surrounding plain. As with ventifacts, this debate has been taken further by the work of Whitney and co-workers (Whitney and Dietrich 1973; Whitney 1978, 1983, 1985; Whitney and Splettstoesser 1982) who have advocated abrasion by the turbulent vorticity of dust particles or even of dust- and sand-free air.

An associated feature are *zeugen* which are perched rocks created by the abrasion of material in the zone a few tens of centimetres high close to the ground in which sand transport by saltation takes place. Although early reports from deserts often featured descriptions of these perched, or 'mushroom', rocks, they are not widespread, and may be as much a consequence of weathering of the rock in the capillary fringe close to the ground as of the action of saltating sand.

Pans

Closed depressions (topographic basins) are a morphological feature which occur in a wide variety of situations, and not all are the consequence of wind erosion. However, there is a large group of features, found in most semi-arid lands, termed 'pans', which are widely believed to be at least partly formed by deflation (Goudie 1991). They are shallow depressions, most of which are periodically filled with water (Fig. 3.5).

The size of pans in south-western Australia is between 0.004 and 100 km^2, with a fairly well-defined modal size at about 0.05 km^2 (Killigrew and Gilkes 1974); in South Africa the range is between 0.05 and 30 km^2, with a mean of 0.2 km^2 (Goudie and Thomas 1985). The biggest recognized pan in eastern Australia is about 45 km across (Bowler and McGee 1978). Pans have a wide variety of plan forms; some are irregular, but most are smoothly rounded, and 'kidney' shapes recur widely. Goudie and Thomas (1985) reported densities greater than 100 pans per 100 km^2 in southern Africa. Pans are commonly developed on soft sediments susceptible to erosion, and are usually sites of closed drainage. They are associated more with semi-arid than with arid or hyper-arid climates.

One of the clearest signs that pans have experienced deflation is that most are associated with a *lunette* on their downwind margin (Fig. 3.6). Lunettes are unmistakably dunes, though composed often of silts and clays (Chapter 5). The size of a lunette is proportional to that of its associated pan: large pans are fringed by lunettes a few kilometres in length and more than 60 m in height.

Early theories of pan formation included animal 'wallows', whirlwinds, and meteor craters, but more recent work has seen deflation as the dominant process and susceptibility to deflation as the major control on their distribution. The closed depression and lunette themselves are the best evidence that there has been deflation of loose sediment from the pan, although much of the eroded material travels beyond the lunette. However, although deflation may be necessary to deepen the pan, and may be the dominant process, others must also occur. The material on the dry lake floor must be loosened, and this is generally thought largely to be the result of salt

Fig. 3.5 A pan in the Kalahari Desert, southern Africa (photo: Jo Bullard).

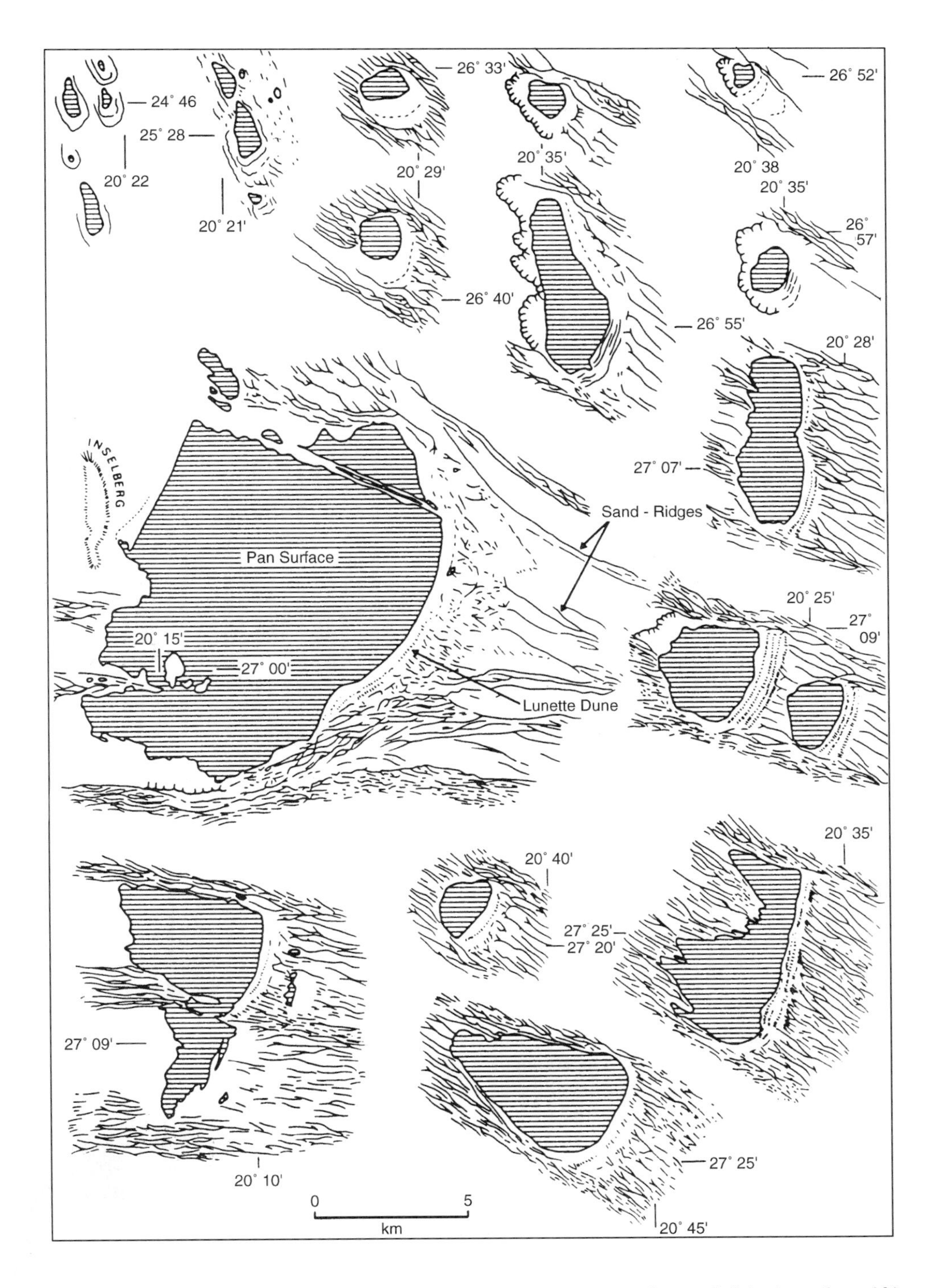

Fig. 3.6 The association of pans with lunette and other dunes in the south west Kalahari, southern Africa (after Goudie 1989).

weathering, itself dependent on the influx of salts in drainage water. In some cases there must also be washing in of sediments of the right size for aeolian deflation, for some pans rest on consolidated material that does not readily deflate or disintegrate easily by salt weathering. In some cases there has been deflation both of the underlying material and of the inwashed material. For example, Lancaster (1978) described a pattern of two parallel lunettes at the edge of some Kalahari pans, one formed of sand-sized material, the other of silt. The first lunette was evidently created with sand from the deepening pan; the second lunette was created with silt which had originated as a deposit from material washed into the pan created by the first episode of wind erosion.

None of these processes could take place unless there was some form of initial closed depression in which water collected, and this is itself associated with a number of processes, such as dunes damming water courses, intermittent flooding, as in semi-arid climates, and in some instances even the gentle tectonic disruption of a drainage pattern (Marshall 1987). In some places, as in Texas, there may also be solution of underlying limestones, both in initiating the depression, and in helping to deepen it (Osterkamp and Wood 1987). Finally, there must be wave action during the occasions when the lake fills, for the smooth elliptical form could not be the result of wind erosion. When the wind itself creates hollows, such as blowouts in coastal sand dunes, they are elongated in the wind direction, but the elongation of pans at right angles to the prevailing wind clearly suggests the involvement of wave processes (Killigrew and Gilkes 1974), for waves apparently swirl round and erode the lateral ends, rather than the one facing the wind (Carson and Hussey 1960). Wave action must also be the explanation for the smoothness of the outline of the pans (as on sandy coastal bays).

There are predisposing conditions which must be fulfilled for pans to occur. They must not be obliterated by the integration of a drainage system nor overrun by aeolian sand, and the environment must be dry enough to encourage deflation. Bowler (1986) believed that semi-arid environments were a necessary condition, for only then could there be the alternation of wet and dry conditions that would give lakes and deflation. Moreover, he believed that pans only formed in certain very specific parts of a climatic cycle, as a wet period was replaced by a dry one.

Notwithstanding Bowler's (1986) observation that pans occur in modern semi-arid environments, there are many other environments in which pan-like lakes now occur, and little doubt that wind erosion played a larger or smaller part in their formation. Some of these may be relics of former semi-arid environments. The best known of these pan-like features are the Carolina Bays, which are elliptical lake basins fringed by Bay trees (hence the name), covering an 1100 km stretch of the Atlantic coast of the USA between Maryland and northern Florida (Fig. 3.7). They are most common in the Carolinas, and may total 500 000 (Stolt and Rabenhorst 1987). Price (1968) listed 15 hypotheses that had been put forward to explain the Bays, including meteorite craters, solution holes, by submarine action, valley damming, and giant schools of fish waving their fins in unison. However, several authors have favoured an hypothesis which invokes deflation (Price 1968). What is undisputed is that the Carolina Bays are inherited from the late glacial period, when, presumably, wind erosion was encouraged by higher winds and less vegetation. Similar aligned lakes, also thought to be inherited from glacial times, occur in the Paris Basin (Matchinski 1962). Small lakes with the same elliptical shape, but where deflation has probably played a much smaller part, and which may be forming today, are found in tundra conditions, as in Alaska, northern Canada and Siberia (Black and Barksdale 1949; MacKay 1956; Carson and Hussey 1962).

Stone pavements

Stone pavements were defined by Cooke (1970: 560) as 'armoured surfaces comprising intricate mosaics of coarse angular or rounded particles, usually only one or two stones thick, set on or in deposits of sand, silt or clay.'

They are also known as *gibber plains* and *stony mantles* in Australia, as *desert pavements* in the USA, as *gobi* in central Asia, and as *hammada*, *reg* or *serir* in Arabic. They occur widely in environments where there is little vegetation, including mountain, arctic, and periglacial regions, but especially in hot deserts.

The deflation hypothesis for stone pavement genesis relies on the particle size selectivity of aeolian transport. The assumption is that the parent soil or sediment consisted of fine, dust- and sand-sized material mixed with coarser material. In this hypothesis, the wind deflates the finer material, leaving behind an increasing concentration of the coarser particles which form a lag (Fig. 3.8). This armour then acts to protect the underlying fine material from further aeolian erosion, which explains

Fig. 3.7 Map of the Carolina Bays, USA, an area of lake basins which may have been formed by wind action (after Prouty 1952).

why well-developed desert pavements come low on the susceptibility list of Gillette *et al.* (1980, 1982) above.

Despite the plausibility of this hypothesis, and its undoubted correctness in some cases, direct evidence of deflation is hard to find, and there are many other processes that can produce coarse-particle concentration on bare surfaces. These include the selective removal of fine particles by surface wash, and the movement of coarse particles from within the soil towards the surface by cycles of heating-and-cooling, freezing-and-thawing or wetting-and-drying or solution-and-recrystallization of salts. These processes are outside the scope of this book, and were covered by Cooke *et al.* (1993).

McFadden *et al.* (1987) recently suggested yet another mechanism for the formation of desert pavements which does involve aeolian processes, but this time depositional ones. Using evidence from the pavements on basaltic lava flows in the Cima volcanic field of the Mojave Desert they suggested that pavements developed in topographic depressions into which fine material was blown and coarse material was washed by colluvial and alluvial processes. The fine aeolian material landing on stony surfaces would be washed off the stones and collect beneath them, actually lifting them up by as much as 0.2 m.

Wind-eroded depressions and plains

The first geomorphologists to visit deserts, struck by the novelty of the landforms, convinced themselves that the wind was very active. Whereas rivers were the main agents of denudation in the humid climates, they believed that symmetry demanded that the wind should be the chief agent in the deserts. The best remembered of these 'eolianists' is Keyes (1912) whose model of *eolation* involved the production of low-angle plains by wind erosion. Another set of features that became embroiled in the controversy about wind erosion were large closed depressions, of which there are many in the desert realm. Keyes' ideas and those of this school (e.g. Passarge 1904; Penck 1905; Walther 1924) were undoubtedly overstated, and the clear evidence that rivers had been active in the best-known deserts, like those of the south-west United States, meant that the very suggestion that the

Fig. 3.8 A 'stone pavement' ('hamada') in central Algeria. The stones are concentrated at the surface over a less stony soil. The deflation of fine material from between the stones is only one among many other possible explanations. These include surface wash, ratcheting up of coarse material through the soil by wetting and drying, or heating and cooling cycles, and dust washing off and accumulating beneath the stones. (Photo: K. Gardner.)

wind could create more than a few yardangs became a target for the derision of geomorphologists like Davis (1905, 1954) and Cotton (1947) who believed that fluvial erosion was the key to desert landscapes. Although Keyes had very little evidence to support his model, and although there are many alternative explanations for both plains and large depressions in deserts, a role for aeolian erosion in both cannot now be dismissed so summarily.

Large enclosed basins, some of them many tens of kilometres across, are the better researched of these large-scale features. The best known are in the northern Sahara, particularly those in southern Tunisia and in the Western Desert of Egypt (Walther 1924; Passarge 1930). The best known of all is the Qattara Depression in northern Egypt, 192 000 km^2 of which is below sea level (Said 1960; 1962). The obvious alternative explanation for a depression such as this is tectonic, but conclusive evidence for either tectonism or wind erosion has been hard to find. Calculations of fluvial inputs, even to basins of fairly well established age in the western United States, show that many basins have apparently far less accumulated material than would have been supposed, suggesting that the wind must have been an important process in removing it (Blackwelder 1928; Langbein 1961). For the Qattara Depression many workers have suggested a predominantly deflational origin (for example, Walther 1924; Ball 1927), although Albritton *et al.* (1990) recently synthesized previous evidence and their own investigations to hypothesize a sequence of events which included fluvial, karstic and mass wasting processes as well as deflation. What can be said of the large North African basins, and of others of a similar nature in central Asia (Suslov 1961), is that the depressions are very old, and that salt weathering, solution and fluvial erosion certainly played large if perhaps subsidiary roles in their formation.

Supposed plains of deflation were the most scorned of the Keyes' proposals (Keyes 1909). Davis (1930) dismissed the notion almost out of hand. But the idea is not dead, and there are some plains where wind erosion seems to have played a major role. The most persistently quoted example is in south-western Egypt and the north-western Sudan (Sandford 1933). Breed *et al.* (1982, 1987) developed a model of the slow destruction of the Tertiary fluvial landscape of this area as it dried up in the Quaternary (Fig. 3.9). The wind abraded the scarps, leaving conical hills, which were themselves eventually levelled. Other candidates for 'eolation' are the 'Serir' north of Tibesti in the east-central Sahara; and the region between the Ahnet and Adrar des Ifoghas in the south-central Sahara (Mainguet *et al.* 1980).

Less level plains, but with rather more convincing evidence of deflation, are found in many deserts. The evidence lies in ridges, whose meanders and braids are

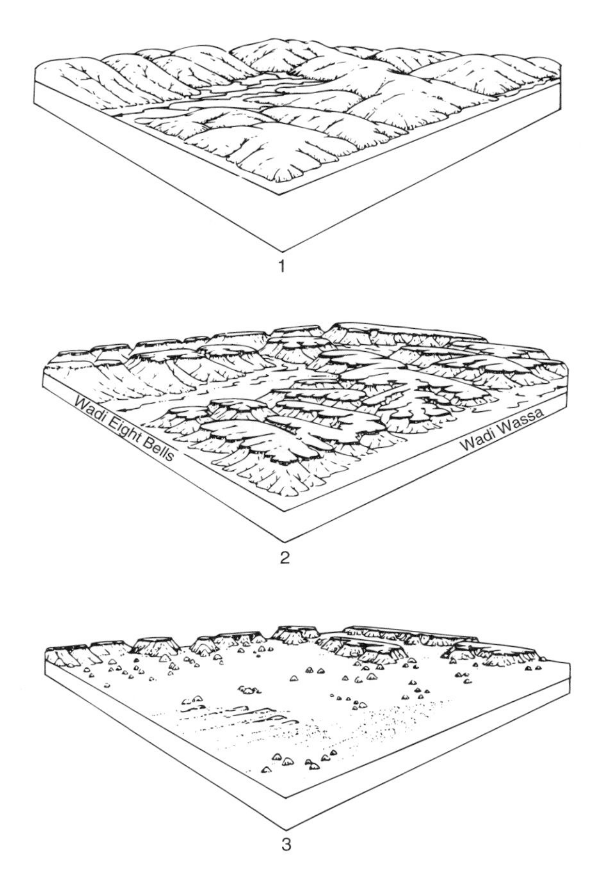

Fig. 3.9 A model of landscape development in south-western Egypt, where a fluvial landscape developed in the late Tertiary has apparently been degraded by wind erosion to form plains and remnant conical hills (after Breed *et al.* 1982).

unequivocal proof of their origins as river channels, but which are now raised up to 20 m above the intervening plain (Maizels 1990). The generally accepted interpretation is that these raised channels are preserved either by coarse channel-bed deposits or by the preferential cementation of material beneath the channel. Finer or less cemented sediments in the inter-channel areas have been deflated, with minor fluvial erosion in places. This explanation follows the pioneering observations of Hörner (1932) round Lop Nor in central China and Miller (1937) in the Nejd in northern Saudi Arabia. In both places yardangs between the raised channels are irrefutable evidence of wind erosion, though elsewhere yardangs are not so evident. The channel systems have been given various ages from Holocene to Pliocene, thereby implying a range of rates of wind erosion (Maizels 1990).

Conclusion

Early in this century there were geomorphologists who believed that the wind, allied with insolation weathering, might have a profound influence in forming erosional landscapes in deserts. This view waned in popularity as work, particularly in American deserts, suggested that water was much more important in landform genesis, even in dry environments. Most recently, however, the erosional work of the wind has begun again to be viewed as important (Goudie 1989), and the extent of wind-eroded landforms has been more widely recognized. Even so, with a few notable exceptions, and in marked contrast to depositional landforms such as dunes, there remain remarkably few studies of wind erosion processes or landform development, and this must surely be an area of research in the immediate future.

Further reading

General discussions of wind erosion have been provided by Breed *et al.* (1989), Goudie (1989), Greeley and Iversen (1985) and Laity (1994). Other reviews have covered specific wind erosion features: stone pavements (Cooke 1970), pans (Goudie 1991), ventifacts (Higgins 1956), and yardangs (McCauley *et al.* 1977a; Whitney 1985; Halimov and Fezer 1989).

CHAPTER 4 Dust

Introduction

Dust is so commonplace that it is in danger of being overlooked as one of the principal ways in which terrestrial sediment is moved and has been moved in the past. It is easy to forget that, on the global scale, the quantities of dust, though probably one order of magnitude less than the vast amounts of sediment taken to the oceans by rivers are very large indeed (Fig. 4.1). The quantities were much greater in the past, as Chapter 8 shows; indeed, because the periods when there was more dust were periods when there was less fluvial erosion, the two modes of transport were probably more comparable in rate than they are today. This Chapter discusses the origin, movement and deposition of dust, both now and in the past.

Even in contemporary north-western Europe, far from the profuse sources, aeolian dust is the sediment that intrudes most into people's lives. Between October 1987 and April 1989 there were nine events at Bochum in Germany when dust was evident on outdoor shiny surfaces, like cars (Littmann 1991a). Although this was rather more frequent than usual, there is at least one and generally more of these events at northern European sites every year. In the dry world, dust is even more mundane. Skies are obscured by dusty haze for much of the year, and shiny surfaces are shrouded many times a day. In parts of southern Israel, where Revelations 6:12 described skies darkened by dust in Biblical times, the problem has now been quantified. People can wipe up as much as $0.25\,kg\,m^{-2}\,yr^{-1}$ (Goosens and Offer 1990), and $8.3 \times 10^{-3}\,kg\,m^{-2}$ after a single storm (Ganor and Mamane 1982). This compares to only $0.07 \times 10^{-3}\,kg\,m^{-2}$ after the much rarer storms at Bochum. Yet there are still dustier places, to judge by the number of times visibility is less than 1000 m. In south-eastern Mongolia, there are over 300 such occasions in the year, compared to only about 27 in the Negev (Middleton 1991).

As well as being mundane, dust is easily defined. It is sediment that travels in suspension in the wind, following the turbulent movements of the air, unlike sand, which falls back to Earth after it has been picked up. In theory, and in practice, there are particles that behave in intermediate ways and deposits of intermediate size (Chapter 2), but the line between dust and sand is usually sharp, because of the way in which the wind rapidly separates them. It must be conceded, notwithstanding, that there are some puzzling size phenomena in dust. Large particles of mica (whose platy shape must help them to remain aloft) and perhaps other large particles generated by aggregation in clouds (Westphal *et al.* 1987) may not be unexpected. But a $75\,\mu m$ diameter particle of quartz, collected north of Hawaiian Islands, which could only have travelled more than 10 000 km from its source in Asia, is much more of an enigma (Betzer *et al.* 1988).

This everyday phenomenon (minor anomalies apart), furthermore, is composed of everyday materials, organic and inorganic. Most of the minerals are common rock and soil minerals and salts; there are also commonplace biogenic materials like diatoms, phytoliths, spores, pollen and ash. In the desert, it is true, there are some extraordinary components of what can loosely be called dust, like the 'manna lichen' which fed the Israelites in Sinai and Alexander's armies in the deserts of central Asia (Donkin 1981), but most of the organic constituents are much more common. Coarse dusts are generally rich in quartz or calcite, which are among the commonest minerals; fine dusts are mostly composed of the almost equally common clay minerals (Pye 1987). As a final measure of its banality, dust and its movements

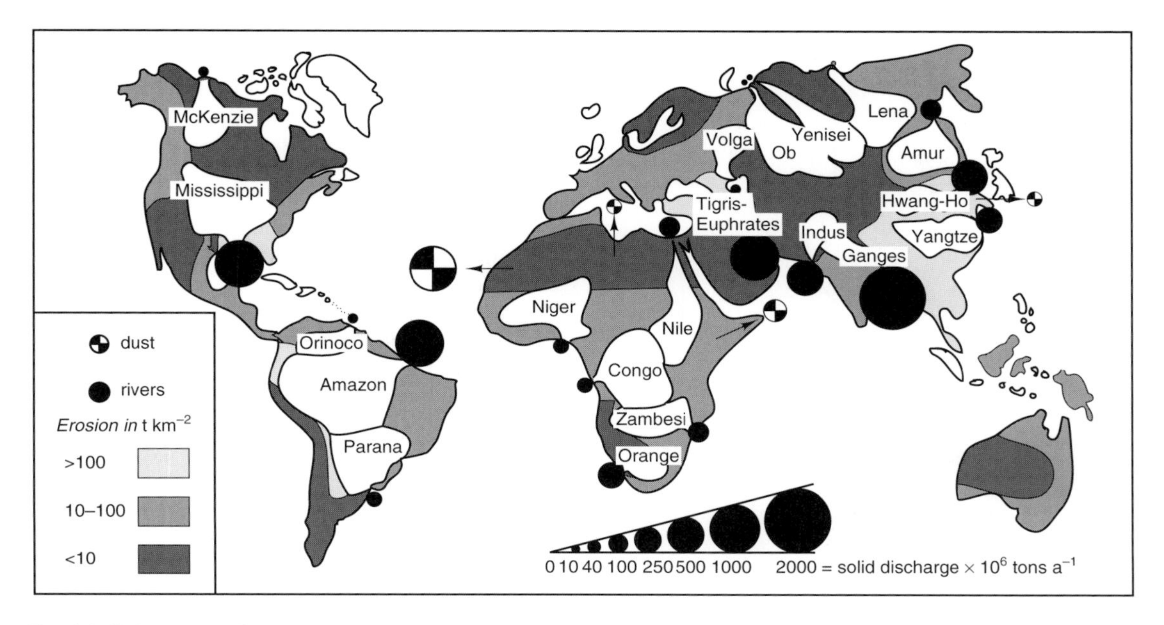

Fig. 4.1 Global rates of dust transport compared to rates of fluvial sediment production.

have been studied by scientists for a long time (Dobson 1781; Darwin 1846).

Mundane it may be, but dust still holds mysteries. Many aspects of its sources, modes of movement and deposition are still open to discovery and debate. There are many unanswered questions about the much dustier periods in the recent geological past, when huge thickness of loess were deposited. Furthermore, dust may have grave implications for the future of the global environment.

The collection and measurement of dust

The collection of dust for sedimentary, mineralogical, chemical or microscopic analysis is an easier procedure than the quantitative estimation of dust entrainment, transport or deposition. McTainsh (1986) provided a useful summary of methods of collection for qualitative analysis. One widely used method is to fly nylon or terylene mesh kites, but not only is the efficiency here thought to be only about 50 per cent, but clay-sized dust can also be missed, and the procedure cannot work at low wind velocities. Pumps with filters have also been used, and where quantitative measurement of flux is not important, these do not need to be too demanding in design, although the filters must be as inert as possible (Adetunji and Ong 1980).

There are three states in which dust can be measured quantitatively: in emission, in motion or as a deposit. These conditions are not always easily distinguished, for dust may rise, move and be deposited at a huge range of scales, from a few millimetres to many thousands of kilometres, resting on a surface for anything from a few microseconds to millions of years. Moreover, the nature of the surface has itself a huge influence on the residence time, rough or sticky surfaces retaining much more than smooth ones. Despite these fundamental problems, a number of devices have been evolved (see the reviews in Knott and Warren (1990) and McTainsh (1986)), but the methodology is quickly evolving (as will be seen below) and reviews become quickly dated. Partly because of this rapid development, there are few standard methods, so that some of the data for dust entrainment, transport and accumulation quoted below may be comparable only very approximately. Recommendations about standardization of measurement have been made (Gillette 1987), but there is still no agreed methodology or widespread network of measuring stations.

Dust emissions can be calculated from measurements of particle concentration above an eroding

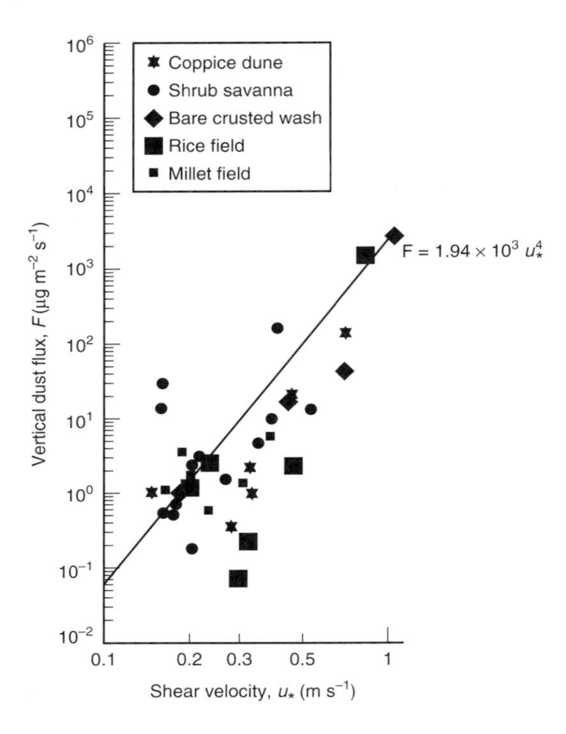

Fig. 4.2 Vertical changes in dust flux rate over different surfaces (after Nickling and Gillies 1993).

surface, collected in a series of samplers. In general, there is an exponential decline in concentration with height above the surface, and Gillette *et al.* (1972) suggested a relation between the concentration curve and the rate of vertical flux. A good relation has often been found between upward dust flux, estimated in this way, and the ambient shear velocity. In Mali measurements showed that $F = au_*{}^4$, where F is the flux rate and the constant a relates to the type of surface (Fig. 4.2; Nickling and Gillies 1993).

Measuring dust in motion can be done in two principal ways. The first is to collect it. This can be done with a simple arrangement of shallow trays, on a pole or tower (Fryrear 1986), or, more commonly, but at much greater expense, with aspirating devices that suck in dust-laden air, capturing the dust either on a filter, in a gravitational trap, or electrostatically. There are several commercial aspirators on the market (for example, Offer and Goosens 1990). The main problem with these devices, when used for quantitative measurements of dust flux, is to ensure that the volume of air that they take in is known, so that concentrations can be worked out. Devices of this type were described by Nickling and Gillies (1993). Trapping the dust can create problems if the filter interferes with the flow.

The second group of methods for measuring dust in motion depend on light-extinction. At their simplest these include measurements of visibility or of the extinction of solar radiation (as on conventional sunshine recorders). Calibrated against dust collections, these can give quite good results very cheaply (McTainsh 1980). Many studies have used visibility records from standard meteorological observations, and although these may include periods when visibility is reduced by smoke or mist, they are generally unambiguous and can be taken from widely available sources of data, although calibration is always a problem (Middleton 1989). There are now relatively cheap hand-held dust meters available from some instrument makers such as Casella.

More sophisticated equipment can be used, but expense restricts frequency. For example, ground-based (or ship-based) sun photometers measure the transformation of solar transmissions at various specific wavelengths (Jaenicke 1979; D'Almeida *et al.* 1983; Ben Mohamed *et al.* 1992). The measurements must be corrected for trace gas interference and must also be calibrated with ground-based measurements (Ben Mohamed and Frangi 1986). Quite sophisticated methods are necessary for areas with low dust concentrations. Dust densities may also be estimated from satellite imagery. Aerosol optical density can be measured from visible and infrared data from Landsat, Meteosat and other satellites (Legrand *et al.* 1989; Holben *et al.* 1991). Estimates are best over the sea or the ocean where the surface albedo is low. Columnar atmospheric densities can be estimated in various ways (for example, Dulac *et al.* 1992).

Measuring dust deposition presents more problems. For a start, it is here that the character of the surface is critical to the interpretation of the results. Moreover, atmospheric dust concentrations and fall-out rates are so highly variable in space and time that it is difficult to obtain representative samples. Dust levels fall off rapidly away from disturbed sites such as roads, and can be raised during the collection of samples. Even without disturbance, dust deposition varies very considerably from place to place, for example round large obstacles such as hills (Goosens and Offer 1990; Holben *et al.* 1991). In Niger, Chappell (1995), using ^{137}Cs (caesium) to estimate long-term patterns of dust-fall, showed that much more dust had collected in groves of trees, than on open ground. Even at the decimeter scale, different plants and the microtopography they create produce 'orders of magnitude' differences in the entrapment of dust (Oldfield *et al.* 1979).

The problem extends to the height above the ground at which collectors are placed, for this can very greatly affect collection rates, as well as the character of the dust, such as its grain size (Goosens 1985). Height can also influence the rate of trapping of contaminants, such as saltating sand and particles splashed in by rain action. There is no standard height of collection, and considering the variety of situations in which dust exists, a standard may not even be desirable. Drees *et al.* (1993) first placed their collectors at 5.0 m above ground, but lowered them to 2.5 m in response to the need for accessibility. Finally, there is the ever-present problem of contamination by vandals or birds.

Various sticky surfaces, for example vaseline or commercially produced sticky labels, have been used to trap dust, but these either clog up quickly, or need elaborate treatments to extract the dust (Clements *et al.* 1963). Bags containing moss, which, when live, is a very efficient dust trap, have been used, although these need watering, and present what may be a very unrepresentative surface (for example, in a desert) (Goodman *et al.* 1979). The simplest collector is a vessel, such as a bucket, but the wind may blow dust over the top, or whirl it around and remove it, so that choosing a size becomes a difficult compromise. The British Standard gauge has a funnel and a mesh to protect it from insects and birds (Goodman *et al.* 1979). Another design was tried by Orange *et al.* (1990). Adding water to the vessel prevents re-entrainment, although the water requires constant attention, evaporating in dry, hot areas and overflowing in wet ones, and provides what may be an unrepresentative surface.

One of the more successful methods is a tray of marbles (Yaalon and Ganor 1980) or polystyrene balls (Drees *et al.* 1993). When dust settles on the spheres, it may be washed down by rain, or may slide down between them. In theory, the spheres protect the dust from re-entrainment (and thus, inevitably, to some ways of thinking, provide an unrealistic trap). It is easy to wash the dust from between the spheres. The biggest trouble here is retaining the marbles against the predation of small children, camels, gophers or other rodents. A field test showed that water traps and trays of marbles had about the same efficiency, but that dust could be re-entrained from between the marbles, whereas it was thoroughly trapped by the water. Both were much more efficient than flat surfaces, wet or dry (Goosens and Offer 1994).

The last and probably the most representative measure of dust fall (considering its apparently high short-term rate of variation and the problem of artificial surfaces) is to estimate rates over long periods of time, as indicated by natural traps. To estimate dust-fall rates over the last few decades, the ^{137}Cs method, mentioned above, is promising, although not problem-free, because of the need to distinguish background caesium fluxes and those due to overland flow and aeolian transport. Slowly accreting deposits have other general problems. The best natural trap may well be snow or ice (Dovland and Eliassen 1976), but, of course, snow and ice have limited distribution, and snow scavenges dust (probably fractionating it as it does so), so that snow-covered areas probably experience greater rates of deposition and of different grain sizes than would otherwise occur (Knutson *et al.* 1977). Rain also has the ability to scavenge, although probably less efficiently. It has been found, however, that rain brings down coarse dust preferentially over the land, but not over the oceans (Buat-Ménard and Duce 1986).

Natural traps, like soils, lakes or oceans may well be efficient, but they also contain other materials that are mixed with and difficult to distinguish from dust, and most of them are subject to post-depositional disturbance by many different processes. In addition, the rate of travel from the surface of a water body to its bed may be very slow and tortuous. The onus in these methods is therefore thrown onto the distinction between dust and other materials and between dusts from different sources. This can be done by grain-size analysis, mineralogy, chemistry, magnetic susceptibility, and the identification of biogenic materials from distinctive environments. Some examples of the analysis of dust deposits are given below.

Dust in motion

Entrainment

The threshold wind speed of entrainment for dust, as explained in Chapter 2, is in theory inversely related to size (Fig. 2.6). In reality, the process of raising dust is not the simple matter of moving a fine particle into the wind that Fig. 2.6 implies, because, as will be explained, very little fine dust is lifted directly, most being created by attrition in saltation and suspension. Wind speed is nevertheless an important control, for more dust is produced by sandblasting and the rate of sandblasting is positively related to u_*. The duration of high winds is also clearly important (Zobeck and Fryrear 1986a). But there are many other controls.

First, there are further aerodynamic controls. Lift (or upward suction imposed by various processes described in Chapter 2), has been shown experimentally to be much more effective for dust than for sand (Iversen 1986a), suggesting that the relationship between particle size and threshold wind speed may not be the same in dust devils (see below) as in other conditions. These observations relate to others, which suggest that dust is also much more susceptible to turbulence than sand (Nickling 1978), and larger scale evidence that turbulence in thunderstorms and cold fronts is important to the entrainment of dust. But wind speeds and even turbulence are only related to dust raising, and not necessarily to its maintenance in suspension. Thus the amount of dust in suspension at a site is poorly related to local wind speeds, since much of it has been raised elsewhere and at other times (Offer and Goosens 1990).

Second there are 'erodibility' controls (to do with the character of the surface). The grain size of the primary particles on the surface is less important than the grain size of aggregates (Nickling and Gillies 1993), and aggregation is related to a number of soil characteristics such as clay mineralogy, salt type and content and organic matter type and content (Breuninger *et al.* 1989). These relationships are explained further in relation to agricultural soils in Chapter 9. Grain size also has an effect on surface moisture, which may be the most critical factor in controlling dust entrainment, for sandy soils dry out at the surface more quickly than loams (Gillette 1988). Soil moisture is often the dominant control on dust emission. At the largest scale, it seems to be the major control of the annual cycle of dustiness of the global atmosphere which is twice as dusty in August than in February, August being the driest month for the northern-hemisphere deserts, which are much larger than those in the southern hemisphere (Joussaume 1990). The same kinds of process appear to control dustiness at a regional scale, for dustiness in the Sahel increases after one or two very hot years (Littmann 1991b). Soil moisture is controlled itself by differences in rainfall and evaporation.

There are yet further complications. The amount of bombardment by saltating sand is probably a major control, for even small amounts of moving sand can liberate far more dust than is liberated from saltation-free surfaces (Gillette 1981). Vegetation and its seasonal pattern is another control. In the West African Sahel it has been found that the seasonal minimum of dustiness comes in October, near the end of the rainy season, when the vegetation has fully responded to the rains (Littmann 1991b). In climates like the United States High Plains, and the Russian and Kazakh steppes, intense frost and snow cover are other strong inhibitors (Zhirkov 1964), though frost may have a longer-term positive control on dustiness, by breaking up clods and preparing them for entrainment after the thaw. The interactions between these controls and land use, which has become even more important to dust production in many areas, are discussed in Chapter 9.

The spatial outcome of all these relationships is that the dustiest places are in dry parts of the world, especially where the surface has a mixture of dust-sized particles with sand (Fig. 4.3), although the relationship is not simple, as will be seen below. There are also temporal implications, for dustiness is associated with drought. This relation, however, is not always direct: droughts may have their maximum effect on dustiness some time after their own climax. In Arizona and dry parts of Australia, dust storms are related to moisture from the previous winter (Brazel and Nickling 1986; Yu *et al.* 1993), but the relationships are not always as clear even as this (Littmann 1991c).

In transport, dust is generally size-sorted by height (Zobeck and Fryrear 1986a), but some workers have observed fine material being held within the saltation layer, from which little escaped (de Ploey 1977). The criteria that determine whether a particle stays aloft are discussed in Chapter 2. Grain size decreases with distance from source in most of the cases where there is evidence (Fig. 4.4; Inoue and Naruse 1991). Tsoar and Pye (1987) calculated that particles coarser than 20 μm travelled close to the ground, and were therefore likely to be trapped by vegetation. Most loesses are of this coarser size range, and have therefore not travelled far. Particles finer than 20 μm travelled throughout a column of dust up to 100 m above the surface, and thus were transported far further (Fig. 4.5). Following these trends, dust deposits, particularly loess (which is discussed later in this chapter) generally become finer away from the source. For example, the Malan Loess, which is the most recent of the Chinese loesses, becomes finer eastward away from its sources in central Asia, changing from a silty to a clayey deposit. At the

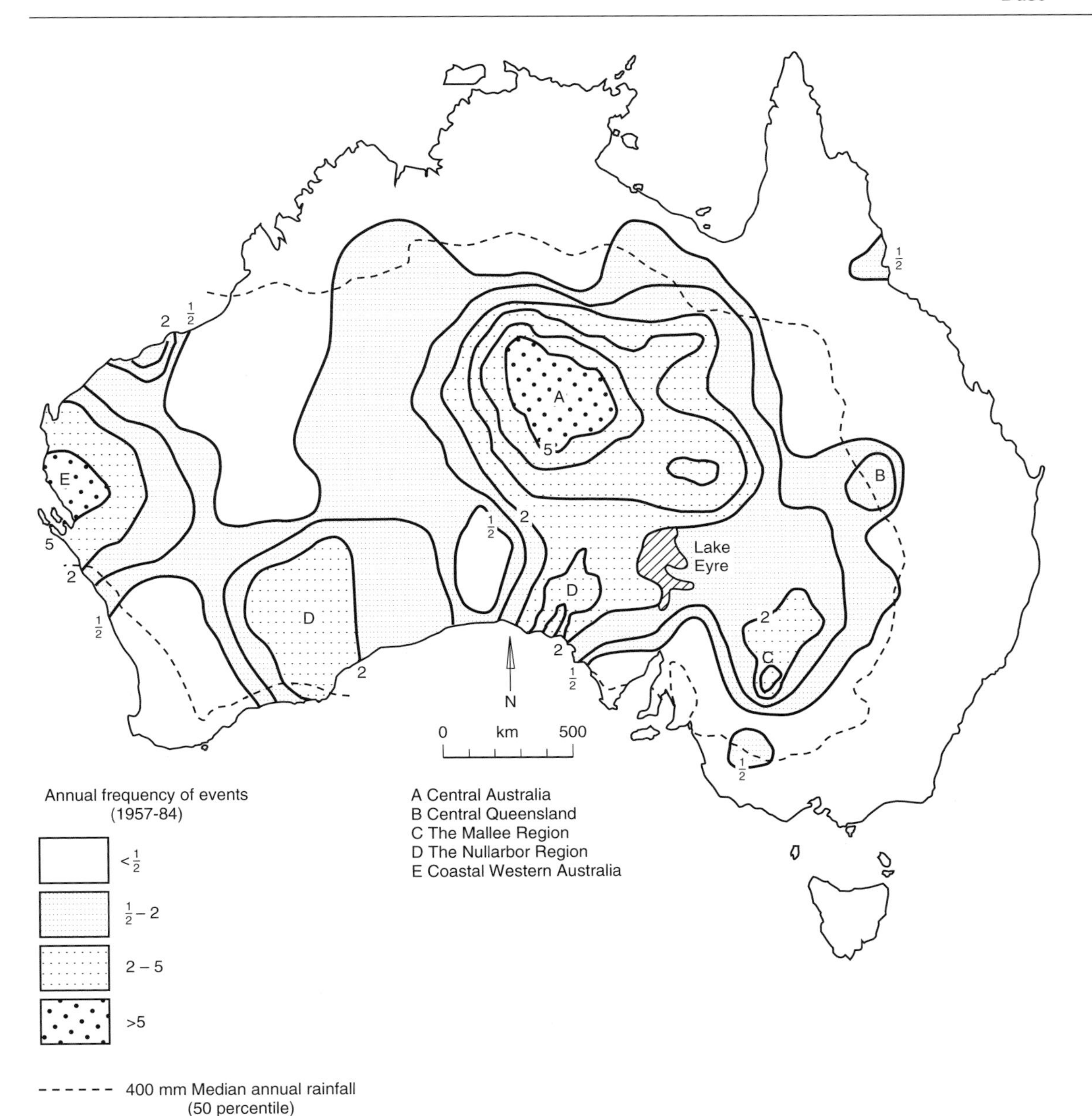

Fig. 4.3 Average frequency of dust storms in Australia, showing the probable importance of Lake Eyre and other central Australian dry lakes as sources of dust (after McTainsh and Pitblado 1987).

same time, heavy minerals are left behind in the western deposits (Eden *et al.* 1994).

Dust-raising and dust-carrying systems

The atmospheric structures that raise and carry dust occur at scales ranging from a small flurry a few millimetres across, to dust storms 1000 km across.

Dust devils, a common dust-raising system at the smaller end of the scale, are thermal vortices, made visible by the dust, which develop when and where there has been intense surface heating (encouraging upward air movement). The need for this heating

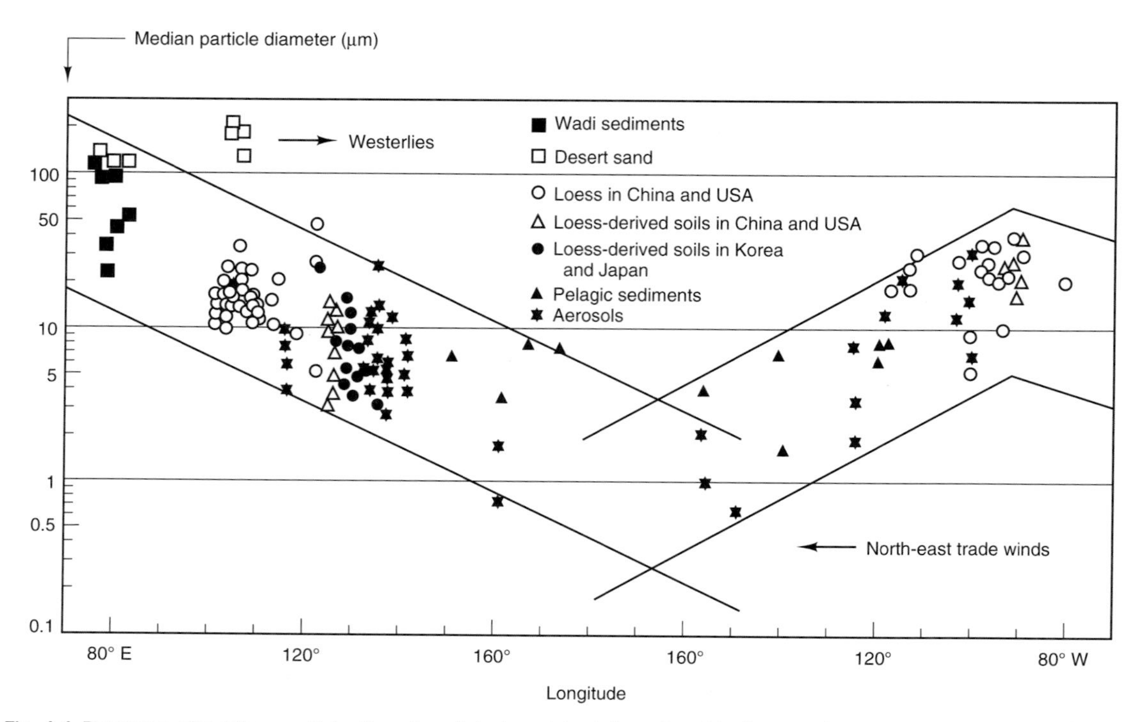

Fig. 4.4 Decrease of median particle diameter of dust and dust deposits with distance from source in the Pacific (after Inoue and Naruse 1991).

confines dust devils in space and time. Thus they are more common on readily heated surfaces, like those of dry lakes (Young and Evans 1986), and at times when the surface is heating up, as in spring (Wigner and Peterson 1987). Dust devils must also have vorticity (Sinclair 1969), and this can come from a number of sources; flow round obstacles such as hills (Idso 1974) and shear in convective, unstable atmospheres (Carroll and Ryan 1970) being two. As the dust devil develops, pressure drops in the centre and wind speed reaches a maximum in a tight ring round it (Hallet and Hoffer 1971). Tangential velocities reach $22\,m\,s^{-1}$, and vertical velocities $3.5\,m\,s^{-1}$.

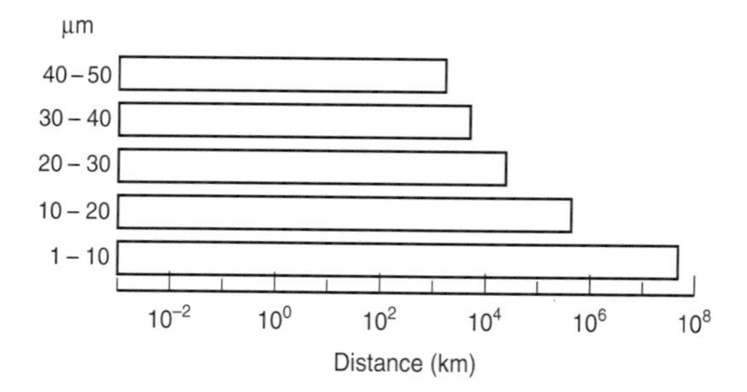

Fig. 4.5 Distances normally travelled by dust particles according to size (after Pye and Tsoar 1987).

At a slightly larger scale, dust is raised by the severe turbulence in thunderstorms (Brazel and Nickling 1986). The term now widely used for thunderstorm-induced dust storms is '*haboob*', a colloquial word from Sudan (Idso *et al.* 1972). It is the strong gusts of density-current downdrafts of cooled air from the cumulo-nimbus cloud, preceding the thunderstorm itself, that create the haboob (Pye 1987). In Arizona, which experiences 41.7 haboobs on average in July and August (Goudie 1983), they raise great quantities of dust. They develop along squall lines associated with incursions of warm humid air into this generally arid area (Brazel and Nickling 1986). Haboobs in the West African Sahel are also associated with squall lines, which move over the surface at about $16\,m\,s^{-1}$, scavenging dust. It is these squalls that raise the dust into the easterly jet over the Sahel (see below) (Tetzlaff *et al.* 1989).

Another organized pattern, spanning a range of sizes, is dust-carrying rather than a dust-raising. It is the 'dust plume' (flowing parallel to the surface). It is a feature early in the life of many dust storms, before the dust cloud merges and obscures its own internal structure. Plumes are presumably formed by horizontal vortex-like flow (Bowden *et al.* 1974; Swift *et al.* 1978; Middleton *et al.* 1986). They may occur singly or in families, and they range from a few tens of metres to many kilometres across. One single plume, seen on satellite imagery, originated in Iraq and streamed 400 km south-eastward in the strong winter 'shamal' wind, spewing large amounts of dust over the Arabian Gulf. It widened from about 10 km near its source to 60 km at its point of dispersal (Middleton 1986). Being narrow concentrations of dusty air, plumes act as jets. It may be that dust dampens turbulence and this should allow wind speeds to increase, so that the dust storm may increase in intensity in a positive feedback fashion, to be controlled ultimately by some other process (Barenblatt and Golitsyn 1974). Another dust-raising weather system at this scale (often itself composed of a series of plumes) is the katabatic wind, sweeping down from mountain massifs (for example, Swift *et al.* 1978). The 'Santa Ana' of southern California is one of these, taking dust from the dry interior of western North America towards the Pacific coast.

Cold fronts, at a still larger scale, account for a very high proportion of the dust raised and carried within high-latitude semi-arid lands. They bring sharp changes in wind direction in zones of high turbulence, a few hundred to 1000 km across, which sweep across country at speeds of about $20\,m\,s^{-1}$, raising huge quantities of dust, if the soil is dry and bare. These systems (and haboobs) produce the horrifying walls of dust like those in Fig. 4.6, which can rise over 2000 m into the sky and travel at rates of $36\,m\,s^{-1}$ (Idso *et al.* 1972). Also like haboobs, cold fronts lift large amounts of dust into higher-level jet streams (see below). Daily particle concentrations may average $1000\,\mu m\,m^{-3}$, but peak values may reach $10^4\,\mu m\,m^{-3}$. Frontal dust storms are so familiar, but such a nuisance, in some parts of the world, that they have vernacular names; in the southern Mediterranean, moving east to west, they are known successively as *gibli*, *khamsin* and *sharav*.

These modes of transporting dust seldom succeed in carrying dust very high. Most of it travels below inversion layers, as can be seen on any aircraft flight over dusty terrain. In central Asia, for example, there are no loess (dusty) deposits above 2000 to 2500 m above sea level (Dodonov 1991).

The mix of mechanisms responsible for dust-raising probably varies very greatly from place to place, depending on local climatic conditions. On the southern High Plains in Texas, which is a zone of unusually strong cyclogenesis, 30 per cent of dust is estimated to be related to well-developed frontal systems; 19 per cent to thunderstorms; and 30 per cent to 'daytime mixing down of high-speed momentum from above', a process associated with variations in the position of the high-level polar front jet, but not with any well-defined synoptic situation on the ground (Wigner and Peterson 1987). In California, more dust is raised by cold fronts and similar disturbances than by haboobs. In Arizona, the reverse

Fig. 4.6 A 'wall' of dust associated with a frontal dust storm in Niger. (Photo: Jake Sudlow.) See also Fig. 1.1(c).

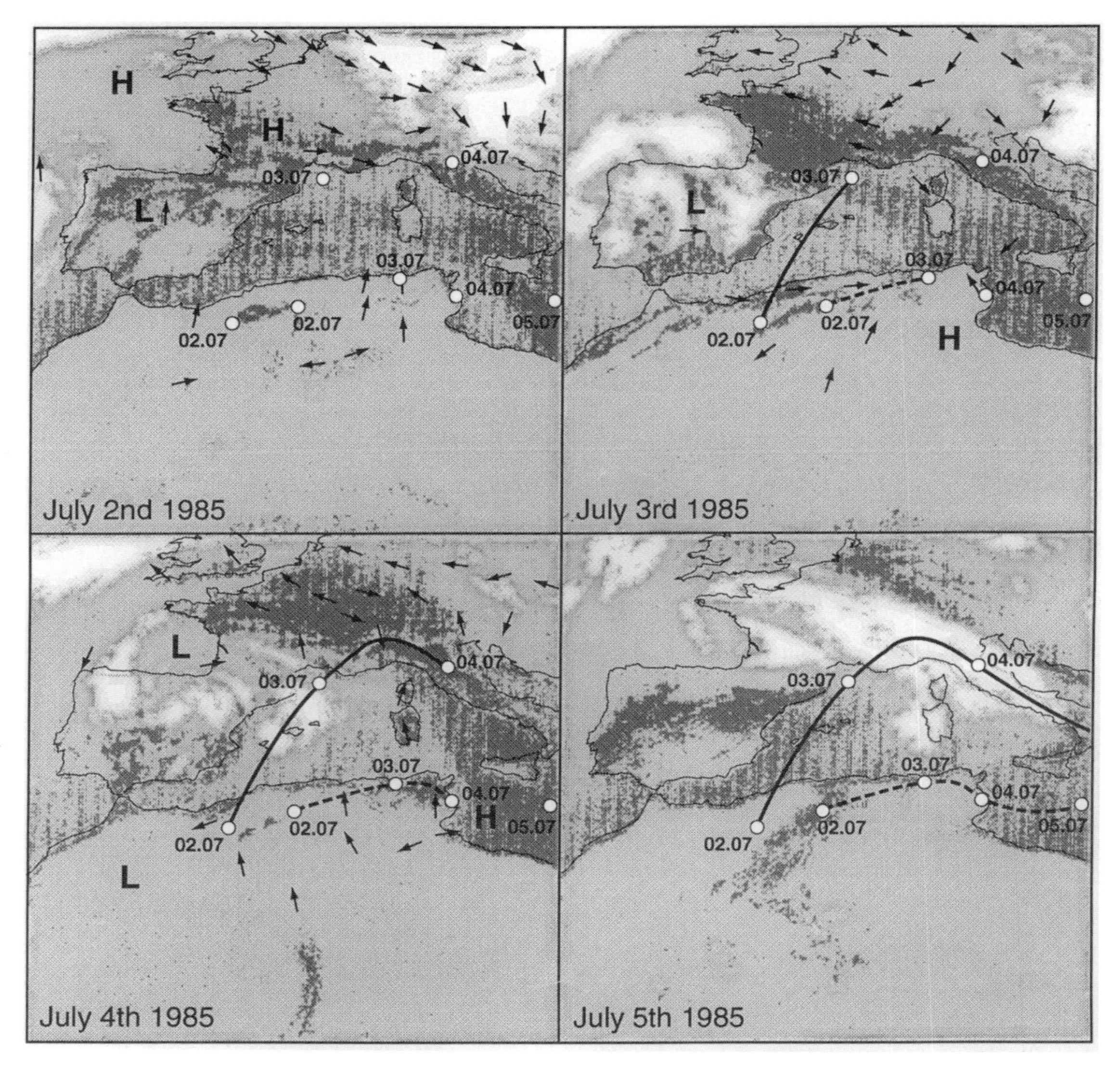

Fig. 4.7 Dust storms over the Mediterranean and Europe, showing a probable source in the Chott ech Chergui in northern Algeria (data from Dulac *et al.* 1992). Arrows indicate direction of wind flow.

is true (Brazel 1989). In other parts of the world katabatic winds, haboobs or any other of the modes described above might be the most important mechanism.

The dust, once raised and carried short distances by one of these mechanisms, merges into much more extensive dust storms (Fig. 4.7). The best known of these is the *harmattan*, the dusty wind that blows down from the Sahara between January and April towards West Africa and ultimately out over the equatorial North Atlantic towards the Americas (McTainsh 1980). Other notoriously dusty winds carrying the dust raised by smaller systems are the *shamal*, which blows from the north-west down the Arabian Gulf, and the *kosa* (dusty outbreaks from China eastward over Japan and the Pacific). Others occur in North America, usually travelling eastward from the south-west of the United States; one of these swept eastward over the Gulf of Mexico in February 1977, at one time covering 2400 km^2 (Breed and

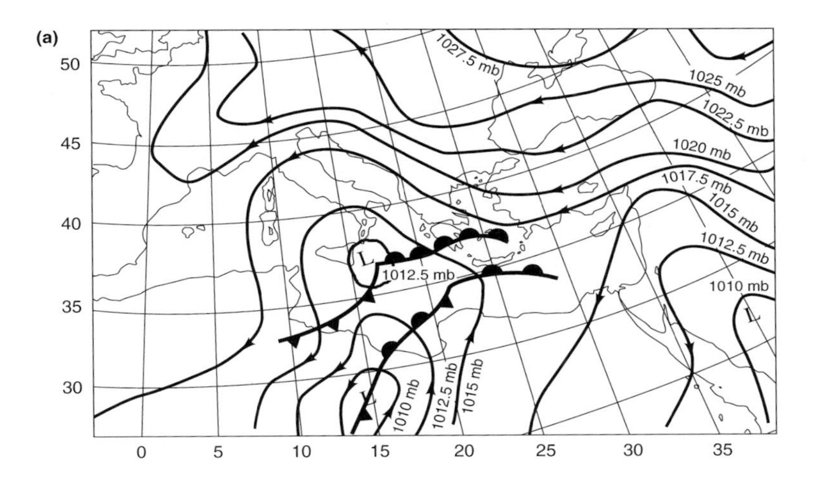

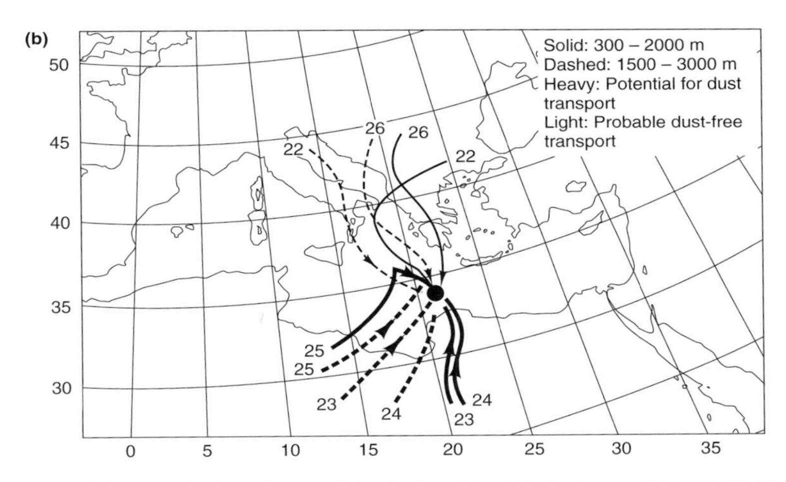

Fig. 4.8 (a) Synoptic situation and (b) calculated back trajectories of dust during a dust storm over the Mediterranean in October 1988 (after Dayan *et al.* 1991).

McCauley 1986). Some of these great clouds of dust move at astonishing speed. Dust raised at Faya-Largeau in northern Chad travels the 1000 km to Kano in 24 hours (McTainsh 1985). These clouds of dust are shepherded by large-scale atmospheric patterns, such as fronts and convergence zones, as shown in Fig. 4.8 (Dayan *et al.* 1991). The dust in these storms is eventually dispersed by rain or extreme turbulence. Rain is of major importance in bringing down dust in many areas, so that rainfall (and rainfall type) are often the best correlated with deposition rate in the short term (Bergametti *et al.* 1989b).

At an even larger scale, dust is sometimes raised high above the surface and travels in major jet streams, such as the African Easterly Jet from northern Africa which takes dust in great waves, with a wavelength of 2500 km, from the Sahara at 2000 to 5000 m above the surface, travelling eventually over the Caribbean (Carlson and Prospero 1972). In the Indian Ocean, the Somali Jet performs the same function, taking dust from north-eastern Africa towards India (Sirocko and Sarnthein 1989). It is the major westerly jet stream that carries most dust out over the Pacific from China.

Sources

Contemporary, proximal sources

Studies of dust just after it has been raised give clues about immediately proximal sources. It consists of two modal sizes: coarse particles (somewhere between 10 and 200 μm) and fine particles (somewhere between 1 and 20 μm); there are fewer particles of intermediate size. The consensus explanation of this size bimodality is that most of the fine dust is created by the attrition of the coarse particles (which are easier to raise, as explained in Chapter 2). Attrition creates fine dusts from two sources: either by breaking down aggregates of fine material; or by blasting fine clays off the surfaces of coarse quartz grains (Gomes *et al.* 1990). The facts that the proportion of fine particles increases with wind speed (more wind energy creating more attrition), and that coarse particles can be seen (in the scanning electron microscope) to be coated with fine ones, both support this hypothesis (Gillette and Walker 1977). Furthermore, a recent study of samples of dust-yielding surface soils found no naturally occurring grains smaller than 80 μm and most populations to range between 100 and 300 μm in size, suggesting strongly that the finer mode in dust can only come from attrition (Bergametti *et al.* 1994). Another study found that only 8 per cent of the fine particles were lifted directly from the surface (rather than being created by attrition) (Gillette *et al.* 1972). Much clay may never be fully disaggregated, for once it has been reduced to dust, it is lifted above the saltation layer, where it suffers much less attrition, and can travel in suspension, still in aggregate form.

By the time dust has travelled several hundred kilometres, its grain size is refined, but its size patterns can still retain clues about its sources. The coarser mode falls out, leaving only the finer mode, which, in transport, becomes yet finer. Some of this dust is unimodal in size, some bimodal. In the Canary Islands, a few hundred kilometres from its Saharan source, the dust has a modal size finer than 5 μm and is unimodal. Some dust collected in Germany, much farther from its source, is also unimodal (mode finer than 1 μm), and being fairly well weathered, may have come from as far away as the Sahel (diatoms indicated a single source either in the Sahara or the Sahel). Other dusts collected in Germany, though fine, are bimodally sized (modes finer than 1 μm and around 20 μm), a characteristic more common on dusts that have travelled only a few hundred kilometres, as in the southern Sahel. In Germany, some of this bimodal dust was found to be little-weathered, suggesting that it came from a hyper-arid source. The model of origin and transport of this bimodal dust, therefore, is that it was brought by a dust cloud that had originally been raised in the Sahel, but had been recharged with dust from the Sahara. This is consistent with meteorological analysis of the paths of dust plumes reaching Europe (Littmann 1991a).

Doubtless mixes of dust from different sources with different travel histories give a range of different grain-size (and other) characteristics. In general, mixing increases with distance from source, increasing also the difficulty of linking dust characteristics to those of the source (Hoogheimstra 1989). For example, the dust that leaves the Saharan coast is already so well mixed that it is practically identical to that which reaches the Caribbean (Glaccum and Prospero 1980).

A very prolific proximal source for contemporary 'natural' dust is dry lakes in arid and semi-arid parts of the world. (The increasing amounts of dust from artificial sources, now amounting to 25–50 per cent of all dust (Hansen and Lacis 1990), is discussed in Chapter 9.) Playas and sabkhas, most of which are at the distal ends of fluvial sedimentary systems, contain large quantities of sediments of appropriate size, both mineral (from alluvium and by the evaporative concentration of salts in solution) and biogenic. Their surfaces are kept free of vegetation by occasional flooding and by salinity, and they are broken up, in readiness for entrainment, by salt crystallization. The northern Chad Basin in West Africa and the Lake Eyre Basin in Australia (Figure 4.3) are two such sources that have been shown to be of major importance. Both are fed with sediments by streams from wetter areas, and both are zones of internal drainage from which no sediment or salt escapes except in the wind (McTainsh 1985). If these deposits are bombarded with saltating sand, as they are in the northern Chad Basin, where the old lake floor is now traversed by a field of active barchans, they are even more likely to produce dust.

There are many smaller dry lakes in the dry lands, and also good evidence that they yield large quantities of dust. The *chotts* in southern Tunisia and northern Algeria (some almost as big as the dry lakes in northern Chad) are the source of the huge quantities of gypsiferous dust that now blanket the surrounding landscape (White and Drake 1993). The Chott ech Chergui, on the high plateau of Algeria, is clearly seen on Meteosat imagery as the source of the major dust storm that moved north into Europe in July 1985 (Fig. 4.7; Dulac *et al.* 1992). In central Nevada, where there are many of these dust-yielding dry lakes, Young and Evans (1986) recorded up to 2.68 $kg\,m^{-2}$ of deposition (of dust and sand-sized dust aggregates) downwind of a playa lake. In Nevada, dry lakes were the main sources of dust during the windier drier periods of the Pleistocene (Chadwick and Davis 1990). Owen's Lake in California, after its desiccation by water-extraction for Los Angeles, is now a source of huge quantities of dust, as explained in Chapter 9.

River alluvium, shortly after deposition, and before colonization by plants or sealing with a crust, is another common proximal source of dust. Alluvium deposited in distal sections of river valleys is most susceptible, for its mode is finer than upstream, and contains smaller quantities of coarse particles which might accumulate as a 'lag' to protect the soil from further attack by the wind. Much of the loess that now blankets parts of central America and central Asia had river alluvium as its proximal source. This is seen in the distribution of the deposits, which thin out rapidly away from river valleys (Fig. 4.9). The 'loess' hills, which follow major valleys such as the Missouri and Mississippi, and the rivers of central Asia such as the Zeraf Shan (Fig. 4.10) are sometimes known as 'loess lips' (loess is more fully discussed on page 57).

The composition of dust is a clue to its proximal source (as seen above in the case of diatom assemblages). Minerals and biogenic sediments are very little altered in transport, and may give good clues about the sources, at least of coarse dusts. But because of the mixing of dusts from different sources, the sources of fine dust are usually much more difficult to identify. Nevertheless, there are some clear relationships. The dust which travels across the Atlantic from the western Sahara, underlain as most of it is by crystalline basement, is more quartzose than the calcite- and dolomite-rich dust that reaches the eastern Mediterranean from the generally sedimentary rocks of northern Sahara (Ganor and Mamane 1982). The dust arriving in Crete from the sandy deserts of Libya is quartz-rich; that from the more calcareous Maghreb is more calcareous (Pye 1992). The clay mineralogy of dusts that reach the Tyrhennian Sea in the western Mediterranean from these and other parts of North Africa is also closely related to source, although there has been some mixing, even over this distance (Bergametti *et al.* 1989a). Neodymium and strontium have been found

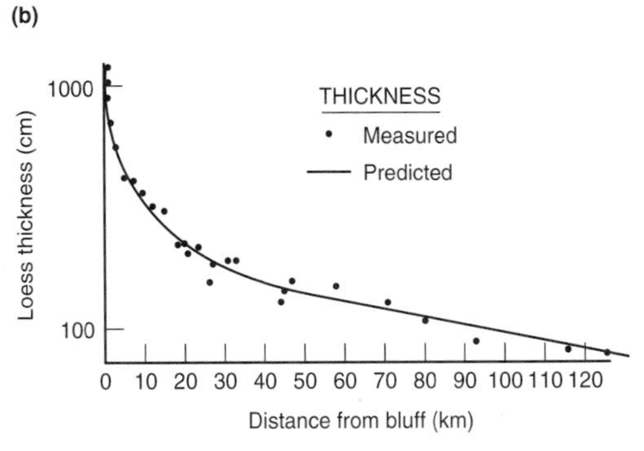

Fig. 4.9 Loess thickness along a traverse downwind of the Missouri River in Illinois (after Frazee et *al.* 1970).

to be useful tracers of these Saharan dusts (Grousset *et al.* 1992).

An important albeit very distinctive source of dust in the atmosphere is from volcanoes. In the 1980s, there were several major eruptions, including that of Mount St Helens, but by far the greatest production of dust came from El Chichon in Mexico in 1982 (Michalsky *et al.* 1990). Eruptions yield material of a great range of sizes, which are carried in a number of ways. Some material of dust size is carried in surges and flows and deposited near to the volcano itself. Great volumes of dusts are also ejected high into the atmosphere and are carried by winds as dust. This component of volcanic dust falls in patterns depending on the height to which it was first thrown by the eruption, its original grain size, its subsequent aggregation, the wind strength and direction at the

Fig. 4.10 Massive cliffs of loess beside the Zeraf Shan River in Tajikistan.

time of the eruption, and atmospheric inversion patterns at that time (Fisher and Schmincke 1994). For example, tephra deposits can be very widespread, and, being also usually of a distinctive mineral composition, provide excellent marker horizons in sedimentary sequences.

The finest volcanic dusts are pumped high into the stratosphere. They consist partly of very finely divided ash, much of which is so dispersed, and so weatherable when it eventually falls, that it is very difficult to trace. Another major component is fine sulphuric acid droplets, travelling as aerosols, which may persist for over a year in the atmosphere; when they are eventually precipitated they fall in even more dispersed patterns and are even harder to detect (Hansen and Lacis 1990).

Primary formation

A major proportion of contemporary dust has not been created from rock in the recent past, but is inherited from breakdown processes in much earlier times. The deposits of late Pleistocene and early Holocene lakes, now dry, are a particularly prolific source, as is shown by their dissection by yardangs (Chapter 3). Many of the deposits in Lake Chad, Lake Eyre and the North African Chotts, which yield such large quantities of dust, are not contemporary. Both the Australian deserts and the Sahara have many smaller dry lakes with ancient deposits, often of Holocene age.

The inheritance of some kinds of dust has caused controversy. The question is best put in Smalley's vocabulary (Smalley and Smalley 1983). In this, a 'P' event is when particles are created *ab initio*; a 'T' event is when they are transported; and a 'D' event is when they are deposited. Most sediments go through many thousands of T and D events, in the sequence P, T1, D1, T2, D2 and so on.

For many dust constituents, there is no great mystery about the nature of these D and T events; neither is there much doubt about their P or formative event. These less problematic constituents include salt particles (gypsum and halite), most of which can only have come from saline lakes, and it is not difficult to see how, as brittle, soft materials, they are broken down to dust-size by attrition. Calcite is another common component of dust that occurs in lake deposits; it is also common in soil horizons. Neither is there much debate about most of the clay mineral dusts. Palygorskite, for example, can be traced fairly unequivocally to high-pH environments as in salt lakes, and is a component of many dusts, and of some peri-desert loesses (see page 60; Coudé-Gaussen 1985). It may be difficult to discover the exact provenance of other clay minerals, like illites, kaolinites and smectites, which are important in many dusts and soils, but there is little mystery about the ways in which they were created as particles of dust size, since they are well known to be the products of the intense weathering of primary minerals.

The main debate is about the P event of quartz dust. The debate is important largely because silt-sized quartz particles are the single most important constituent of the loess deposits that cover large parts of the globe as it is of many contemporary dusts. Loess is such an extraordinary and extensive deposit that its aeolian origin was hard to accept at first. Early debates about loess included some wild ideas such as that it was of cosmic, lacustrine, volcanic or marine origin or had formed by *in situ* weathering. Biblical fundamentalists even suggested that it was a remnant of Noah's flood. Most nineteenth-century geologists believed it to be fluvial, citing the many fluvial structures (thought now to represent reworking of aeolian dust deposits), but all these theories had been dismissed by the early twentieth century in favour of an aeolian origin, except by a handful of geologists (Pye 1987).

However, to acknowledge loess as aeolian is only to start another debate. Quartz is very common in igneous and metamorphic rocks, but at a much greater mean size than in dusts and loesses. In gneisses and massive plutonic rocks the mean size is 720 μm, with 50 per cent and 20 per cent being mono-crystalline respectively (Blatt 1987). Thus quartz dust and loess have experienced considerable reductions in size. Blatt noted, however, that quartz in schists had a mean size of 440 μm (40 per cent mono-crystalline) and he used $\delta^{18}O$ values from source rocks and sediments to conclude that most of the finer quartz was derived from the finer crystals in these rocks and in slates. If correct, this finding may settle some of the debate: at least part of the explanation of the size of the quartz dust occurs at source. There is, notwithstanding, a need to link quartz dusts specifically to these metamorphic rocks, and to explain how such large quantities of fine particles have been fractionated to form loess and other dust deposits.

Two other ways in which silt could have been produced are comminution in streams and weathering. Although fluvial action is capable of reducing quartz crystals to silt size (which could then be blown off alluvial deposits), it is generally thought not to have produced sufficient quantities to account for loess. It is probably effective only in high-energy mountain streams where there are boulders in transport. There is more debate as to whether intense weathering, perhaps in the seasonally wet tropics where weathering is intense, could produce sufficient quantities of silt; it could have produced silt-sized quartz particles that were later concentrated in sediments by rivers, and these deposits might then have provided silts for reworking by the wind, but the argument is unresolved (Pye 1987).

Attention has focused more on two other groups of possible size-reducing processes. These fit categories which are as old as Obruchev's (1945) discussion of quartz loess: those of 'cold' environments and those of 'hot'. The 'cold' or 'mountain' dust-producing processes are said to be frost action and glacial grinding. Smalley (1990), having sustained the argument for over two decades, still insisted that glacial grinding was an important mechanism, but he added a new twist to his argument (following Blatt): it could only produce significant quantities from the stressed quartz derived from metamorphic rocks. Smalley had come round to the belief that frost action was an additionally important mechanism (working on the same parent material), but that it could only have proceeded at a great enough rate on massive uplands, as on the huge Tibetan Plateau. The building of the Plateau since the Tertiary, moreover, would have stressed many quartz crystals. This source, in Smalley's view, provided the silts that became the Chinese loess.

Smalley's belief in the importance of frost action was not shared by Pye (1987), who, though acknowledging that it might produce silt-sized quartz particles, did not believe that it could produce sufficient parent dust for loess. He later conceded that frost action might be very effective in cold wet 'Icelandic' type environments, but still maintained that it had not been very active in cold dry environments like Siberia or the Antarctic (Pye 1989). Smalley's position on glacial grinding is also not wholly secure, for although large quantities of dust are certainly associated with some glacial environments, and silt-rich deposits have been found beneath glaciers, experiments and scanning electron microscopy of sub-glacial grains throw doubt on whether glacial grinding could produce large enough amounts (summarized in Pye 1987); but then it may only be able to do so given the shattered quartz in the parent rock that Smalley's case demands. In any event, most authorities agree that, if not the sole mechanism of silt formation, glacial grinding was nevertheless a major source of dust-sized quartz. Most acknowledge also that the seasonally bare expanses of dusty sediment left by glacial outwash channels, and the strong winds associated with glacial times and glacial margins were additionally important to fractionating, moving and depositing 'glacial' loess.

The close relationship between the largest accumulations of quartzose loess and recently glaciated areas does not in itself clinch the argument. It has been repeatedly suggested, for example, that the Chinese loess is more a desert than a glacial product, for it is close both to an intensely glaciated and to a large, arid area (Zhang Linyan *et al.* 1991). The thickest North American loess, in Nebraska, is also in a somewhat eccentric position in relation to the Wisconsin glaciers and ice-sheets and the wind patterns that flowed around them. It may be that it was simply older silts carried in highly seasonal meltwater streams, deposited on their vegetation-free flood plains, and then exposed to the high winds of glacial times that were responsible for much of the glacier-dust association.

These long-running arguments about the formation of 'cold' loesses were, until recently, conducted in ignorance of the extent of quartzose silty deposits all around the Sahara and some other deserts. These are now acknowledged as loess by most of those who have studied them. One reason that these deposits were slow in being recognized as loess is that they are coarser than the European, Asian or North American loesses. But the deep accumulations of silt in the Atlantic off the coasts of Senegal and Mauritania, downwind of the Sahara, which have come to light in recent ocean-floor explorations, and which are widely acknowledged to be of aeolian origin, are very strong evidence that the Sahara has produced large quantities of dust. The marine deposits have now been joined, in the literature, by major terrestrial deposits in the West African Sahel, Tunisia and southern Israel, all of which are now widely conceded as 'peri-desert' loess (McTainsh 1987). Although there was a Permo-Carboniferous glaciation in what is now the Sahara (a possible, but very remote glacial 'P' event), there is no direct link between its deposits and the dust now being produced and that which was produced in the late Pleistocene and early Holocene.

There are two processes that might produce quartz dust in deserts (Obruchev's (1945) 'hot' loess and Coudé-Gaussen's (1991) 'peri-desert' loess): salt weathering and attrition. Dust-sized particles can undoubtedly be formed by the salt weathering of quartz, but, following Blatt's argument, it may be that this can happen only in quartz which has micro-fractures (Smith *et al.* 1987). This is suggested by the experimental finding that salt barely attacks texturally mature quartz dune sands (which are supposed already reduced to a resistant core), but does attack fresh grains from granite regolith (Pye and Sperling 1983). Silt particles have also been produced experimentally by the detachment of siliceous cements or overgrowths on sedimentary grains (Smith *et al.* 1987). This may be because overgrowths are easier to detach by salt weathering than are pieces of shattered quartz crystal. In the experiments of Smith *et al.*, magnesium sulphate and sodium sulphate were effective at producing silt, whereas sodium chloride was not.

Given the right kind of quartz and the right kind of salt, salt weathering would be most intense in areas where salt can be carried to the sites of weathering by water (at landscape and grain scales). These would be semi-arid rather than arid. Salt weathering may not be intense enough to produce large quantities of dust in hyper-arid deserts. This may be the explanation of the confirmation by McTainsh *et al.* (1989) of an earlier finding by Goudie (1983) who had discovered that dust storms were most frequent in areas with about 200 mm mean annual rainfall and less frequent in drier and wetter areas. The combination of salt and frost weathering, as might happen in central Asia, may help to explain the vast quantities of loess in China (Pye 1987).

The second possible way in which quartz dust might be produced in deserts is by attrition in saltation, which though widely acknowledged to be the source of clay-mineral dusts (see page 52), is less certainly able to produce quartz dust. The attrition model for quartz dust is very old, and is now being revived. Despite Goudie's and McTainsh's findings about the association of dust with semi-arid rather than arid climates, there is an undoubted association of some dust deposits with some very arid areas. These are particularly areas downwind of some large sand seas, as in Tunisia and the Sahel (Coudé-Gaussen *et al.* 1983; McTainsh 1987). Although the spatial associations are strong, conclusive field observations of the process in operation are virtually impossible, and even experimental proof is difficult. In the laboratory, quartz dust with characteristics close to those of peri-desert loess has been produced by the attrition of sand in machines in which air is blown through quartz sands, causing innumerable impacts, but how like the real thing these machines are is open to debate (Whalley *et al.* 1987). Quartz dust of loess size has even been produced by abrasion in the laboratory of local Miocene and Plio-Pleistocene sandstones from the Hungarian Plain, suggesting that even some of the so-called 'cold' loesses may be of this origin (Smith *et al.* 1991).

Gross spatial patterns of production and removal

The processes of production and transport combine to create well-developed spatial patterns of dust. The main areas of contemporary production are undoubtedly the Sahara and the central Asian deserts. Subsidiary sources are Australia and south-western USA, southern South America and south-western Africa (Fig. 4.11). The Saharan source has been known since at least the times of early Arab navigators, who called the Atlantic in this area the 'Dark Sea'. Darwin (1846) collated some earlier accounts and noted a maximum rate of dust deposition on ships near the Cape Verde Islands. Contemporary estimates of total transport from the Sahara vary from 600×10^6 to 700×10^6 t yr^{-1}, with 190×10^6 t yr^{-1} travelling westward over the Atlantic; estimates of transport to Europe are between 7.6×10^6 and 10×10^6 t yr^{-1} (D'Almeida 1989; Littmann 1991a). The Chinese deserts send vast amounts of dust out over the South China Sea and the western Pacific; even out in the central northern Pacific, it has been (conservatively) estimated that 6×10^{12} to 12×10^{12} t yr^{-1} of dust is added to the Ocean surface (Uematsu *et al.* 1983). The average dust flux to the Arabian Sea over the last 8000 years is estimated at 100×10^6 t yr^{-1} (Sirocko and Sarnthein 1989).

It is very hard to estimate the total present global rate of dust production, for the origin of most dust is very difficult to pinpoint, as is clear from what has been discussed so far. One estimate is 1800×10^6 to 2000×10^6 t yr^{-1} (D'Almeida 1989).

Deposition

Gross patterns

The amount of dust fallout at the regional scale is a function of two sets of variables. First is the rate of production in the source region, itself a function of rates of weathering, moisture content and vegetation cover, and of disturbance. Second is the efficiency of transport from source to sink, which is a matter of lifting efficiencies, transporting power, and circulation patterns.

Only where there are high rates of production in the source area, wind systems of some constancy and

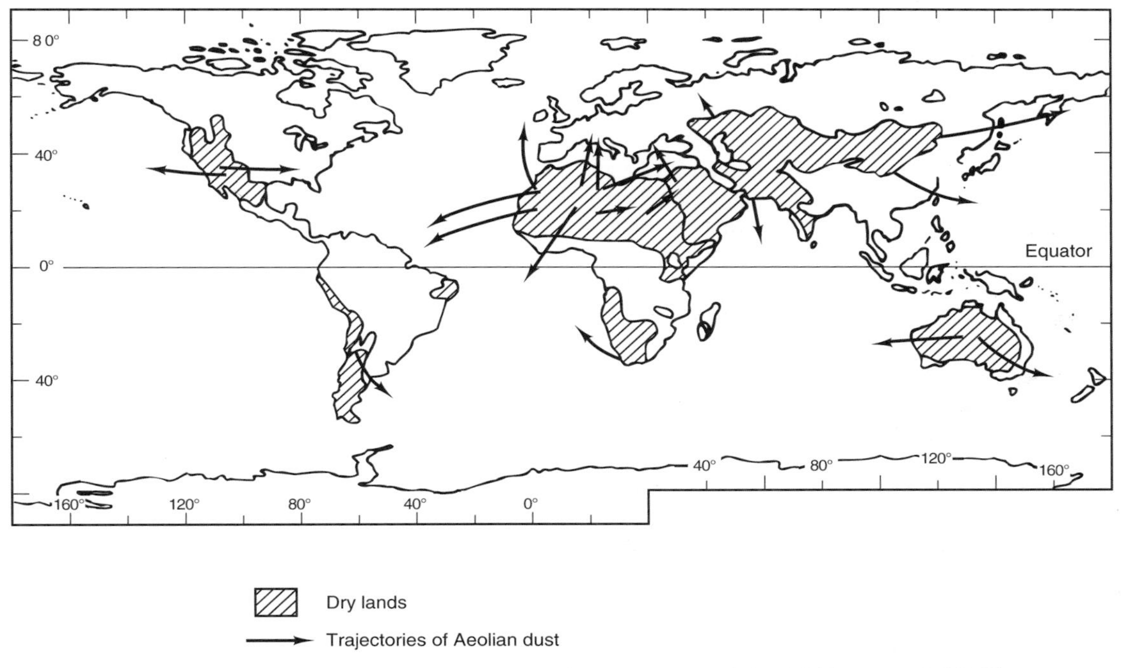

Fig. 4.11 The principal global dust sources and pathways of dust movement (after Middleton *et al.* 1986).

efficient traps, as in the Atlantic Ocean off Senegal and Mauritania, are distinct patterns likely to be discernable. In these situations, there is undoubtedly a rapid pattern of decline in transport and fallout away from a source of dust, at least close to source (Jaenicke 1979). Similar exponential patterns can be seen in the pattern of decrease in loess thickness away from the parent river valleys in North America and central Asia as explained above (Fig. 4.9). At a smaller scale the exponential decrease is seen in the pattern of dust deposition downwind of sources like playa lakes (Young and Evans 1986).

Dust in crusts, soils and geomorphological processes

The most obvious manifestation of dust falling in cooler, wetter areas is 'red snow' events, which are quite frequent in the Alps, Pyrenees and Scandinavia (for example, Lundqvist and Bentsson 1970), but on a soil surface the quantities of dust that reach humid areas in present conditions are generally too small to form identifiable deposits, being incorporated by weathering and mixing with other materials. Pye and Tsoar (1987) modelled the relationship between deposition rate and weathering rate (Fig. 4.12); Pye (1992) used the model to calculate that one would need $0.325\,\mathrm{kg\,m^{-2}\,yr^{-1}}$ for there to be a discernable deposit of loess accumulation 'under humid conditions', whereas in Crete he found only 0.01–$0.10\,\mathrm{kg\,m^{-2}\,yr^{-1}}$, and where therefore the dust was incorporated into other soils. In any event, the addition rates to many Mediterranean soils are quite appreciable. In Corsica, where red rain and red snow are quite common, it has been estimated that there is about $10\,\mu\mathrm{m\,yr^{-1}}$ accretion (Loÿe-Pilot *et al.* 1986).

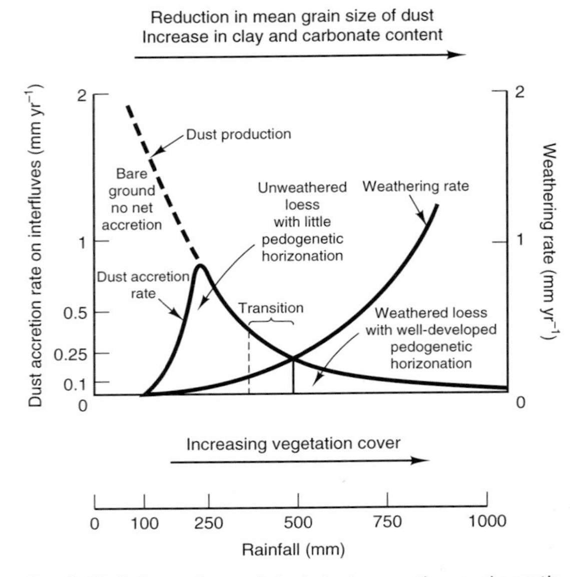

Fig. 4.12 Schematic model of dust accretion and weathering in relation to mean annual rainfall (after Pye and Tsoar 1987).

Dust contributes to several types of surface formation in semi-arid and arid areas, close to its sources. A large proportion of the elements in desert varnish (a dark coating on stones and rock surfaces in many deserts), particularly of the iron and manganese, comes from dust. Not only does the dust itself contain these elements in abundance, but varnishes are often far richer in some elements than the rocks they coat, suggesting that the source is external. The same arguments and observations apply to 'desert glaze', a shiny silicious coating of rocks in deserts, and to case hardening, which is toughened carapace on rocks in deserts and other areas. Dust may also carry the iron that reddens many desert sands (Cooke *et al.* 1993). Finally, some dust remains long enough on the surface of desert soils, particularly ones with rough surfaces like alluvial fans and lava flows, to be washed into crevices and to create a dust-enriched upper horizon (Amit and Gerson 1986; Wells *et al.* 1987).

That dust is a major contribution to semi-arid soil formation is not surprising, given the high rates of dust deposition in many of these areas. The massive calcrete (or *caliche*) horizons in many arid soils have now been shown fairly conclusively to have derived from calcareous dust, deposited on the surface and then washed down the profile. The most conclusive evidence is that calcrete overlies rocks which could not possibly have delivered so much calcium when weathered (Machette 1985). At one site in the southwestern United States, only 1 per cent of the soil carbonate can have come from bedrock; modern carbonate dust flux at the site was estimated at $5 \times 10^{-3}\,\mathrm{kg\,m^{-2}}$ (Mayer *et al.* 1988). Dust has also been a major mechanism for the transport of gypsum to many semi-arid soils. In Wyoming, gypsum dust is deposited at a rate of 2.6×10^{-4} to $6.0 \times 10^{-4}\,\mathrm{kg\,m^{-2}\,yr^{-1}}$, well within the rate required to produce gypsum-enrichment in soils, which was estimated to be between 1.1×10^{-5} and $20 \times 10^{-5}\,\mathrm{kg\,m^{-2}\,yr^{-1}}$ (Reheis 1987).

Even when incorporated into soils, dust can still sometimes be detected mineralogically and in other ways, and in some cases this shows that it has made an important contribution. In much of the Mediterranean, early pedologists believed that the red soils that overlay limestones (so-called *terra rossas*) had been derived by weathering and accumulation of the small quantities of silicious material in the limestones. Weathering and accumulation undoubtedly play a role, as on Crete (Pye 1992), but Danin and Gerson (1983) calculated that it would take 200 000 years for a terra rossa soil of the thickness of those near Jerusalem to accumulate in this way (and the weathering of 20 m of limestone). It would take only 10 000 years for the soil to accumulate if it were being created by dust at present rates of fallout and local redistribution (and rates were probably much higher in the late Quaternary as is shown in Chapter 8). Moreover the terra rossas near Jerusalem rested on limestone surfaces which had clearly been pitted by endolithic lichens, indicating that the surfaces had once been bare, rather than that they had been weathered to produce a soil.

Other humid areas surrounding present-day deserts also have significant rates of accretion. The rate of deposition over the Senegal and Gambia river basins in West Africa is about $10\,kg\,m^{-2}\,yr^{-1}$ or about $200\,\mu m\,yr^{-1}$ (Orange and Gac 1990). In South Australia, near Adelaide, measurements suggest accretion rates of the order of $5–10\,t\,km^{-2}$, or 2.5–5 mm in 1000 years (Tiller *et al.* 1987). Dust, derived from China, is being appreciated as an important contribution to soils in Japan and Korea. In Japan the dust flux is said to be of the order of 4–7 mm per 1000 years (Inoue and Naruse 1991). In parts of the rain forest, where leaching is fast, dust may provide the main source nutrients, as in Ghana, where the ecosystem appears to derive all its potassium from harmattan dust (Tiessen *et al.* 1991). Dust seems to have been an extremely pervasive, if minor, contributor to soils the world over, for example distributing biogenic components far from their sources (Jones and Beavers 1964). Dust may even have been a major contributor to lateritic and bauxitic deposits (many of which are commercial resources), where analysis reveals that a significant contributor may have been dust blown out of the world's deserts and their margins (Brimhall *et al.* 1988).

In the semi-arid Negev of southern Israel, it is notable that the rates of dust accretion are in some cases closely matched by the rates of fluvial erosion: the amount of sediment taken out of present-day catchments by streams is roughly equivalent to that which is added to them in dust. Whatever the significance of this coincidence, it is plain that dust can have a profound effect on fluvial processes in this, and probably other areas. Yair (1994) noted that a deposit of dust is more permeable than rock, so that dust accumulation can reduce runoff, and hence the rate of fluvial erosion. If, as some authorities believe (see Chapter 8), loess accumulated in wetter periods of the Pleistocene and Holocene, then this process might have created a decrease in runoff during wet periods, not, as might have been expected, an increase. Fluvial erosion might decrease even though rainfall had increased.

Where the dust is no more than a thin covering over rock, however, it is redistributed by runoff very shortly after leaving the aeolian system. In these cases the dust is redeposited as colluvium at the base of slopes; and much dust ends up occupying shallow depressions. These dust-filled dry lakes are very obvious on volcanic terrain, as in parts of Saudi Arabia and Syria, where their pinkish hues contrast vividly with the dark colour of the rock. Dust deposited on hilly desert terrain is quickly washed off and is only retained in valleys, as in Sinai (Dan 1990). In the past some of these deposits, evidently derived from dust, have accumulated as lacustrine terraces in confined valleys, producing very distinctive landforms (Fig. 4.13; Rögner and Smykatz-Kloss 1991). The landforms of loess in wetter climates, where there have been phases of fluvial dissection of the terrain, are discussed below.

Loess

Loess is a material that originated as aeolian dust, and is thus composed mostly of particles in the 10–50 μm range. Quartz is a varying but usually dominant component of the mineral suite, averaging between 60 and 70 per cent (Pécsi 1990). There is also a high percentage of carbonate in most loesses. The origin of the dusts that are the parent material of loess has been discussed above, but loess must not be seen simply as loose dust. In loess the dust has been transformed by diagenesis, in which it is lightly cemented, usually by carbonates dissolved out of the dust itself, and by local reworking by sheetwash or slumping. Some loesses have been thoroughly reworked and redeposited by streams.

It has been claimed that loess covers 10 per cent of the terrestrial globe (Pécsi 1990), although there are lower estimates (Fig. 4.14; Pye 1987). Clearly, figures for cover must vary according to the definition of

Fig. 4.13 A loess terrace in the Wadi Firan, an extremely arid part of southern Sinai. See also Fig. 1.1(c).

what is or what is not loess. Thickness is not an easy criterion, since thin covers of dust are incorporated in soils at varying rates (see above). Grain size is another problematic area for definition, because there is a gradation from sandy aeolian deposits to loess in many parts of the world. In the Low Countries of north-western Europe, the transition may often be sharp (Lebret and Lautridou 1991), but in other areas there is a gradation over great distances. The Blackwater Draw Formation on the southern High Plains of New Mexico and Texas, for example, was apparently blown north-westward from the Pecos River and exhibits a gradually diminishing grain size on this trajectory (Holliday 1989). Some authorities

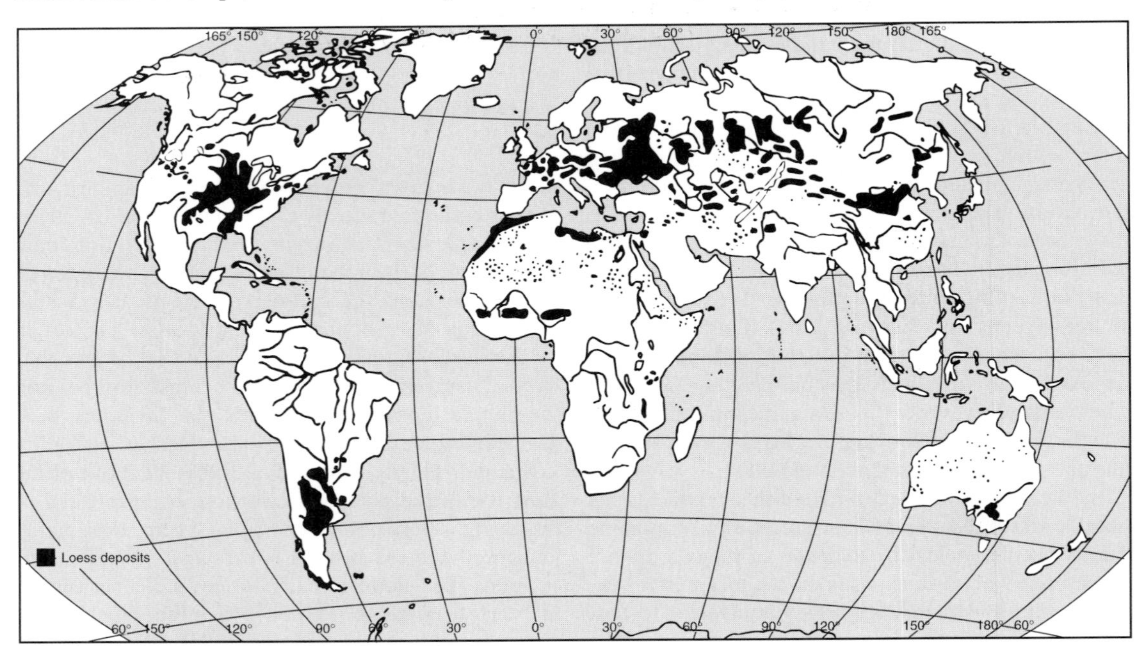

Fig. 4.14 Principal loess-covered areas in the world.

adopt rigid attitudes to the origin of the material, excluding silts reworked by slope or fluvial activity from the loess category, and here the extent of reworking becomes an issue. Thus any definition of loess necessarily has a large arbitrary component, and estimates of its extent can only be very approximate.

Following the discussion above, loesses can be divided into two types, though the distinctions are not always very clear: (1) high-latitude loess, related either to mountains, deserts or glaciers; and (2) peri-desert loess in lower latitudes. The origins of the dusts that created these two loess types has been discussed above.

High-latitude loesses

The high-latitude loesses are by far the more extensive, and until recently attracted the lion's share of attention from sedimentologists and Quaternary geologists. This is because they provide the best terrestrial record of climate changes over the last two million years or so. The only records that exceed the loessic record for detail and completeness are on the floors of some ancient lakes and on the bed of the oceans (where the material is also dust largely of aeolian origin).

In spite of the intense research into the stratigraphy of loess, it is still open to great controversy. One reason is that, although it has the potential of an enormously valuable source of information, it is only recently that dating methods, like palaeomagnetism, and particularly optical dating (Chapter 8), have allowed a good chronostratigraphy. Mineral analysis is an older analytical tool, but one that also still contributes strongly to the understanding of stratigraphy. This is because, in general, loesses of the same age have been shown to have very similar mineralogy, just because the wind distributes them so widely and mixes them so thoroughly (unlike fluvial sediments with a much more localized range of sources and sinks).

The oldest loesses, both high-latitude and peri-desert (apart from the dubious 'loessites' discussed in Chapter 8) occur in China, where they began to accumulate some 2.5 million years ago. In parts of China and Tajikistan the sections that reach back this far are over 200 m thick (Goudie *et al.* 1984) and near Lanzhou in China about 300 m has been recorded (Derbyshire 1983). In these high-latitude situations, most of the loess was deposited in cold, windy and dusty periods; soil horizons developed at contemporary surfaces in warmer, perhaps wetter and less dusty interludes (Fig. 4.15). In the 2.5 million year history of the Baoji section in China, which is nearly complete, there were 37 major cold/warm cycles (Ding Zhougli *et al.* 1992). The warmer, wetter periods were apparently episodes when the south-west monsoon penetrated further west; in the cold dry periods it retreated and was replaced by the north-east monsoon.

The deposition of these loesses (and the peri-desert loesses) intensified during the Pleistocene, the thicker members being towards the top of the sequence. This

Fig. 4.15 Peorian Loess (last glaciation) separated by the Sangamon palaeosol (interglacial) from the Loveland Loess (penultimate glaciation) in Nebraska, USA.

corroborates a picture of intensifying global aridity (Chapter 8), but loess deposition in China and central Asia may have been even further intensified by the aridification consequent on the uplifting of the Tibetan Plateau during the Quaternary. This produced not only the thickening of successive layers of loess, but its progressive extension over new areas further east in China (Zhang Linyuan *et al.* 1991). During the Pleistocene the periodicity of loess accumulation and intervening periods of soil formation can be closely fitted to the Milankovitch rhythms in some sections, as in Alaska (Béget and Hawkins 1989). These rhythms are discussed in Chapter 8.

Even in Britain, where the loess is thin (though defined only as a deposit thicker than 0.3 m), there is a record of Late Devensian, Wolstonian and Anglian loesses in many parts of the country, some interbedded between glacial tills (Catt 1979). Some of the British loesses are interglacial, belying a simple correlation of loess with cold phases. The last of the British loesses, whose deposition peaked in some sections at between 14 000 and 18 000 years ago, seem to have come in easterly winds from the North Sea Basin, then an outwash plain with deposits from the Thames and the Rhine, and has a comparable mineralogy throughout (Eden 1980; Gibbard *et al.* 1987).

Peri-desert loesses

Peri-desert loesses are generally much thinner and sandier than the high-latitude loesses. They have only recently been widely accepted as aeolian, and are now attracting more and more attention, as the review above shows. In the Negev in southern Israel, peri-desert loesses reach only about 12 m thick (Dan 1990), and they are also much less complex than the middle Asian or North American loesses, having far fewer palaeosols (Bruins and Yaalon 1979). But in other respects they are loess-like, for they have 30–60 per cent quartz and up to 70 per cent calcium carbonate (Coudé-Gaussen 1990). The distinctive clay-mineral palygorskite, derived only at very high pH conditions, and therefore undoubtedly of desert origin, can make up between 20 and 70 per cent of the clay-mineral fraction of loesses in Tunisia (Coudé-Gaussen 1987). Many peri-desert deposits of dust are so thin and dispersed that they have been incorporated very thoroughly into soils, as described above.

Australian pedologists identify a distinctive form of clayey peri-desert loess which some of them term 'parna', others calling it more simply 'loessic clay' (Dare-Edwards 1983). Similar clayey loesses have been found in Israel (Bruins 1976). Most of these deposits form a blanket-like covering of the landscape, or 'sheet-parna'. These sheets are best developed downwind of ephemeral lakes or alluvial spreads. The particle size of loessic clay sheets in Australia is bimodal with peaks in the clay and silt or sand fractions. Clay content varies from 30 to 60 per cent, increasing away from the source (Butler and Hutton 1956).

Distribution of loess in the landscape

The distribution of loess is controlled by a number of processes. The most obvious pattern of distribution is deposition in a tapering plume downwind of source. This can be seen very well in the plumes of Pleistocene dust deposits downwind of small playa lakes, for example in Nevada (Chadwick and Davis 1990). It can also be seen in the distribution of loess on the western European coast between Brittany and Holland, where the loess occurs in basins downwind of the outfalls of large rivers (like the Seine) onto the exposed bed of the English Channel (Lebret and Lautridou 1991). Many of the thickest loesses are associated with linear sources along major river valleys, such as the Pecos River in Texas (Holliday 1989) and the Missouri (Fig. 4.9).

Another important control is vegetation cover. If an aerodynamically rough cover of vegetation is essential to the long-term accumulation of dust, as many authorities maintain (Tsoar and Pye 1987), its distribution would be an important control on the pattern of deposition. Chappell's (1995) finding that dust enriched with ^{137}Cs from the period of nuclear bomb-testing in the 1960s had collected more round groves of vegetation than on open ground has been referred to. In the Negev in Israel, mosses seem to have been a major factor in both trapping the dust and stabilizing the surfaces on which it accumulated as loess (Danin and Ganor 1991).

Many less direct pieces of evidence show that little dust is retained on open desert surfaces, there being little vegetation to hold it. Because water is still a major constraint on the distribution of vegetation in semi-arid areas, its distribution is patchy, and this may be one of the main reasons for the patchiness of dust deposition. In the Negev in Israel it is claimed that more dust is deposited on lower slopes where there is more vegetation (Dan 1990) (an alternative explanation is discussed below). The central Asian and western Chinese loesses may be confined almost exclusively to piedmont zones (Dodonov 1991), perhaps for the same reason. A similar explanation has been applied to the loess-edge ramps in Saxony and other parts of northern Germany. In these, the thickness of loess increases markedly away from the ancient position of the ice front, suggesting that they were the result of deposition in areas of increasingly denser vegetation in the glacial period when they were formed (Leger 1990). The importance of vegetation can also be seen in the apparent association of the peri-desert loesses of Tunisia with times of greater rather than lesser vegetation. There may not have been enough vegetation in the drier periods to trap the dust (Coudé-Gaussen 1990).

Bedforms and penecontemporaneous fluvial landforms in loess

Even coarse dust, strictly speaking, is too coherent to form into dunes, and so usually adopts a smooth surface. Occasionally, as when there is a high proportion of intermixed sand, for example close to the source, dune forms can be detected (Leger 1990), but they are usually localized, and fade out downwind. Striations are the only other common 'free' bedform in loess. They are a prominent feature of the central European loess lands (Leger 1990) and of central North America and the Palouse in the Northwest of the United States (see Cooke *et al.* 1993).

In areas with little surface relief, as in south-eastern Australia, loess blankets the landscape as a nearly uniform sheet (Dare-Edwards 1984), but bedforms are created where loess accumulates around topographic obstacles, and these have been the subject of some debate. One school maintains that most deposition occurs in the lee of the obstacle, where wind speeds are low (Leger 1990; Yaalon and Dan 1974). Another school points to observations in the field and in a wind-tunnel that suggest that deposition may occur preferentially on the windward sides of hills (Goosens and Offer 1990), and this is somewhat confirmed by the distribution of loesses in Belgium (Goosens 1988). It is known that wind speeds fall immediately upwind of some obstacles (Chapter 5), and this might have more of an effect on the deposition of dust than of sand. The field evidence is equivocal. For a start, the direction of the winds that laid the dust is not always clear. Thicker deposits upwind of hills may also be formed as dust-sized material is washed or soliflucted off the hill after deposition as dust. The patterns downwind of large hills may also be complex, for in some cases there may be 'hydraulic jump' phenomena in these lee areas, creating complex patterns of dust deposition (Queiroz *et al.* 1982).

Loess was seldom, if ever, deposited into landscapes with no fluvial or slope activity. Indeed, the need for vegetation to trap the dust, explained above, means that many of these were landscapes with enough runoff to sustain active slope and fluvial processes. Thus in many places the dust, once deposited from the atmosphere, was quickly reworked and redeposited at the surface by runoff, and a complex landscape developed, which was composed of the accumulating loess and contemporaneous fluvially cut valleys with terraces and landslides (Fig. 4.16; Zhang Linyuan *et al.* 1991).

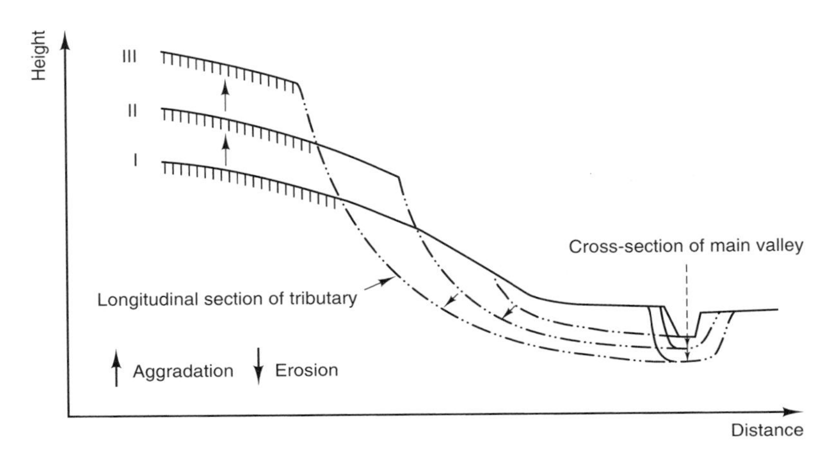

Fig. 4.16 Loess accumulation and penecontemporaneous valley development (after Zhang Linyuan *et al.* 1991).

In these landscapes the distinction between primary aeolian loess and reworked material is difficult if not impossible.

Dust deposited in the deep oceans, contemporary and ancient

The massive accumulations of dust off the West African coast (derived from the Sahara), in the western and central Pacific (from the Asian dust plume), in the South Pacific off Australia, and in the Mediterranean (again from the Sahara, but also from other parts of North Africa) have been repeatedly alluded to above, as some of the best evidence for the global transport of dust. The Saharan plume is best developed between 12° and 25° N; the North Pacific plume between 35° and 42° N; and there is another plume eastward from Australia over the South Pacific. There are further, thinner continental dust deposits in the North Atlantic (derived from North America) and the South Atlantic (from Argentina). Table 4.1 gives some data on present rates of the flux of dust to the oceanic surface.

Dust is thus a major contributor to submarine sedimentation, comparable with inputs from fluvial sources, even in seas with quite large inputs of fluvial sediment. It is almost the only source of mineral sediment in places like the central Pacific and on the Mid-Atlantic Ridge, but even on the bed of the western Mediterranean basin, as much sediment derives from Saharan dust as comes from the River Rhône (Loÿe-Pilot *et al.* 1986). Saharan dust also dominates over all other sources in the North Atlantic Trade Wind belt. But fluvial inputs are, of course, much more important in seas like the Bay of Bengal or the China Sea close to the coast (Chester 1990).

The character of the sediments derived from these inputs is one of the best indicators of their origin as dust. For example, clay minerals in dusts and oceanic sediments are dominated by kaolinite in the lower latitudes (where kaolinite is the main product of continental weathering), and by illite in mid-latitudes (where less kaolinite is produced on the continents) (Chester 1990).

The actual mechanism for accumulation is poorly known, for it has been calculated that dust would take many decades to reach the ocean deeps. It may be aggregated in faecal pellets near the surface, and descend in this way at a greater velocity than it would as single grains. Once on the ocean floor, there is considerable bioturbation, so that time-reconstructions can seldom be finer than 500 years (Tetzlaff *et al.* 1989).

Table 4.1 Flux of mineral dust in 10^9 kg yr^{-1} (after Chester 1990).

North Atlantic (north of the trade-wind belt)	12
North Atlantic (trade-wind belt)	100–400
South Atlantic	18–37
Indian Ocean	336
Western North Pacific	300
Central and eastern North Pacific	30
South Pacific	18

Considering the vast increases in global dustiness during periods of the Pleistocene, it is not surprising to find that submarine deposits of dust, referred to above, were being much more actively accumulated during these times. In the Arabian Sea, for example, where the dust is derived from Africa, the glacial-age maximum of dust production was 160×10^6 t yr^{-1}, compared to the Holocene rate of only about 100×10^6 t yr^{-1} (Sirocko *et al.* 1991). One of the best known of the deposits is off the northern African coast between 12° and 25° N; it extends as a tongue of thick aeolian sediment some 200 km from the coast

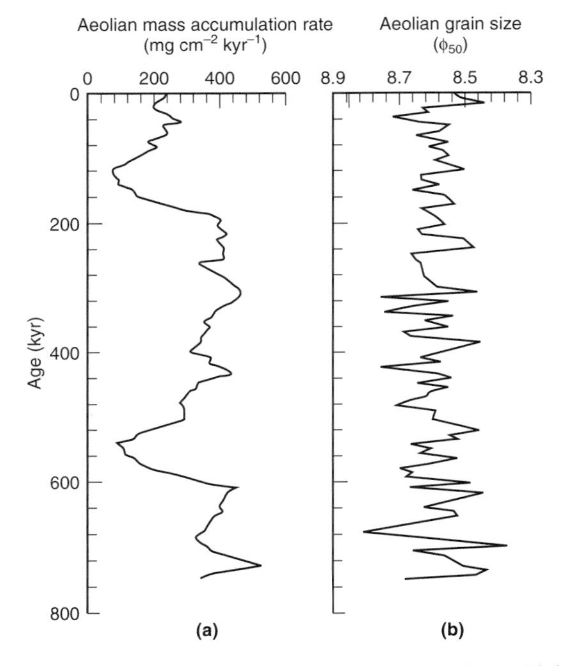

Fig. 4.17 The Pacific dust record, showing variations of (a) accumulation rate, and (b) grain size over time (after Rea 1990).

(Tetzlaff and Peters 1986). The dust stream from the southern Sahara to this deposit is calculated to have carried about 2.5 times more dust at 18 000 years BP than it does today (Sarnthein *et al.* 1981). The dust record, both in terms of mass input and grain size in the Pacific is shown in Fig. 4.17 (Rea 1990).

Submerged, terrestrially deposited loess is far less extensive than the deep-sea dust deposits, but it is interesting for its testimony of conditions when sea level was lower. Loess must have been submerged in many places but, like dune sand (Chapter 8), it may also be too fragile to survive an active wave environment, except in small fragments. Evidence for submerged loess has been cited in the Bohai Sea off the north-eastern Chinese coast (Li and Zhou 1993).

Conclusion

This chapter has shown that the landforms of dust are distinct from those of sand (except in some marginal situations). This is doubtless a function primarily of rapid grain-size fractionation in motion, whereby dust moves quickly away, leaving sand to travel much more slowly. But the separation extends much further: to the mechanics of movement, the mineralogy, the character of the landforms, their location, and to the place of the deposits in the geological record. The distinction between the two types of aeolian landform has required their almost totally separate treatment in this book.

The importance of dust in global geomorphology can be seen in several facets. The first is the extent of dust-covered (loessic) landscapes (which are arguably economically much more important than the sandy lands of semi-arid and arid regions); secondly there is its importance in ocean sedimentation; third is the value of the loessic record of recent climate change; and finally is the importance of dust in contemporary pedology, geomorphology and climatology.

Further reading

The best general and very comprehensive source of information on dust is Pye's book (1987). A good, more recent review on dust was provided by Middleton (1989). Loess in general is also covered in Pye's book (1987) and in a number of recent collections of papers, such as those by Pécsi (1987) and Pécsi and Lóczy (1990). McTainsh (1987) and Coudé-Gaussen (1987) have reviewed the position with reference to desert loess. There is a burgeoning literature on deep-sea dust deposits; possibly the best recent collection is in Leinen and Sarnthein (1989). Chester (1990) provides another useful review.

CHAPTER 5 Dunes

Introduction

Of all aeolian landforms, dunes are the most impressive and the most unambiguously aeolian. Research into dunes over more than a century still leaves great uncertainties, as will be plain in what follows, but considerable progress has been made very recently, and this will be the main basis for this chapter.

An aeolian dune is an accumulation of sand-sized sediment deposited by the wind and shaped into a bedform by deflation and deposition. Dunes are built of rock-mineral sand, sand-sized aggregates of clays, salt crystals or ice. The terms used to describe dunes are illustrated in Fig. 5.1. The definition requires some provisos, first about scale. At the lower end dunes must not be confused with ripples, which have a fundamentally different mode of origin (Chapter 2), and they must also be distinguished from agglomerations of dunes, such as dunefields and sand seas (Chapter 6). Dunes are therefore defined as features between about 0.3 m and 400 m high and between about 1 m and 500 m wide (although there is still some overlap between these dimensions and both ripples and dunefields). Second, to distinguish dunes from yardangs, which are aeolian landforms at the same scale, dunes are defined as features on which the sediment is loose, and is moved grain by grain; the material of yardangs is coherent and is generally eroded by abrasion (Chapter 3).

Dune processes

This section examines fundamental processes shared by all dunes, namely: initiation, replication, early growth, short-term adjustment of windward slopes, slip-faces, and elementary types of movement. Most of these processes can be viewed on a two-dimensional cross-section; in general, it is three-dimensional shape that distinguishes dune types, and this is examined later in this chapter.

Initiation

The initiation of dunes has seldom been documented, and even theory is fragmentary. Dunes could begin to accumulate in many ways. One of the very few studies of initiation, on Padre Island on the Gulf Coast of Texas, found that most began as accumulations around slight indentations in the surface, changes of roughness, or small obstacles like plants, and despite the distinctive conditions on the island, this may well be how most dunes everywhere begin (Kocurek *et al.* 1992). Others have been reported 'calving' off bigger dunes as in Fig. 5.2, though the reasons are obscure. Yet others probably originate where streams of sand-laden winds converge, as round an obstacle, or where secondary flow patterns converge. Others again may be formed by 'ground-jets' (sudden gusts of wind at the surface), which might occur in various ways, one of which is the break-up of waves on an early-morning atmospheric inversion, which Knott (1979) observed at In Salah in Algeria. These jets should be capable of sweeping sand from limited areas, and depositing it when they dissipate. The resulting patches of sand, like the ones accumulated in other ways, could be templates for dunes (Warren and Knott 1983).

The population dynamics of a group of patches of sand involves the rapid dissipation of most, leaving only a few to survive and grow into free, moving dunes. Survival probably depends first on the achievement of some minimum size, the evidence being that dunes smaller than a few metres across are

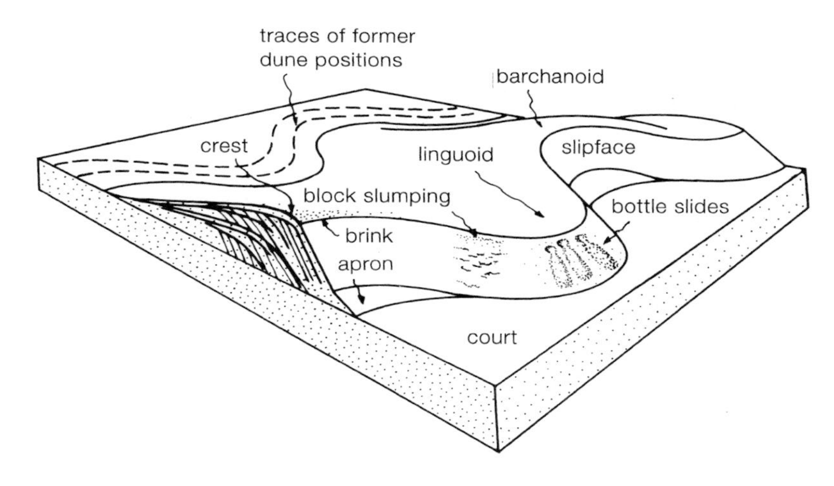

Fig. 5.1 The major features of a dune.

rare. Bagnold (1941) observed fluctuations in sand-carrying capacity when a wind passed over the leading edge of a patch (Chapter 2), and suggested, on this basis, that the minimum size was a patch that was big enough to accommodate these. This gave a minimum of between 2.5 and 6 m (in the wind direction). Bagnold's observations have now been greatly extended, as reported in Chapter 2, but Butterfield (1993), whose work is also discussed in Chapter 2, believed that turbulent fluctuations in the wind meant that any form of equilibrium state was unlikely ever to be reached. There are clearly many more features to be discovered about the nature of flow adjustment and its relation to the minimum size of stable sand patches.

However, there are probably other controls on minimum size, as shown by observations of two dunes within the same ambient wind regime in Oman

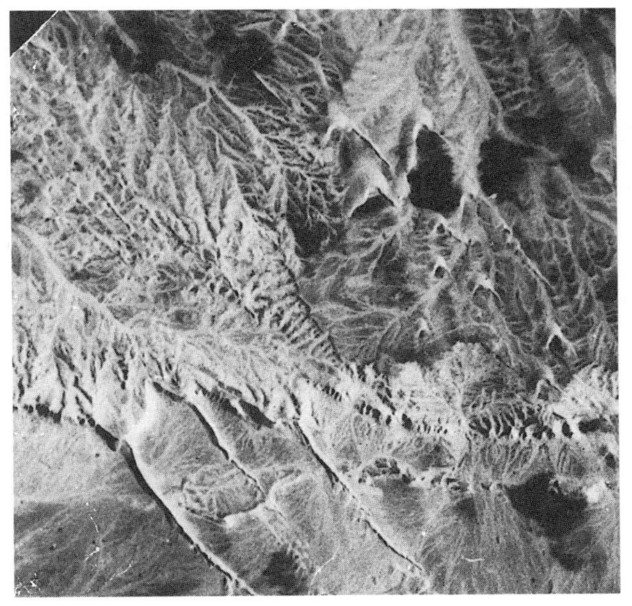

Fig. 5.2 The 'calving' of small barchans downwind of a linear lee dune in central Sinai.

Fig. 5.3 A small barchan dune, Oman.

(Fig. 5.3). The smaller dune, about 2 m long (smaller than Bagnold's supposed minimum), was a mobile, natural and apparently stable barchan. The larger dune, which was about 3.5 m long, was built with a dump shovel on an open, otherwise dune-free plain. The small, natural dune outlived the larger, artificial one, which though becoming a recognizable barchan, lasted only three weeks. It appears that the critical factor in survival was sand supply, for the smaller, natural dune was well supplied, while the larger, artificial dune was starved of sand; the constant and inevitable loss from the lee side of this dune was not being made up.

Replication

The term 'replication' is used here to describe the development of successive dunes all of roughly the same form, and with a regular spacing (Fig. 5.4). This property, though a basic and common characteristic, is still mysterious. Early geomorphologists dabbled with the idea that dunes were formed by gravity waves at the boundary between the less dense atmosphere and the denser cloud of saltating particles, or even the denser bed itself – ideas that were also applied to ripples (Chapter 2). These models were rapidly dismissed (see discussion in Cooke *et al.* 1993). Nucleation round regularly spaced obstacles must also be dismissed as a general explanation, for this simply transfers the problem of explaining regularity to the obstacles, and, short of regular faulting (an explanation that has been used), this is even more difficult to explain.

Three main groups of hypotheses survive (although they are not mutually exclusive, and may even be complimentary). Each is better developed for dunes beneath water, though most authorities believe in

Fig. 5.4 Replication – the repetition of the same basic landform – is a striking characteristic of dunes.

fundamental similarities between these and aeolian dunes. All the models require an initial disturbance or some discontinuity on the bed (perhaps the 'first' dune initiated in a manner described above), though the character of the bedforms thereafter is independent of the discontinuity.

The crucial feature of the 'delayed response' or 'kinematic instability' model is a 'delay factor' which magnifies the initial disturbance to the flow until it reaches some equilibrium configuration which is propagated downstream. Kennedy (1969) believed that the factor reflected two kinds of delay: first, in the flow properties, and second, in the sediment transport (processes similar to those described above in the context of dune initiation). More recently, McLean (1990) endorsed the delay factor as necessary for dune formation in flows of infinite depth, but pointed out that real dunes form rather differently (see below). Both Yalin (1977) and Raudkivi (1976) were sceptical about Kennedy's factor, but his model, though not fully substantiated with empirical observations, is still widely discussed.

The 'organized turbulence' model assumes a pre-existing pattern in the flow, dunes being created when the pattern is fixed by a discontinuity on the bed (Yalin 1977). Yalin's argument was based on the observation that wavelengths (though not heights) of dunes appeared to be already determined before

Fig. 5.5 Incipient dunes, Oman. The dunes are only a few centimetres high, but many metres across.

growth began. Moreover, the wavelengths of subaqueous dunes were strongly related to the depth of flow, suggesting that they related to the largest possible eddies, these being of order of the flow depth. Other 'turbulence' models link subaqueous dunes to surface waves (Hammond and Heathershaw 1981; Allen 1982). Yalin (1977) saw little problem in extending his model to aeolian dunes.

This model has many supporters among those who have studied aeolian dunes, although the evidence, either morphological or meteorological, is sparse. The morphological evidence includes 'spontaneously' developing low, regularly spaced dunes in snow (Kobayshi and Ishihara 1979) and on beaches (Bourman 1986); Fig. 5.5 shows another example. The meteorological evidence is of wave-like motion in the lee of hills and in other meteorological situations, some of which have been thought to be related to dune patterns (Kolm 1985).

The 'flow response' model is related to the organized turbulence model. It requires an initial, fully-formed dune, which produces regular disturbance of the downstream flow. Beyond the first brink, flow takes off and reattaches to the bed downwind, enclosing a 'separation bubble' or 'lee eddy'. Intense turbulence at the reattachment point allows no accumulation of sand, but downwind, in the expanding internal boundary layer, turbulence rapidly diminishes to a point where deposition can occur. This produces a second dune at a position depending on the energy of the turbulence, itself dependent on ambient flow conditions (McLean 1990). Several authorities have noted that in aeolian transverse dunes, the wind does not recover its full pre-dune characteristics until distances of between 10 and 15 dune heights downwind (although this figure must depend also on ambient velocities) (for example, Lancaster 1989a).

The windward slope

Bagnold (1941) developed the classic explanation of how a patch became a dune. Large quantities of sand could be carried over a stony surface, because of the effectiveness of the rebound. The deceleration of this densely charged flow, when it encountered the edge of a patch of sand, induced deposition, and the patch grew to a dune. The cross-sectional form of this new dune would quickly adopt an asymmetric form. Exner (1927, quoted by Graf 1971) explained this process in the following way: as the new dune grew upward, the upper parts were subject to faster winds than the lower parts, and therefore moved forward to produce the asymmetry. The observations on Padre Island produced a further model of this kind of growth: dunes begin as irregular patches; become rippled 'protodunes', over which the flow contracts, and in the lee of which it expands again; further growth induces flow to separate in the lee and sand to fall out into the zone of separation; these dunes

Fig. 5.6 Low, regularly spaced transverse dunes.

then become fully fledged barchans with slip faces; and finally these, in turn, merge with, cannibalize or link laterally with other dunes (Kocurek *et al.* 1992).

The windward slopes of fully developed dunes have three distinct zones: toe, main slope and crest (Burkinshaw and Rust's (1993) phases 1, 2 and 3). Most authorities who have studied these complain of three major problems (Mulligan 1988; Burkinshaw and Rust 1993; Wiggs 1993; Frank and Kocurek 1994; Lancaster *et al.* 1994). As mentioned in Chapter 2, the available methods of measuring shear are very inadequate where, as on these slopes, wind velocities manifestly do not conform to the log-height rule of Kármán and Prandtl. Second, the wind is rarely constant enough for real dunes to reach equilibrium with it (which, as will be seen, may itself explain some of the characteristics of windward slopes). Third, there is strong interaction between form and process (termed 'morphodynamics').

The toe is a place where slope angle, shear, roughness and sand discharge all change abruptly. Most observations and models show that velocity and, more important, u_* (shear velocity) decrease at the toe (Fig. 5.7; Howard and Walmsley 1985; Jensen and Zeman 1985; Tsoar 1985, 1986; Tsoar *et al.* 1985; Livingstone 1986). Theory suggests that this is because flow is backed up against the slope. If the wind were saturated with sand as it approached the dune (which it usually is), sand would then be deposited at the toe, and most mathematical models of dune formation do indeed develop an accumulation of sand here. But real dunes do not experience this accumulation, for, if they did, they would grow backwards into the wind, which they manifestly do not.

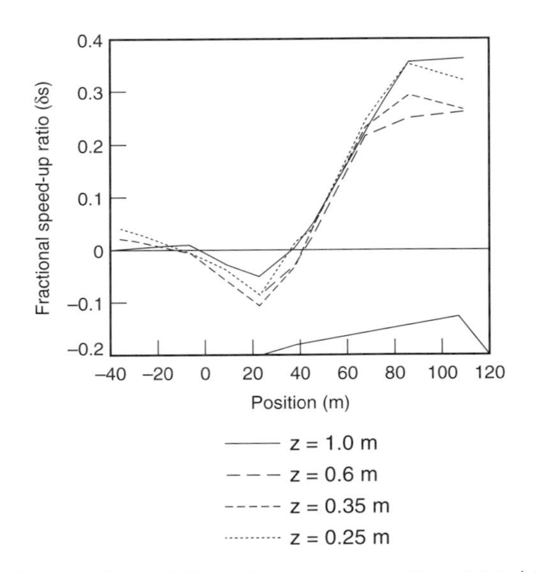

Fig. 5.7 The variation of wind speed with height (z), expressed as a ratio against the upwind speed, over a dune cross-profile, showing a decrease in speed (and apparently of u_*) at the dune toe.

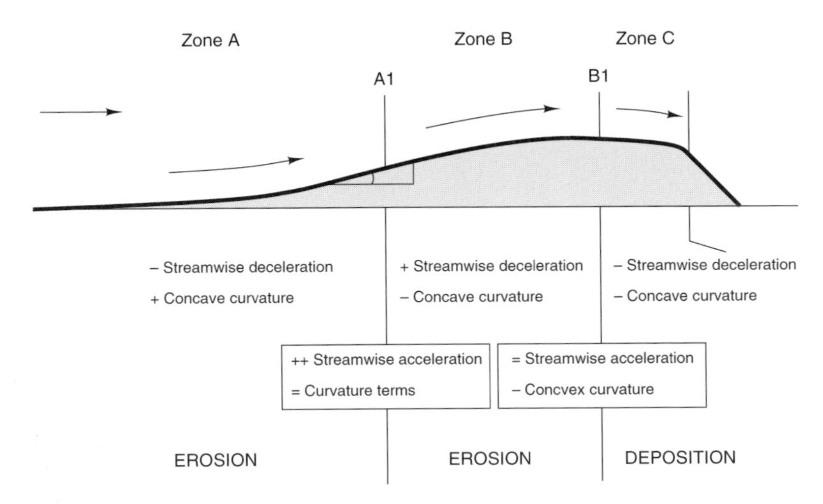

Fig. 5.8 A model which accounts for flow curvature (after Wiggs *et al.* 1995).

Recent wind-tunnel observations suggest that this apparent discrepancy of theory and reality arises from methods of calculating shear which rely only on assumptions about the nature of the velocity profile (Chapter 2), for surface shear stress is not controlled solely by velocity, but also by turbulence. When measurements of Reynolds stresses (indicators of turbulence) at the surface of a model dune are made, they do indeed show additional surface stresses at the toe, despite the decrease in velocity (Wiggs *et al.* 1995). These additional stresses may partly be attributable to concave streamline curvature (zone A in Fig. 5.8) (Finnigan *et al.* 1990). The opposing effects of velocity and curvature must be balanced to maintain transport of sand, suggesting that the toe is a zone of delicate and constant adjustment to the wind.

The main slope is a zone which Bagnold (1941) showed must be the zone of greatest erosion, if the shape of the dune is to be preserved as it advances. It is also a zone in which shear must increase upslope 'such that the increasing volume of sand eroded from the... slope can remain in transport' (Lancaster 1987a: 519). Measured differences between sand flux at the toe and the top of the main slope can be in the ratio of 1 : 42 and more (Lancaster *et al.* 1995). The form that delivers these conditions is apparently the straight slope at 5° to 10° which is found on most natural dunes. For all these observations, the nature of the adjustments of flow over this kind of slope is still obscure, since such slopes induce ground jets, in which flow close to the surface is speeded up relative to flow immediately above (giving a velocity reversal with height) (Hunt *et al.* 1988). The depth of the jet is directly related to the size of the hill or dune, and on most dunes that have been studied, which are small, it has been impossible adequately to measure flow parameters within this narrow layer (Wiggs *et al.* 1995).

Using models and observations of flow and sand transport equations (Chapter 2), the main slope can be modelled mathematically. Although there are problems with many of the models, they do give valuable insights into dune behaviour. Howard and Walmsley's (1985) and Walmsley and Howard's (1985) pioneering models showed that erosion and deposition were sensitive to minor variations in the shape of the slope, and especially to the value of roughness height (z_0) and its spatial variation. This is confirmed by field observations (Reid 1985; Warren 1988a). Wippermann and Gross's (1986) model, though depending on a simple flow model, produced a recognizable barchan from a conical pile of sand in the equivalent of eight days with $u_* = 7\,\mathrm{m\,s^{-1}}$ (Fig. 5.9), a situation not dissimilar to the field observations on the dune in Fig. 5.3. Jensen and Zeman's (1985) model showed that the fluid forces were always trying to steepen the dune, a tendency the authors tried to counter (unsuccessfully) by introducing a slope correction for the transport rate, or a lagged relation between wind speed and sand flux.

The relation between wind speed and slope angle is disputed. Lancaster (1985b) predicted steeper slopes at higher wind speeds, although the models of Howard *et al.* (1978) and of Wippermann and

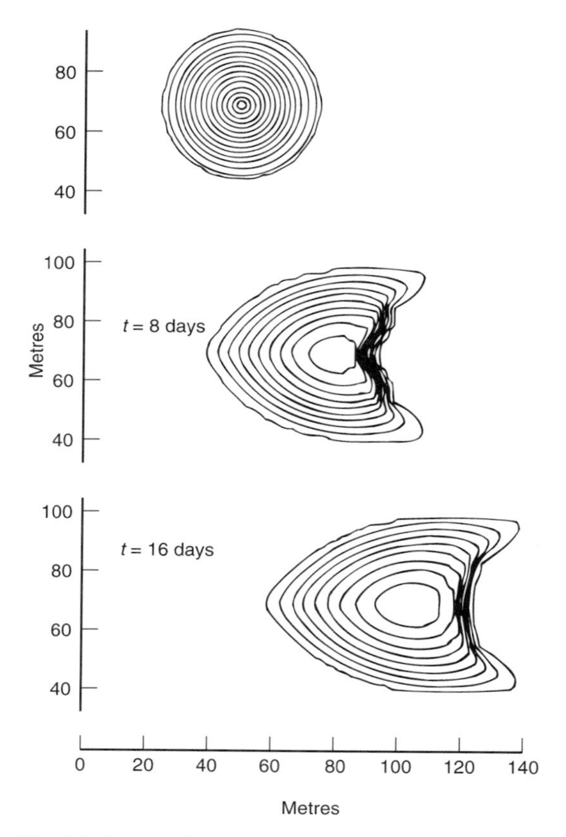

Fig. 5.9 A numerical model of dune development (after Wipperman and Gross 1986).

Gross (1986), and observations in strong wind environments, as in mountain passes, indicate low angles (Gaylord and Dawson 1987).

The crest

The crest presents further puzzles. The difference between dome-shaped crests and those in which the windward slope leads straight to the brink (both common forms) has not been resolved (Fig. 5.10). There are three main groups of theory.

The most widely held explanation, though it is only partial, is evolutionary change. The most prevalent view is that small dome dunes (with marked separation of crest and brink), evolve to larger dunes with less crest–brink separation and finally to dunes with straight slopes to the brink (Lancaster 1987a). Yet many dunes with crest–brink separation, including some dome dunes, seem to be equilibrium forms (Breed *et al.* 1980), and some dome dunes (even up to

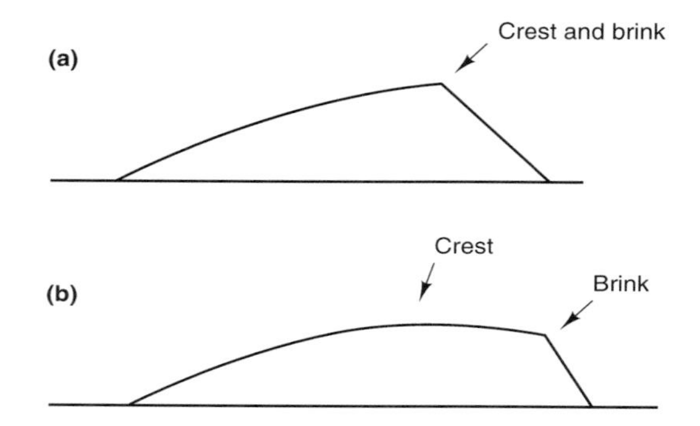

Fig. 5.10 The relation of a dune's crest (the highest point) to its brink (the top of the slip face). In (a) crest and brink are together while in (b) they are separated.

6 m high) are closely juxtaposed to dunes with slip faces (McKee 1966). Furthermore, the reverse evolutionary sequence was suggested by Capot-Rey (1963) and Verlaque (1958), and Fryberger *et al.* (1984) observed changes in both directions.

A second set of explanations is aerodynamic. If flow were to decelerate before the crest, in one argument, deposition would occur and build up the crest. Yet there is disagreement as to just where the point of maximum wind speed occurs. Mulligan (1988) found it to be well before the crest, but Lancaster (1987a) found it at the crest. There could be many explanations for the anomaly, the most likely being that the two sets of observations were made at different stages in the development of the slope. Moreover, theory suggests that the position of the peak value is sensitive to the shape of the slope, the u_* value, the wind direction and the character of separation in the lee (Jensen and Zeman 1985). A related suggestion is that crest–brink separation is a consequence of lag between changes in shear and the rate of sand transport (Bagnold 1941). The point where deposition takes over from erosion, the brink, would therefore be downwind of the point where shear begins to decline, which is assumed to be at the crest. Because the lag depends only on wind speed and sand size and not on dune size, crest–brink separation would become progressively less marked as the dune grew, and this prediction was confirmed by Hastenrath's (1978) field observations. Finally, in the aerodynamic category, is a theory involving flow curvature, for unlike the toe, flow is curved in a convex fashion over the crest, and, by analogy, this should complement any other mechanism that decreases shear and so sand transport (Fig. 5.8; Wiggs *et al.* 1995).

A third explanation for dome-shaped crests also has its origin with Bagnold (1941). In varying wind speeds, the point of maximum shear would be in constant migration, and the point of maximum deposition would also migrate. Crests are undoubtedly subject to greater variations in sand transport than lower slopes, for when winds are light, the dune crest may be mobile when the base of the slope is static. Lancaster (1985b) found that, in light winds, there was a ratio of 158 : 1 between the sand transport rate on the crest and that at the windward base; at higher wind ambient velocities the ratio dropped as low as 13 : 1. Thus light winds erode the crest while the base of the dune is inactive, and so create a convex profile, if not crest–brink separation. In this explanation, therefore, the crest of the dune readjusts to changing winds, and never achieves equilibrium. If, as is likely, the period of adjustment is of the order of a day, the crest would be constantly readjusted (Hunter and Richmond 1988).

For whatever one (or more) of these reasons, there is certainly deposition on the crests of many dunes, for they are covered with a 'cap', about 0.5 m at its thickest, consisting of low-angle sets of ripple strata, separated by many truncation surfaces, dipping downwind (Fig. 5.1; McKee 1966; Embabi 1970/71).

The lee slope

Discussion here of the lee slope is confined to the processes in the lee of dunes that have brinks and steep lee faces (and ignores the poorly understood processes in the lee of zibars and dome dunes). As most dunes grow upward, there comes a point, depending on ambient flow characteristics, at which the wind can no longer follow the downwind slope, and the flow separates from somewhere near the crest. The flow here is highly erratic, with major gusts and reversals of flow, but its existence is easily detected with smoke (Fig. 5.11) and in the patterns of ripples and small shadow dunes (Hoyt 1966; Sharp 1979; Hunter and Richmond 1988; Sweet *et al.* 1988). The velocities of this return flow are very rarely great enough to have any material effect on the lee slope itself.

A few grains, even of quite coarse grades, travel far beyond the brink (Howard *et al.* 1978; Hunt and Nalpanis 1985), but by far the greater quantity of sand, even in high winds, reaches only as far as the prevailing saltation length, and then falls into the relatively calm air of the upper separation bubble. Since the saltating grains have begun their flights shortly before they pass over the brink, one might expect an exponential decline in deposition on a horizonal plane projected from the brink. This does not mean, however, that there is an exponential decline in the rate of deposition on the lee face. Indeed, Anderson (1988) found a bulge on the lee slope, 0.2 and 0.4 m downwind of the brink, created by more intense grainfall than nearer the brink. He attributed this to two effects: first, the slip face falls away at an angle which is greater than the general

Fig. 5.11 A lee-side eddy visualized using smoke. Overall wind flow is from right to left, although surface flow is from left to right in the lee-side eddy.

falling flight angle of grains; and second, the drag on particles falling into the quieter conditions beneath the brink. He predicted that coarse grains would fall near the top of the slope, and finer ones further out.

Bagnold (1941) predicted that slip faces could not be smaller than a saltation jump length, and this conforms to observation: as a slip face declines in height, there comes a point below which the height rapidly falls off. Anderson's model predicts that particles reach no farther than about 1 m, when u_* is $0.5\,m\,s^{-1}$ and grain size is 250 μm (common values), and this should give a minimum height of the slip face of 0.6 m. If so, dunes over about 1 m in height would have slip faces in most conditions.

When grain fall has built up the lee slope beyond a critical angle, it fails, and sand flows down it to form a 'slip face'. The position of the failure is the 'pivot point', which is at about the crest of Anderson's bulge. After failure, a tiny 5–10 mm scarp cuts rapidly back upslope from the pivot point towards the brink (Fig. 5.12; Hunter 1977a). It is active for many minutes, and feeds a sand-flow avalanche. The flow narrows or 'bottlenecks' through the bulge, and then expands downslope until it reaches a constant (or slowly expanding) width, where it halts. The flow is fastest at the position of the bulge and slows downslope. McDonald and Anderson (1994) measured a rate of $0.2\,m\,s^{-1}$ at the bulge. The avalanche 'tongues' are a few centimetres thick (Lowe 1976) and about 0.2 m wide (Hunter 1985). In the avalanches, turbulence mixes the sand, taking coarser and more platy grains (mica, shale, shell or bark) to the surface, and to the sides, where they may form 'levees', and finally to the toe (Fryberger and Schenk 1981, 1988; Sneh and Weissbrod 1983).

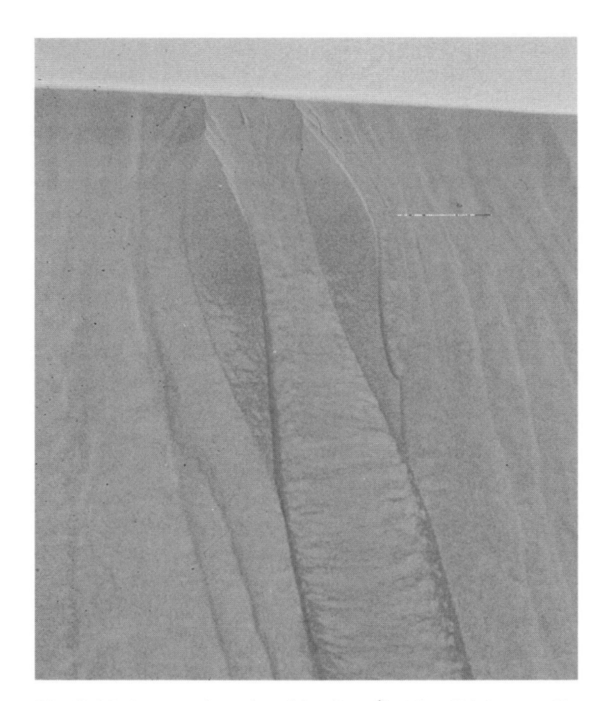

Fig. 5.12 Dry sand avalanching in a 'bottle slide' on a slip face to maintain the angle of repose.

Not all flows reach the base of the slope, for on some the supply of sand runs out before the base is reached, and in others large amounts of sand are immobilized in the levees (McDonald and Anderson 1994). When and if the flow hits the base, a wave travels back up the avalanche at between 0.05 and $0.10\,m\,s^{-1}$ (Allen 1971). The wave may mark the limit between settled, stationary sand downslope, and the more fluid, active avalanche upslope. Avalanche deposits are at almost the minimum possible bulk density (Allen 1971), and include sand grains imbricated with axes pointing directly downslope (Ellwood and Howard 1981).

Slope failure reduces the angle of initial yield to 'the angle of repose' (or the 'residual angle after shearing'). The difference between these two is about 2.5° (Bagnold 1966; Allen 1969; Carrigy 1970). The angle of repose (the angle of most of the slip face) alters little with sand grain size, but is sensitive to grain angularity, moisture, salt content, and perhaps static electricity (Van Burkalow 1945; Allen 1969, 1970; Carrigy 1970). It is commonly between 30° and 33°.

Bagnold (1966) suggested that if some force, such as someone jumping onto the slip face, were to increase the descent velocity of the avalanches above its normal speed, then an oscillating dilation and compaction might take place, and this might be the mechanism of sound production in booming dunes. His model of this process produced frequencies close to those of acoustic sands measured in the field.

When there is cohesion in the slip face, as from early morning dew or in salty environments, whole blocks of sand may slowly slump without mixing (Fig. 5.13; McKee 1979a). When they are only weakly coherent, the slumps soon disintegrate, and movement on most lower slip faces is almost wholly by sand-flow (Schenk 1983), but rain or dew may create very coherent slumps that take sand all the way to the base of the slip face.

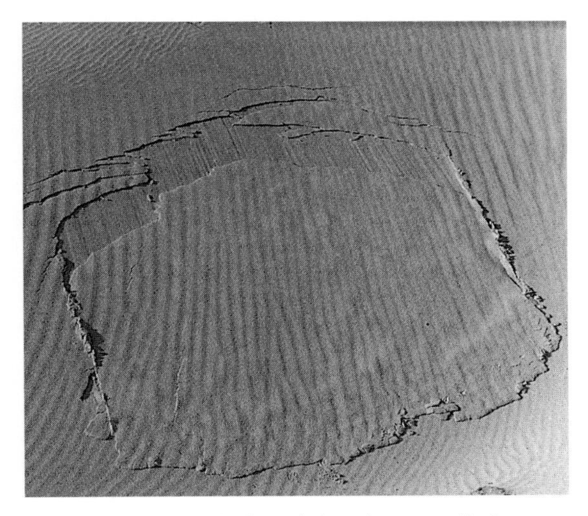

Fig. 5.13 A wet block of sand slumping on a slip face.

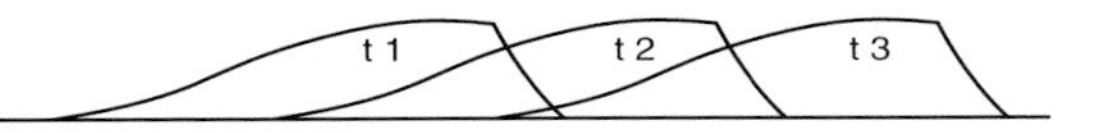

Fig. 5.14 The downwind progress of a transverse dune.

Movement

Mobility is the most striking and alarming property of dunes. A dune that moves a metre in a day during windy seasons is not uncommon (for example, Hunter and Richmond 1988), and this can be a considerable nuisance: dunes repeatedly bury roads, gardens and buildings (Busche *et al.* 1984). Capot-Rey (1957) described how the dune that had covered the camel market at Faya-Largeau in 1935 had moved a kilometre by 1955 (equivalent to $20\,\mathrm{m\,yr^{-1}}$). Haynes (1989) discovered that a barchan, at the base of which Bagnold had camped, had moved relentlessly at $7.5\,\mathrm{m\,yr^{-1}}$ in the 57 years since.

Movement occurs in all un-anchored, unstabilized transverse dunes (Fig. 5.14). It is the consequence of piecemeal deflation on the windward slope, and of deposition in the lee. The rate of movement, V_d, can be expressed as follows (Simons *et al.* 1965):

$$V_d = q_b / kH\rho_{pb} \qquad (5.1)$$

where q_b is the transport rate of the sand trapped by the slip face. $q_b + q_{th} = q$, the total transport rate, where q_{th} is through-going (untrapped) transport. In dunes in near-unidirectional wind regimes, most of the sand arriving at a slip face is trapped (i.e. $q_b \gg q_{th}$). Where winds are more variable in direction, q_{th} approaches q_b, and dune advance is slowed relative to the total drift potential, DP (Chapter 2). In equation (5.1), $k = A_c/LH$ where A_c is the two-dimensional cross-sectional area of the dune, and L is its wavelength (distance between neighbouring dunes downwind). For a simplified, triangular, continuous dune, $k = 1/2$ (Rubin and Hunter 1982); H in equation (5.1) is the height of the dune; and ρ_{pb} is the bulk density of the sand in the dune. Variations of this basic formula have been produced for different purposes (Bagnold 1941: 204; Wilson 1972a; Greeley and Iversen 1985: 185). The argument, of course, refers only to transverse dunes; the growth (rather than movement) of linear dunes is discussed below.

An assumption of equation (5.1) is that dunes maintain their shape in movement. This can only apply to mature dunes, supposedly in equilibrium with their wind and sand-supply environment, for 'younger' dunes are accumulating some of the sand that arrives on their windward slopes, and sand supply conditions do change. None the less, the broad predictions of the formula are adequately confirmed by observations (Fig. 5.15). Doubts remain, however, on two counts. First there is doubt about the size–movement curve at either extreme of the size range. Finkel (1959) found that a linear formula, as above, was a good predictor of dune movement for dunes between 2 and 7 m high, but was wildly out for small dunes of the order of 1 m high. This may have been due to the higher bulk densities of low dunes (Hastenrath 1978), or to a more variable wind environment for smaller dunes, sheltered by large ones. At the other end of the size range, Fig. 5.15 shows that in many cases the relationship should be exponential rather than linear (Sarnthein and Walger 1974). Larger dunes seem to reach a plateau in their rate of movement beyond which size makes no difference. This may be due to greater wind speed-up. Second, there are very few observations of continuous dunes, where much more of the sand is trapped (where $q_b \gg q_{th}$), although some observations in this condition do confirm the predictions of the formula (Warren 1988c).

Bulk transport is the transport of sand by dune movement (the rolling over of sand in the dune), as opposed to transport across the desert floor by saltation (discussed in Chapter 2) (Lettau and Lettau 1969). Hunter *et al.* (1983) developed a formula for the amount of sand in bulk transport (Q_b):

$$Q_b = kHV_d$$

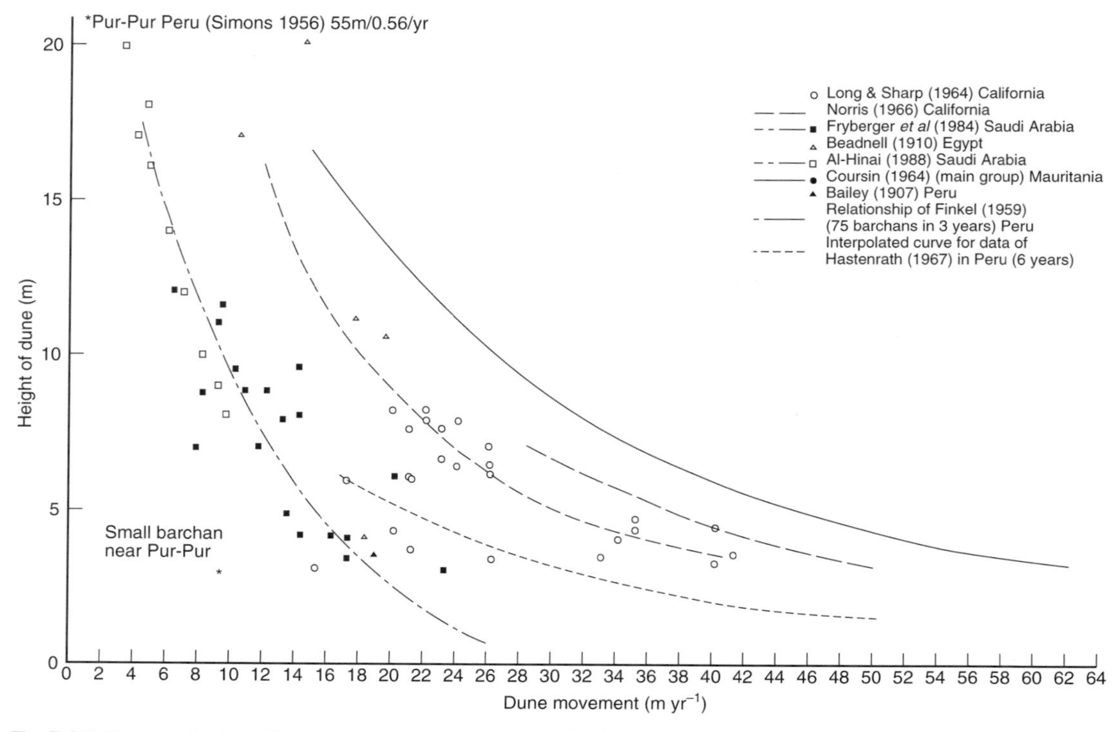

Fig. 5.15 Measured rates of transverse dune movement (after Cooke *et al.* 1993).

where k, the form factor $= A/LH$; A is the cross-sectional area of the dune; L is dune spacing; H is dune height and V_d is dune migration speed (Hunter *et al.* 1983).

Bulk transport can be calculated from data on the shape (volume), bulk density, and rates of movement relative to size in a field of dunes. In Baja California, Inman *et al.* (1966) found bulk transport in a field of low, but continuous transverse dunes to be 23 m^3 (m-width)$^{-1}$ yr^{-1}. In a field of dispersed barchans in Peru Lettau and Lettau (1969) found bulk transport to be 5×10^3 m^3 (m-width)$^{-1}$ yr^{-1}. In another field of barchans, in Mauritania, the figure was calculated at only 1.1 m^3 (m-width)$^{-1}$ yr^{-1} (Sarnthein and Walger 1974).

Classifying dunes

Although the fundamental processes, described above, are common to all dunes, there are a multitude of dune forms. The first stage in explaining this variety is to classify it. There have been many classifications (e.g. Melton 1940; Hack 1941; Petrov 1976; McKee 1979b; Mainguet 1984; Thomas 1989a; Pye and Tsoar 1990; Cooke *et al.* 1993), which differ for two main reasons. First is the very variety itself, the range being different in the different areas with which the various authors were familiar. Second is the difference in the purposes of classification, some being for practical purposes (like a classification of dune trafficability); others being for geomorphological or meteorological interpretation. The plethora of the terms that have now been used makes it very difficult to compare the different systems (see, for example, the lists of synonymous terms in Breed and Grow (1979)).

The aim of the classification suggested here (Fig. 5.16) is to simplify the explanation of dune formation. But, because it is as yet impossible to provide a classification based on consensus about the formative processes, the system is based only on elemental forms. It is a modification and extension of the system of Cooke *et al.* (1993), but unlike it, stabilized forms are not included; because of their palaeoenvironmental significance they have their own chapter (Chapter 8). The classification recognizes a fundamental difference between dunes which

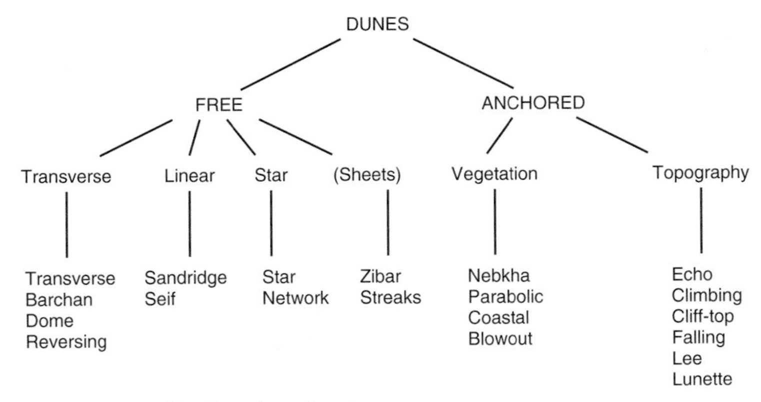

Fig. 5.16 A classification of aeolian dunes.

develop because sand is immobilized by vegetation or topographic obstructions and those which develop without these obstructions. This latter type is termed 'free' here and 'self-accumulated' by Pye and Tsoar (1990).

Free dunes types

Transverse dunes have slip faces that all face roughly the same direction and are characterized by net sand transport normal to their crest. Some are lengthy continuous ridges (Fig. 5.6a); others are discontinuous, in which case they are termed 'barchans', which are crescentic dunes isolated on firm, coherent desert surfaces, such as desert pavement (Fig. 5.17). Some are little more than 0.5 m high, in which case they migrate very quickly and can be re-orientated by any change in wind direction lasting more than a few hours. Others, are over 100 m high, as in the Idehan Mourzouk in Libya, and probably very stable in shape although slowly migrating. The curvature, and wavelength of the curves, in transverse dunes is very variable, but one undoubted control is the height of the dune, the radius of curvature being bigger in higher dunes.

Although they have no slip faces, dome dunes are included here because they have an orientation and pattern of sand transport akin to transverse dunes. Equally, reversing dunes are included with transverse dunes, because they experience net sand transport normal to the crest; in these dunes slip faces develop on opposite sides of the crest in response to a bimodal wind regime with diametrically opposed modes.

Transverse dunes cover some 40 per cent of active and stabilized sand seas (Breed *et al.* 1979a). Small barchans, which re-orientate quickly to changing winds, are very common, but barchans over 2 m in height are rather rare; it has been estimated that they contain less than 1 per cent of aeolian sand (Wilson 1973). Barchans with heights of the order of 100 m are confined to very constant annual wind-regimes; some stabilized ones in the Nebraska Sand Hills were described by Warren (1976a). Some extraordinary collections of small dunes in barchan-shape have been termed 'mega-barchanoids' by Kar (1990).

On *linear dunes* net sand transport is parallel to the crest (Fig. 5.18). They frequently have slip faces on either side of a central crest line, although only one of these is active at any one time, the activity alternating seasonally, or more occasionally, daily. Linear dunes

Fig. 5.17 Classic barchans in Chad.

(a)

(b)

(c)

Fig. 5.18 Different types of linear dune: (a) a seif 3 m high, Namib Desert; (b) a large linear dune 80 m high, Namib Desert; and (c) sand ridges, Kalahari Desert.

are often divided into sharp-crested '*sayf*' forms and more rounded sand ridges (even though the origin of these two may not be greatly dissimilar). The term 'linear dune' is used here generically to cover all dunes which have been commonly termed *seif* (also *sayf* and *sief*), 'longitudinal' or 'sand ridge' because this avoids genetic connotations.

Linear dunes, which are found in all the world's major sandy deserts, vary in shape and size even more than transverse dunes. They reach from less than 2 m high to around 150–200 m in the Namib Desert, the Great Eastern Erg of the Sahara and the Rub' al Khãli in Arabia. Some extend for tens of kilometres, and some exhibit considerable parallelism, although it is a mistake to see all linear dunes as great sheets of corduroy. Many are highly sinuous with very varied spacing, and they may join at what are known as 'Y' or 'tuning-fork' junctions. Much has been made of the regularity of form of linear dunes, but, though there is remarkable regularity in some linear dune fields, there is very little in others, such as those of the south-west Kalahari (Bullard *et al.* 1995).

Dune networks (Fig. 5.19) and *star dunes* (Fig. 5.20) are patterns in which there is a confused set of slip faces pointing in several directions (though not all are active at any one time). Dune networks occur in continuous sand cover, and contain individual dunes that are no more than a few metres high, spaced in the order of 100 m apart. They are very widespread. In plan form, star dunes have a number of arms radiating from a central peak. At the peak, slope angles may be quite steep (15°–30°), but the body of the dune rests on a plinth with much lower angles. Lancaster (1989b) reported star dune spacings from 150 m to more than 5000 m. They are reported to be up to 400 m high in the Ala Shan, China, and the Lut of Iran, and over 300 m high in the Namib and the Great Eastern Sand Sea of Algeria. Although much less common than transverse or linear dunes, star dunes are found in many

Fig. 5.19 Network dunes, Wahiba Sand Sea, Oman.

of the world's major sand seas, but the only sand sea where they cover a large area is the Grand Eastern Sand Sea of which they cover 40 per cent. These striking features have many local names including demkha, ghourd, rhourd, ogrhoud and sand pyramid. In both networks and star dunes, sand makes little overall progress.

Zibar are dunes with no slip faces which are formed from coarse sand and have hard surfaces (Fig. 5.21; Nielson and Kocurek 1986). Modal sand sizes of 2000 μm have been described for the sand in zibars in the Selima Sand Sheet area of north-western Sudan (Breed *et al.* 1987). They are very widespread in some deserts, and certainly cover a more extensive area in total than do barchans or star dunes (even together). Most zibar are of low relief, though some of the zibar in the Ténéré Desert are higher than nearby sayf dunes and reach over 5 m (Warren 1972). Zibar are common upwind of sand seas in zones from which finer material has been winnowed (Warren 1972; Lancaster 1983a; Breed *et al.* 1987). They occur both in extreme deserts such as the central Sahara (Fig. 5.21) and in lightly vegetated areas such as southern California. Individual sheets of zibars are extensive: Monod's (1958) *mréyé* in Mauritania cover 10 000 km^2. Most are straight in plan form, and transverse to the wind, although linear and parabolic forms have also been described (Anton and Vincent 1986; Gaylord and Dawson 1987; Goudie *et al.* 1987; Haynes 1989). Very little is known of their dynamics.

Equilibrium, morphodynamics, hierarchies and complexity

Three fundamental concepts are essential to understanding the controls on the form of dunes: equilibrium, morphodynamics and hierarchy. A vocabulary for complexity is another useful preliminary.

Two properties in subaqueous and subaerial dunes suggest some kind of *equilibrium*: the first is the way in which they maintain a stable configuration as they migrate; the second is the repetition of form in large groups. However, though intuitively necessary, equilibrium is a rather difficult concept to define. Yalin (1977), for example, suggested that an equilibrium shape was one that produced the smoothest flow and thus equalized energy loss in the stream-wise direction, but this is a difficult property to specify or measure.

At a small spatial and temporal scale, as discussed in the first section of this chapter, 'dynamic equilibrium' can be no more than an abstraction, for real dune slopes are in constant change. Readjustment to changed wind conditions takes time (the reaction and relaxation time), during which the slope is not fully in equilibrium with any wind (Allen 1974). Readjustment to an new wind velocity or a new wind direction is rapid at first and then slows down, probably following some kind of exponential law. Winds that persist long enough to establish equilibrium forms are very rare, for the lag between a change in the wind and

(a)

(b)

Fig. 5.20 Star dunes: (a) Namib Desert, Namibia, (b) Great Eastern Erg.

a change in the dune probably exceeds even the length of the semi-diurnal cycle of speed and direction in desert winds. Thus equilibrium may be a useful concept, but can only be modelled, being most unlikely to occur in the field.

The ideal equilibrium conformation of a transverse dune would reflect its response to an ambient wind regime that is fairly constant in direction, if not speed, and the quantity of sand available. It can be hypothesized that all other dune types are moving towards this end-point, but are arrested before they reach it by changes in wind direction. In a changeable regime, only very small dunes can respond as they have short enough relaxation times to be able to adjust their form. To survive, and reflect a changeable regime, therefore, dunes have to be large, their size allowing them to retain some of the characteristics of a wind from one direction after the wind has changed to another, and to retain some imprint of the second set of conditions when the wind again changes, and so on. For them, equilibrium is dynamic, and could be defined by the range of shapes they adopt over perhaps a decade. These arguments apply particularly to linear and star dunes.

This introduces the second essential concept: *morphodynamics*, being the interplay or feedback between form and process. Four examples of morphodynamics will be explained below: the

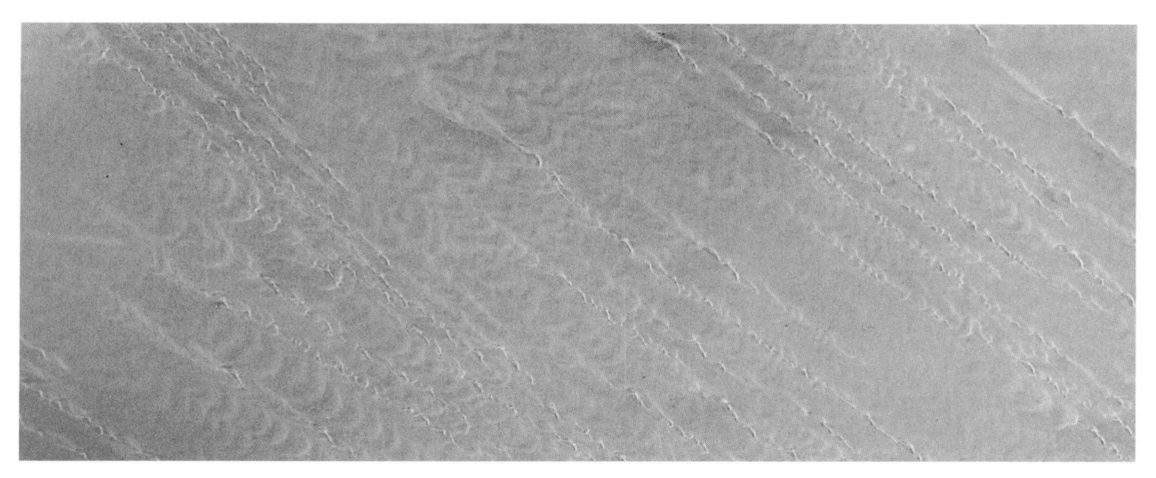

Fig. 5.21 Zibar (the light and dark stripes transverse to the wind), overlain by linear dunes, Ténéré Desert.

three-dimensional barchanoid–linguoid shape of transverse dunes; the process whereby linear dunes increase the near-surface velocity of the wind in their lee, above that of the oncoming wind, if the wind approaches the dune at angles of about 30°; the ways in which star dunes create wind conditions that encourage the accumulation of sand, and the maintenance of their form; and finally the morphodynamics, involving aeolian and littoral processes, of coastal fore-dunes. Morphodynamic responses, it should be noted, are somewhat size-dependent. Only features above a certain threshold size can create fully effective morphodynamic responses, the threshold size being different in different environmental circumstances, the most important of these being the wind regime.

The third fundamental concept in dune morphology is *hierarchy*. Some workers have suggested that dunes are grouped in discrete size categories, this being termed a hierarchy (Wilson 1972b). Empirical support for this view comes from a number of sand seas (Lancaster 1988a), but the explanation is more elusive. Wilson (1972b) believed that the hierarchy was a response to a hierarchy of atmospheric eddies, but there is little evidence of this. An alternative offered by Cooke *et al.* (1993) was that hierarchies were a manifestation of the differing lag time of dunes of different sizes (which they termed 'dune memory'). Small dunes around 1 m high might be responding to daily cycles of wind regime; larger dunes were responding to the annual cycle; the size of the largest dunes, which reached hundreds of metres in height (which Cooke *et al.* termed 'mega-dunes') might be controlled by Milankovitch cycles of climate change.

Finally, a useful *vocabulary* of complexity was introduced by McKee (1979b) and Breed and Grow (1979): 'simple' patterns are ones in which there are no superimposed dunes (they tend to be small dunes); 'compound' patterns are ones in which two dunes of the same type are superimposed or coalesce, almost always involving a large basic pattern overlain by smaller dunes; 'complex' patterns are where two different dune types coexist (as where large linear dunes, adjusted to a complex wind regime, are overlain with small transverse dunes adjusted to the wind of the last few days).

Controls on free dune type

The proliferation of dune forms is the result of complex interactions between a number of factors. For free dunes, the factor that has the greatest explanatory power is the wind regime. Using his system of classifying wind regimes that was explained in Chapter 2, Fryberger (1979) confirmed the findings of many other authorities, in showing that (in general) transverse dunes were associated with unimodal wind regimes, linear dunes with wide unimodal or bimodal regimes, and star dunes with obtuse bimodal or complex wind regimes. He attributed the exceptions to poor wind data. This information can be combined with information about sand transport (Table 5.1).

Explanations based on wind regime can be taken a little further. Work in Australia showed that the amount of sand available for dune building might be a second important control on dune type (Wasson and

Table 5.1 Fundamental desert sand dune types (after Livingstone and Thomas 1993)

Dune type	Wind regime	Mode of activity
Transverse	Unimodal	Migrating
Linear	Bimodal	Extending
Star	Complex	Sedentary

Hyde 1983). Building upon Lancaster's (1994) recent re-examination of Wasson and Hyde's approach (Fig. 5.22a), it has been extended here, as shown in Fig. 5.22b. In the figure, wind direction variability is an index calculated by dividing resultant drift potential by total drift potential (Chapter 2). The amount of sand available was calculated by Wasson and Hyde from an estimate of the depth of sand if all sand in dunes was spread in a layer of even thickness. This parameter has been criticized because it fails to take account of the potential of sand currently in interdune corridors to become incorporated into dunes, and because it does not allow for other controls such as the nature of the desert surface (Mabbutt 1984; Rubin 1984). For Fig. 5.22b, the sand supply axis can be regarded as some unspecified measure of the sand available for dune formation. The positions of the various dune types on the diagram is more speculative than in either Wasson and Hyde's or Lancaster's studies (although it accounts for their data).

On the diagram (Fig. 5.22b) transverse dunes occupy the whole of the lower portion of the directionality (x) axis. This acknowledges the widely observed behaviour of dunes where there is little sand: small dunes re-orient themselves into transverse ridges with any new wind if it blows even for only a day. Examples of these kinds of observation come from Nielson and Kocurek (1987) in respect of star dunes, and Warren (1988a) in relation to network dunes, both discussed below. The diagram extends the area occupied by linear dunes from its restricted area on the Wasson and Hyde plot, following Lancaster's discussion. The originally restricted position of linear dunes seems to have reflected only Australian experience, for in the African and Arabian deserts there are many linear dunes in deep sand. The diagram also introduces network dunes as a thin-sand equivalent of star dunes.

Further help with an explanation of the relationship between wind regime and dune trend was provided by some recent flume experiments (Rubin and Hunter 1987; Rubin and Ikeda 1990). Although

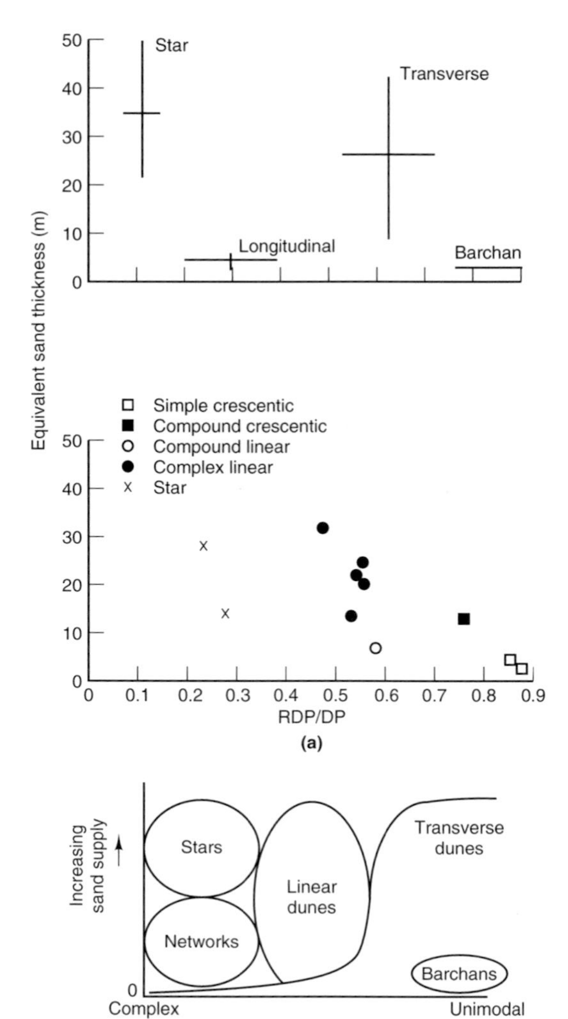

Fig. 5.22 The relationship between wind direction variability and sand supply in controlling dune types: (a) Wasson and Hyde's (1983) original diagram with additional data from Lancaster (1994); (b) a speculative model.

working under water, their results are highly applicable to subaerial situations (Fig. 5.23). A sandy bed was subjected to unimodal and bimodal flows. In the bimodal flows the angle between the modes (the divergence angle) was varied, and the balance between the time that the flow came from each mode was also varied (the transport ratio). Unimodal flows (0° divergence) produced transverse dunes, while obtuse bimodal flow regimes (135° with a

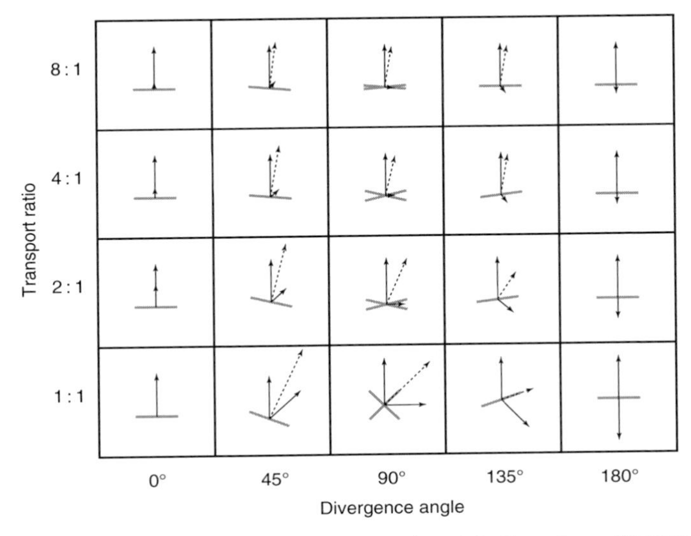

Fig. 5.23 Dunes formed in a flume experiment in flows from different directions. Shaded bars are bedform trend, full arrows are the vectors of flow, and dotted arrows are resultants (after Rubin and Ikeda 1990).

transport ratio of 1 : 1 and 180° divergence, all cases) produced linear dunes; intermediate conditions produced dunes which were transverse, linear, a combination of transverse and linear, or oblique to the resultant transport direction. The conclusion, that transverse dunes formed when the divergence angle was less than 90°, is not fully supported by experience with aeolian dune systems, but flume experiments cannot fully replicate natural systems. In general these experiments offer some useful pointers for future research, especially when allied to computer simulation of dune development (Werner 1994) and experimental studies of ripple patterns (Chapter 2; Goossens 1991).

Free dune processes

Because the fundamental free dune type is a *transverse dune*, their two-dimensional configuration has been discussed in the section above on dune processes. They apparently move downwind according to the relationships discussed above. Figure 5.22 shows that transverse dunes are associated with fairly unidirectional winds, though directional variability can be quite wide before a recognizably transverse form is destroyed (Fryberger 1979; Wasson and Hyde 1983). As the wind regime moves towards greater variability in direction, basic transverse patterns are overlain with network dunes.

The basic form of transverse dunes is the barchanoid–linguoid configuration shown on Fig. 5.1. The explanation of this form must come by analogy with subaqueous ripples and dunes (Allen 1968), for there is no good, empirically verified explanation of it for aeolian dunes. Wilson's (1972a) model for aeolian dunes was based on Allen's ideas. He believed that the linguoid sections of the ridge were swept forward by longitudinal zones of faster flow (the roll-vortices that have been invoked by some authorities to explain linear dunes; see below). Wilson explained the general misalignment of linguoids on successive transverse ridges by a morphodynamic response, in which the roll vortices were displaced sideways as they passed over the brinks of the successive dunes (see Cooke *et al.* 1993, for a fuller explanation of this notion). An alternative model, proposed by Yalin (1977) for subaqueous ripples, invoked spheroidal eddies in the flow. If the analogy with subaqueous dunes is accepted, greater sinuosity should be associated with stronger winds (Allen 1968; Rubin and McCulloch 1980).

The maintenance of the barchanoid–linguoid pattern contains a number of morphodynamic feedbacks (Howard *et al.* 1977, 1978). First, all parts of the dune must migrate at the same rate, implying constant rates of erosion on slopes at different angles to the horizontal and to the oncoming wind. This must be associated with morphodynamically controlled spatial variations in shear. Second, the height of the barchanoid sections (almost always the highest parts) must be controlled by a self-limiting process, for higher dunes mean more divergence, and this means that less sand reaches the crests. Third, the width of the barchanoid element also controls divergence of flow, and thus the height and shape of the dune, although how this morphodynamic process works is obscure. Fourth, the varying height, curvature and angle to the oncoming wind of the slip face must be delicately adjusted to allow the whole slip face to move forward while maintaining its three-dimensional shape. There are probably also adjustments to the linguoid elements, but little research has been applied to these.

Barchans are the simplest form of transverse dune (Fig. 5.17), consisting only of the barchanoid element (with no linguoid attachments). Isolation on a firm surface accentuates the barchanoid shape, for near-surface wind speeds are higher over the pebbly desert surface on either flank, than over the dune itself, and this helps to sweep the 'arms' of the barchan forward (Bagnold 1941). Some of the sand arriving from upwind is channelled round the flanks, and leaves from the wings, trailing downwind. The barchan 'court' (Fig. 5.1) is kept partly clear of sand by the intensely turbulent wake in the lee of the slip face (Knott 1979). Courts may be up to 5 km long on barchans no more than 10 m high. The description of the experimental barchan above (Fig. 5.3) shows that a barchan can only survive if the unavoidable loss of sand from the wings is replaced by sand from upwind. Thus barchans are continuously renewing stores of sand. In a Peruvian barchan field, it was estimated that an average 3 m high barchan gained and lost some $18\,m^3\,yr^{-1}$, which meant that its sand was totally renewed in 64 years, while the dune had travelled 1.7 km (Lettau and Lettau 1969).

There have been a number of observations of the allometry of barchan form. Wings apparently lengthen relative to the body as a barchan grows (Capot-Rey 1957; Verlaque 1958), width between wings and crest height, as well as windward slope angle and height, also change (Finkel 1959; Hastenrath 1967). In a field in which there are many sizes of barchan (as is common) the smaller ones move more quickly than the bigger ones (by equation (5.1)). Most of these small dunes end their lives by colliding with and being absorbed by larger neighbours.

Dome dunes, as the term is used here, are dunes with no slip face, and are built of fairly fine-grained unimodal sand (Breed and Grow 1979). Most are only 1–2 m high, though some are as high as nearby barchan and transversal dunes. Dome dunes, defined in this way, may not be completely distinct from zibars, and indeed Tsoar (1986) found that low, 'flat' dunes in Arizona and California could have a wide range of grain sizes; in other words, they could be dome dunes or zibars. The dynamics of dome dunes are very poorly understood.

Reversing dunes could be regarded as transverse dunes that reverse, or as a special case of dune networks (below) in which the winds of two seasons are diametrically opposed. They occur in distinctive wind regimes, as for example in the Great Sand Dunes of Colorado where they respond both to anabatic and katabatic winds generated by the nearby Sangre de Cristo Mountains, and to the regional Westerlies (Andrews 1981). Reversing dunes are quite common, though in general one of the seasonal elements is much smaller than the other, giving rise to a pattern of small 1–2 m high slip faces on top of the larger slip faces associated with the stronger wind (Cornish 1897; Hedin 1903; Burkinshaw and Rust 1993). The process of erosion on the erstwhile slip face is probably accelerated by thresholds of movement that are significantly lowered at high angles, and by significant speed-up on slopes that are distinctly out of adjustment with the reversed flow (Burkinshaw *et al.* 1993). Reid (1985) found a complex morphodynamic system of grain sorting by the two winds on a reversing dune.

The origin and dynamics of *linear dunes* have been the object of considerable contention, much of which is reviewed elsewhere (Cooke *et al.* 1993; Pye and Tsoar 1990).

Nineteenth-century explanations of linear dunes included a submarine origin, proposed by Sturt (1849) in Australia and Stapff (1887) in the Namib, and a link with tectonic horst-and-graben patterns in the Thar (Frère 1870). Others argued that the dunes were the consequence of interdune deflation rather than deposition (Blanford 1876; Aufrère 1928; Capot-Rey 1945; Mainguet 1983), a theory which was termed 'windrift' (Belknap 1928) and may apply to some lineations in loess (Chapter 4). Where there was no deposition of eroded material on the residual ridges this process would form yardangs rather than dunes. There is some evidence from Australia that deflation of fine-grained alluvium is responsible for the formation of some linear dunes (King 1956, 1960; Bowler and Magee 1978; cf. Mabbutt and Sullivan 1968), and it is also claimed as the origin of linear-like features in sandy material in Hungary (Borsy 1993), but most authorities now believe most linear dunes to be predominantly aeolian depositional features. In some places linear dunes have been regarded as the greatly extended arms of parabolic dunes (see below) (Cornish 1908; Verstappen 1968, 1970) and 'Y' junctions have been interpreted as advancing blowouts (see below) (Madigan 1936, 1946; Folk 1971a). Others have seen them as the result of asymmetrical extension of the arms of barchan dunes (Bagnold 1941; Tsoar 1974; Lancaster 1980), although there is some dispute over the nature of the response to an asymmetrical wind regime.

The most vigorous discussion has been between those who suggest that linear dunes are mainly the product of bimodal wind regimes, and those who hold that 'roll-vortices' may have a part to play in their formation.

The 'bimodal' hypothesis has the support of field studies that have monitored processes on linear dunes, these being a major advance in the recent past. Both Livingstone (1986, 1989a, 1993) and Tsoar (1978, 1983a) worked on active linear dunes that were responding morphodynamically to bimodal wind regimes, and strongly argued for the bimodal hypothesis, although they differed over the exact mechanism of near-surface wind-flow modification (Fig. 5.24). Both showed that the crest of the linear dune migrated laterally in response to seasonally bimodal regimes, but that net sand transport was along the dune. In the Namib, lateral migration was around 15 m back and forth per year (Livingstone 1989a) and subsequent surveys confirmed a highly active crest on a relatively stable plinth (Fig. 5.25; Livingstone 1993). The downwind tips of the Namib dunes were found to be advancing at up to $1.8\,\mathrm{m\,yr^{-1}}$ (Ward 1984).

Based on his work in Sinai, Tsoar (1983a) argued that the intrusion of the dune into the boundary layer affected both speed and direction of the flow over the dune. At the crest a lee-side separation bubble was created, just as with a transverse dune, but because flow in a bimodal regime was oblique to the crest, the reverse flow on lee slope had a strong along-dune element. Tsoar also found that in the lee of sinuous seif dunes wind was accelerated above the velocity of the oncoming wind, causing erosion on some lee-slope elements. Where the dune bent round, this lee-slope wind was decelerated, and this caused deposition in some places, giving rise to a succession of cols and hills (Fig. 5.24a). Fluctuating seasonal winds caused erosion and deposition in different areas, and in general moved the cols and hills along the dune. In general also, the accelerated flow carried

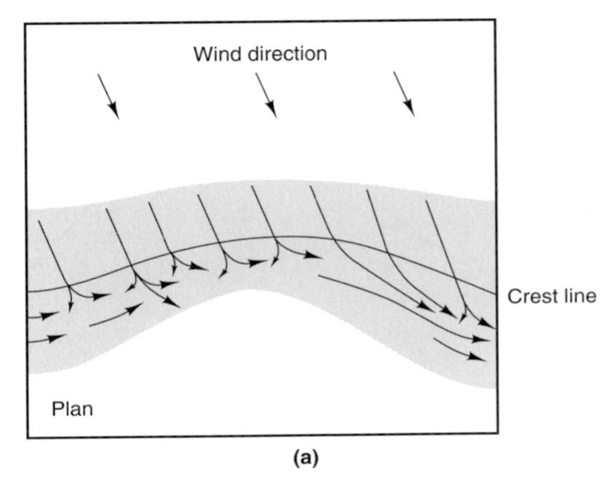

(a)

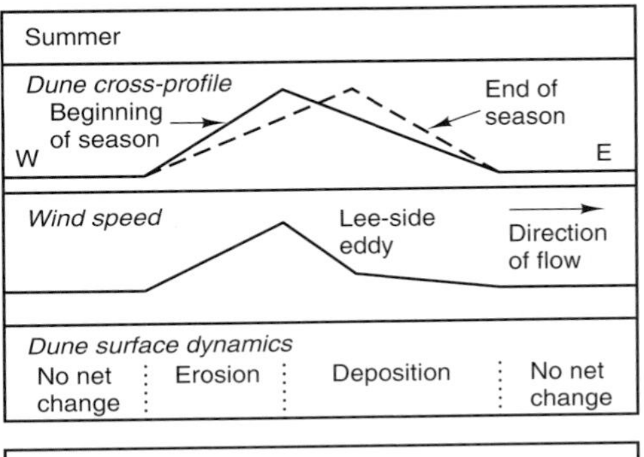

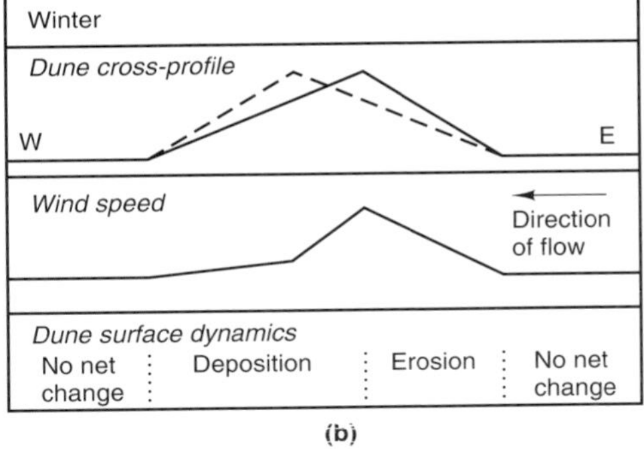

(b)

Fig. 5.24 Bimodalist explanations of linear dunes: (a) a model relying on deflection of flow on the dune's lee slope (after Tsoar 1983a), and (b) a model relying on the distribution of surface wind velocity over the dune cross-profile (after Livingstone 1988). In both cases patterns of erosion and deposition are reversed seasonally.

sand rapidly along the dune, extending it downwind by many metres per year.

In the Namib, Livingstone (1986) showed that the lee-side acceleration which Tsoar saw as so crucial could not control the entire lee slope of large linear dunes. Livingstone argued that the pattern of wind

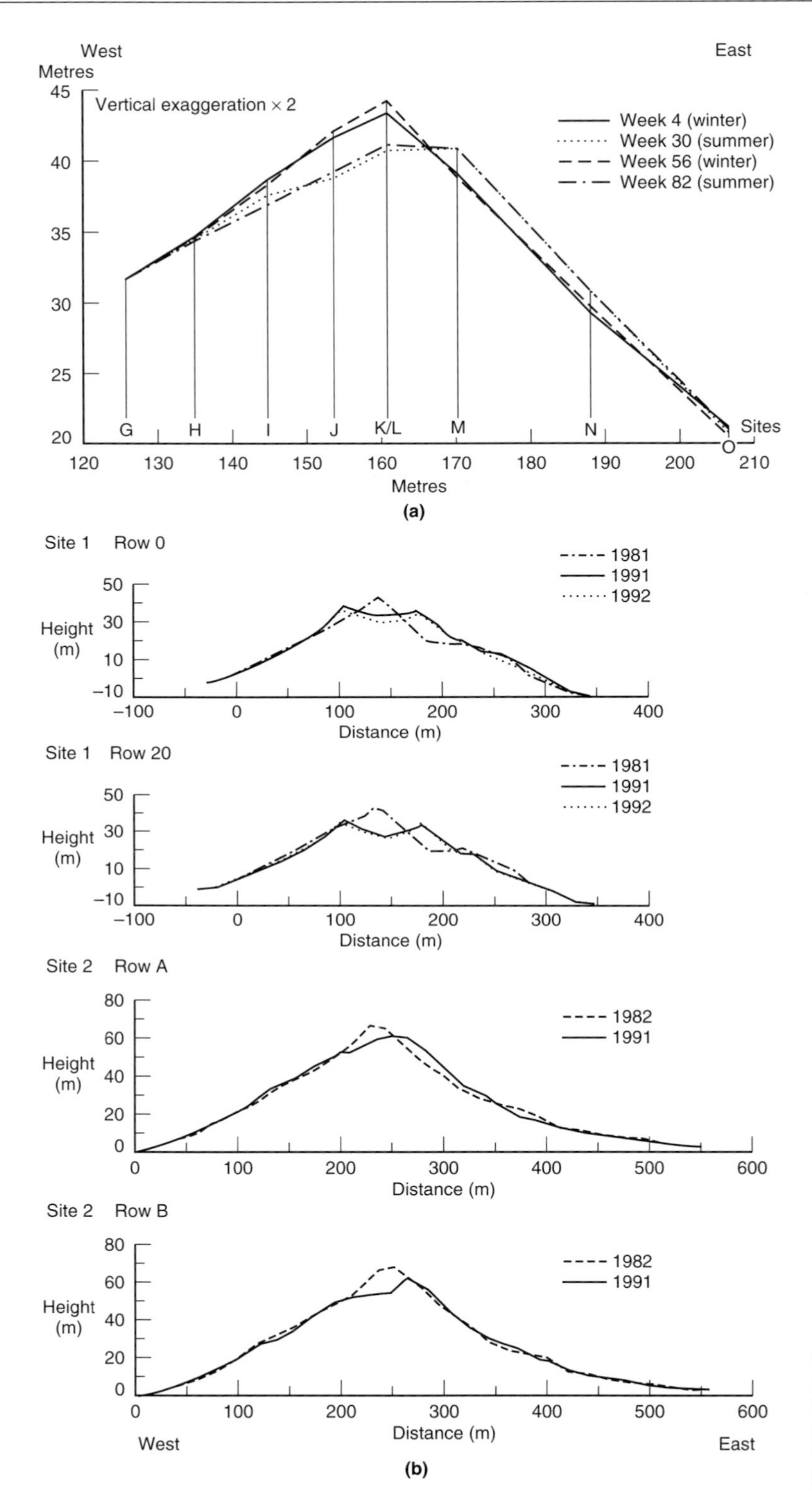

Fig. 5.25 Measurements of cross-profile change on a Namib linear dune: (a) seasonally over two years (after Livingstone 1989a), and (b) for four profiles over 11 years (after Livingstone 1993).

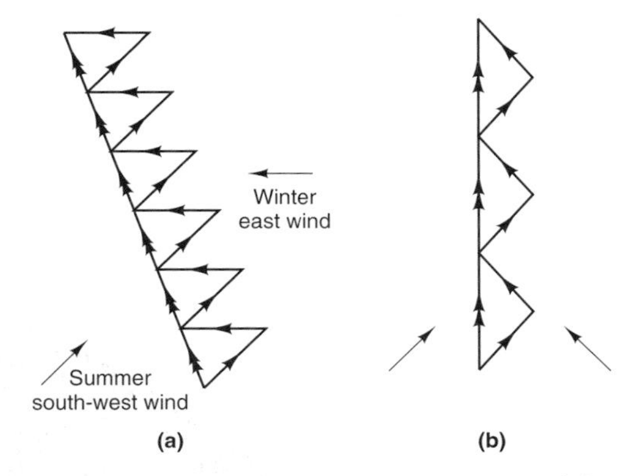

Fig. 5.26 Patterns of advance of linear dunes in (a) an obtuse bimodal wind regime, and (b) an acute bimodal regime. Resultants, marked with double arrows, represent the dune extension in one annual cycle (after Livingstone 1986).

velocity, used as a surrogate for surface shear stress, which followed the pattern of flow over transverse dunes (explained above), was sufficient explanation for the maintenance of the dune's form. He also argued that the angle between the modes of the wind regime might, in part, control the rate of advance and the height of the dune. In acute bimodal regimes, as in Sinai, the throughput of sand along the dune would be rapid and low dunes would be formed, while in obtuse regimes, as in the Namib, elongation would be slower and sand would pile up and form higher dunes (Fig. 5.26). The Namib dunes, which are 50–150 m high, do indeed exist in an obtuse bimodal or complex regime (cf. Fig. 5.23), whereas the Sinai dunes, which are low, exist in a narrow bimodal regime (Tsoar 1983a).

The 'roll-vortex' hypothesis of linear dune formation (Fig. 5.27), which might also explain regularity of spacing, relates linear dune to helical or roll-vortices created in the atmospheric boundary layer. Bagnold (1953) based his version of this hypothesis on laboratory experiments by Brunt (1937) and argued that intense thermal convection in a desert combined with a strong geostrophic wind would create paired roll-vortices which might sweep sand into linear dunes). Despite a very limited basis of empirical findings, the roll-vortex theory continues to find supporters. For instance, Corbett (1993) argued that trains of barchans in the southern Namib were controlled by roll-vortices and could be viewed as proto-linear ridges. Tseo (1993) also argued quite forcefully, largely on the basis of observation of roll-vortices outside dune fields or in wind tunnels, that roll-vortices could indeed be responsible for linear dune formation. His most compelling field evidence came from tethered kites, which appeared to show that winds blew in opposite directions on opposite flanks of the dunes (Fig. 5.28; Tseo 1990). Tseo interpreted this evidence as proof of the roll-vortex theory, although a similar flow pattern might be explained by invoking separation bubbles as Tsoar

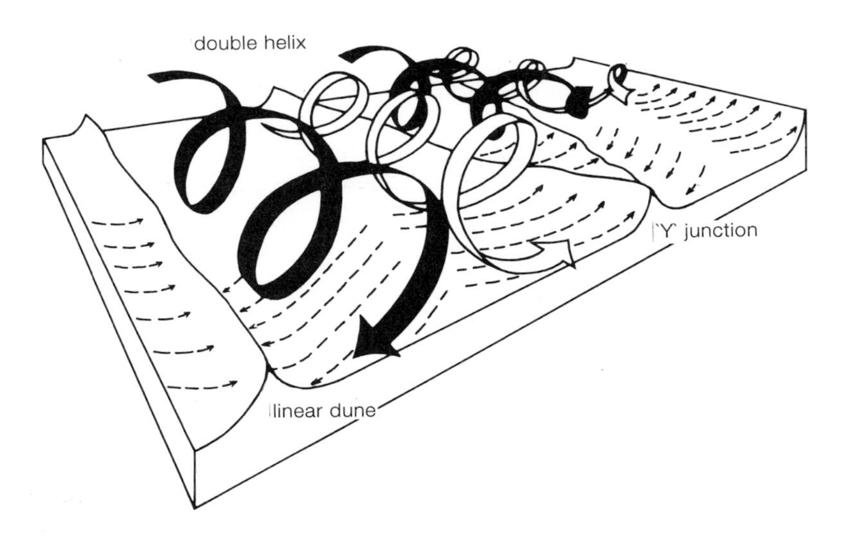

Fig. 5.27 Bagnold's hypothesis for linear dune formation in which a pair of thermally induced roll-vortices sweep sand from the interdune corridors onto the dune.

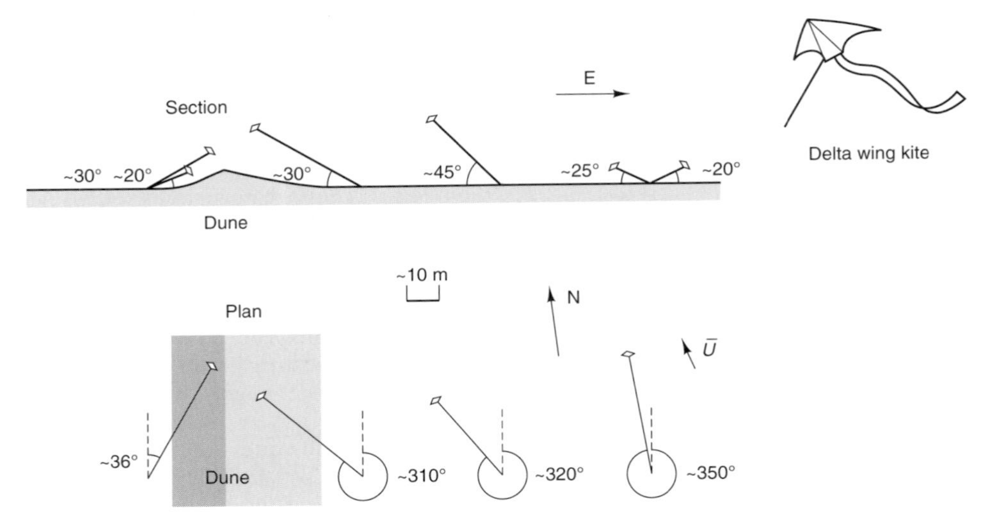

Fig. 5.28 Evidence from tethered kites that winds blow in opposite directions on either side of Australian linear dunes (after Tseo 1990). The pattern could be accounted for by both a roll-vortex model and a flow modification model.

(1983a, 1990b) did, and it could also be argued that, once a linear dune had been formed, such a flow pattern would be inevitable.

Another area of recent contention has been the distinction between sharp-crested, sinuous, unvegetated linear dunes, often termed 'seifs', and convex-crested, straight, vegetated linear dunes, sometimes termed 'sand ridges'. Tsoar (1989) and Tsoar and Møller (1986) argued that these were dynamically distinct types, the former developing in bimodal regimes and the latter in unimodal regimes. It may be that they are only differentiated by the amount of vegetation. The vegetation on the sand ridge cuts the supply of sand to the crest from the windward flank so that the crest is deflated and rounded. It is also possible that the vegetated dunes are remnants of sharp-crested dune, degraded following a change in climate. More studies are required to resolve this issue.

The patterns of *dune networks* and *star dunes* are associated with complex or at least obtuse bimodal wind regimes where net (or resultant or overall) sand transport rates are low. Dune networks are the result of the overlap of a number of transverse dune systems, each aligned to a different wind in a complex annual regime. The huge range of types of regime gives an almost infinite mixture of network types. One network in Oman (Fig. 5.29) is the result of a strong summer wind which creates a basic framework of transverse dunes. This framework survives, albeit much modified throughout the year, and in winter it is overlain by a series of smaller, less long-lasting systems of transverse dunes each aligned to a wind from a different direction, fickle, shifting winds being characteristic of that season (Warren 1988a).

Because of their size and the complexity of their dynamics there are fewer studies of star dunes than of the other types, although some recent work, reviewed by Lancaster (1989b), has brought some understanding. Lancaster (1989c) and Nielson and Kocurek (1987) showed that the major arms of a star

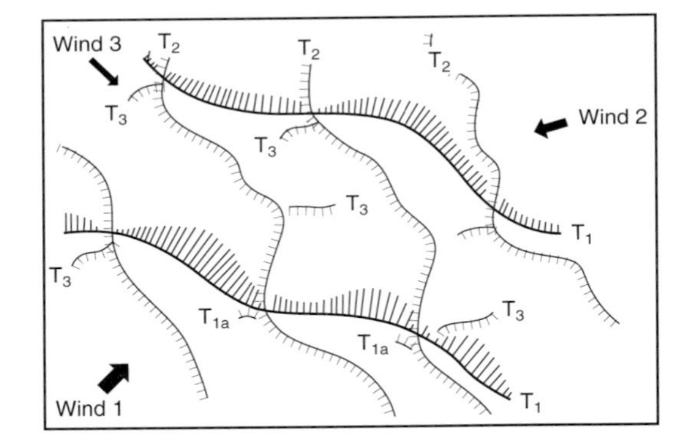

Fig. 5.29 A model of network dune formation, in which the strongest wind (wind 1) creates the most prominent set of transverse ridges (T_1), onto which other ridges are superimposed as a result of the action of less strong wind systems from different directions (after Warren 1988a).

dune, like those in dune networks, were aligned roughly transverse or slightly oblique to the locally dominant wind directions (Fig. 5.30). Nielson and Kocurek (1987) argued that there was a threshold size of dune, below which seasonal shifts of wind would cause wholesale realignment of dunes (as in a dune network). They believed the threshold to be about 100 m diameter and 20 m high for the Dumont dunefield in California. When the star dune reached this critical size, there would be a high degree of morphodynamic feedback, which would control near-surface wind regimes, so that sand arriving at the dune tended to remain there (Lancaster 1989b).

Nielson and Kocurek (1987) believed that star dunes were characteristic of areas where there were frequent shifts of direction, with seasonal winds close to perpendicular to each other. If this were so, star dunes might be virtually 'sedentary', although this is not confirmed. Others have argued that no wind regime is so completely balanced as to give zero net movement (Nielson and Kocurek 1987; Lancaster 1989c), although the morphodynamics of star dunes might itself contribute to virtual stasis. It is likely that star-like dunes exist in a continuum from linear features where there is some directionality in the annual pattern of sand movement, to virtually stationary features where the regime is equally balanced.

In addition to the dune types covered above, attention has recently turned to *oblique* dunes. The term was first used by Cooper (1958) when describing dunes of the Oregon Coast (which he supposed were not aligned with the sand transport direction although he never calculated it), but the term has subsequently been used by several other authorities, and Hunter *et al.* (1983) argued that it represented a morphogenetic type lying between transverse and linear.

Some of the confusion about the term results from a lack of precision in its use (Cooke *et al.* 1993). The term 'oblique' may refer simply to a condition in which sand is transported obliquely across the crest of a dune, but it may also refer to a situation where the entire dune is moving laterally. Sometimes dunes are termed oblique simply because they are asymmetric (in which case most linear dunes are oblique) or because their inferred internal structure is asymmetrical.

The most unambiguous use of the term oblique is when it refers to dunes where crest alignment is neither normal (75°–90°) nor parallel (0°–15°) to the calculated resultant sand transport direction (Hunter *et al.* 1983). However, even here, there are major

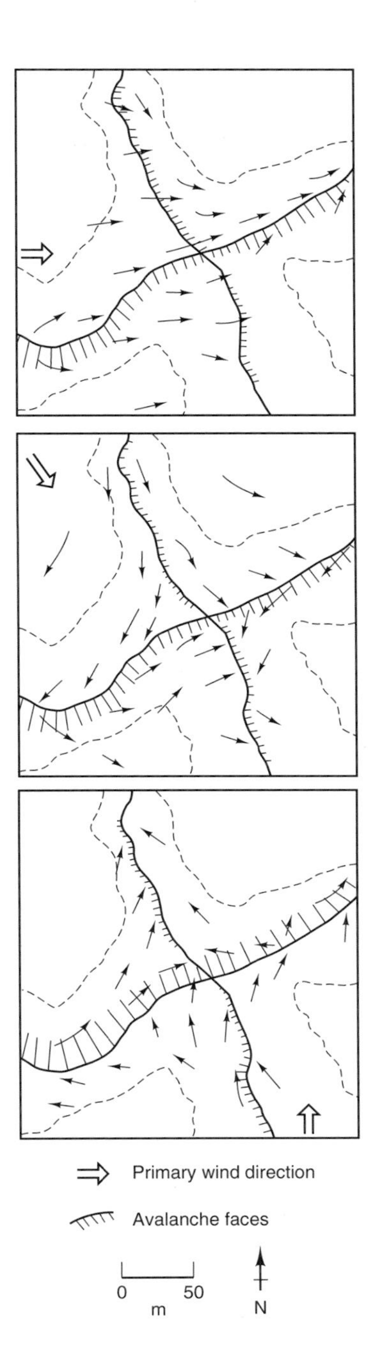

Fig. 5.30 A model of surface winds on a star dune. Winds from different directions create slip faces on different parts of the dune and lead to the growth of the dune (after Lancaster 1989a).

practical problems associated with data about the local wind regime, and with the dune's relations to the current wind regime.

Because of the diversion of flow by the dune, dunes in bimodal regimes of very unequal energy may develop a primary form related to the dominant mode but have some sand transport associated with the secondary mode (Rubin and Ikeda 1990). Thus some transverse dunes exhibit some crest-parallel sand transport and some linear dunes display some lateral movement (Rubin and Hunter 1985; Hesp *et al.* 1989; Rubin 1990; Nanson *et al.* 1992).

Anchored dunes

Anchored dunes accumulate in various positions around fixed obstacles, such as hills or bushes (Fig. 5.31). They are associated with patterns of

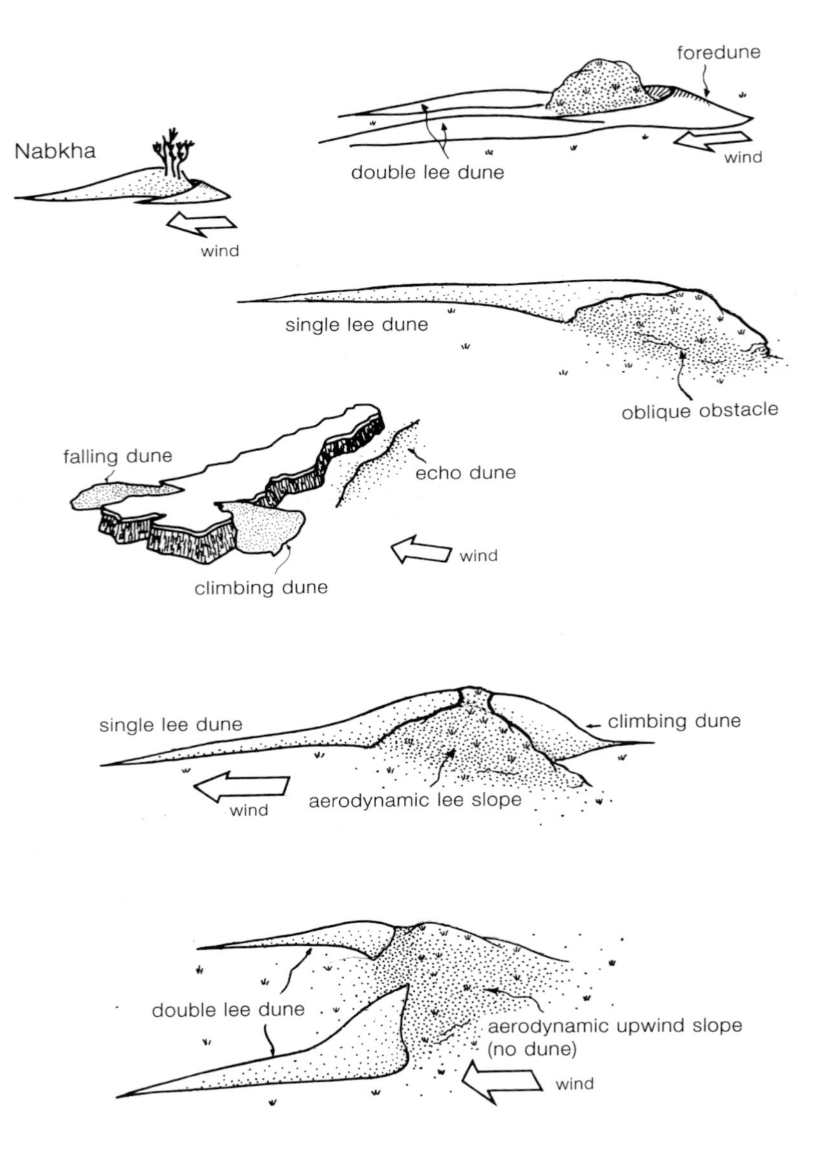

Fig. 5.31 Schematic diagram of dunes anchored by topography and vegetation.

flow separation and acceleration around the obstacle. All types of anchored dunes, with obvious exceptions, occur almost regardless of the size of obstacle, whether it is a major hill massif or a small bush.

As with flow towards a free dune, described above, wind approaching a fixed obstacle is modified. The deposition rate has been found to be a function of the Froude number ($Fr = u_*/(gh)^{1/2}$, where h is the height of the obstacle) (Iversen 1983, 1986a). The volumetric concentration of sand in the wind and the wind regime are other factors (anchored dunes do not survive great changes in wind direction, so that large ones cannot accumulate where the regime is very variable in direction) (Howard 1985).

Wind-tunnel studies have helped to explain something of the patterns of deposition. If the upwind slope of the obstacle is less than about 30° sand is transported up and over it, but above 30° it is trapped in a *sand ramp* or *climbing dune*. These dunes grow to an equilibrium shape after which sand is transported up and over them (Tsoar 1983b). When the upwind slope is greater than about 50°, an *echo dune* is formed, detached from the scarp by an upwind distance of about three times its height. Flow separates at the base of the scarp, and the reverse flow within the separation bubble prevents deposition. The equilibrium height of echo dunes appears to be 0.3–0.4 times the height of the obstacle (Howard 1985).

Just beyond the crest of a scarp there is a zone of slightly reduced wind velocity where *cliff-top dunes* may accumulate, both in deserts and on coasts (Carter and Wilson 1993; Marsh and Marsh 1987). Flow is accelerated round the flanks of obstacles that are narrow across the wind, sweeping sand to zones further out, where it may form *flanking dunes*.

In the lee of wide obstacles, where there is calm air, *falling dunes* may lie up against the lee slope. Where the obstacle is narrower, and flow takes sand round the flanks, *lee dunes* may extend some distance downwind (Fig. 5.32). The simplest lee dune is a single ridge extending downwind of a narrow obstruction, best developed in unimodal wind regimes. More complex patterns include two parallel ridges extending from either side of wider obstructions. Although lee dunes have attracted research because of their occurrence round craters on Mars, they are still the least well understood of anchored dunes (Greeley and Iversen 1985). Beyond their obvious association with patterns of turbulence in the lee, themselves related to the height, width and shape of the obstacle and the character of the approaching flow, there is still a lot to be learnt.

Fig. 5.32 Vertical air photograph of lee dunes, Bilma, Niger.

Downwind of a small lava cone in southern California, for example, the surface shear stress in the wake was 23 per cent greater than its ambient level, creating a zone where there was no deposition (Greeley 1986). Higher and wider obstacles create lee dunes of differing character, adjusted to the pattern of horizontal- and vertical-axis vortices shed from the flanks. There may be one, two or three of these lee ridges. There is probably some kind of feedback between the developing lee dune and the flow, which reinforces the dune-forming mechanism, for some lee dunes extend tens and even hundreds of kilometres downwind of their parent hills, well beyond the zone in which the hill, on its own, can have an effect. Examples are Draa Malichigdane in Mauritania, which is 100 km long (Breed and Grow 1979), and others stretching 72 km downwind of plateaux in the Western Desert of Egypt (El-Baz 1984a).

Plants as anchors

Plants are a distinct type of anchor or focus for the development of dunes, because of the permeability of plant structure, and because of the interaction between plant and dune growth. Dunes in which plants are merely anchors need to be distinguished from those in which vegetation has had wider-reaching effects, as in blowouts and parabolic dunes. This latter category is discussed separately.

The most widespread type of dune anchored to plants can be termed '*vegetated sand mounds*', otherwise '*nabkha*' (or *nebkha*), shrub dunes, coppice dunes or hummock dunes. Vegetated sand mounds, as the term is used here, describes a range of features from isolated single mounds to extensive vegetated ridges, although where these occur on coasts, and are subject to a distinctive interaction between beach and aeolian processes, they are given separate treatment later in this chapter. The distinction between sand mounds formed by aeolian and non-aeolian processes can never be exact, since many, if not most, mounds are produced by combinations of processes, in varying mixtures. These other processes include rainsplash, runoff erosion, animals or even seismic activity (Cooke *et al.* 1993). Vegetated sand mounds are very common indeed; they cover huge parts of semi-arid areas, occur universally in valley floors in very arid areas, and are found on most coasts. Even in the interiors of some quite humid and cold parts of the world, like Ireland (Wilson 1988, 1989), Scotland (Ballantyne and Whittington 1987), and Colorado (Thorn and Darmody 1980), some aeolian activity occurs in upland and mountain environments, creating dunes in association with the sparse, low-growing vegetation.

The development of vegetated sand mounds is an interactive process. First, there is the trapping of sand by the bush or clump of grass (*qua* obstacle); second, there is the growth of the parent plant in the favourable environment created by the accumulating sand; and third, there is the colonization of the mound by yet further plant species. To act as an initial trap, the plant must be at least 0.10–0.15 m high (Hesp 1979), but must also have a branching pattern that is conducive to trapping. Plants which spread laterally create undulating sheets, while those that form tussocks create individual mounds. The most successful plant species at forming vegetated mounds are those that can grow up through accumulating sand. Some trees may even be able to do this, and in such cases the mounds can be many metres high and even kilometres long (Warren 1988b). The longevity of the plant is also important, for only long-lived species can accumulate large mounds.

Small vegetated sand mounds behave much like hills in the way that they create foredunes, echo-dunes, flanking dunes and lee dunes (Fig. 5.31). The lee dunes attached to vegetated sand mounds, which are usually single, are their most striking features (Clemmensen 1986).

Though anchored, many vegetated sand mounds are ephemeral. This must clearly be the case with annual plants, but even round perennials the parent plants may suffer changes in water-table, rainfall, sediment supply or land use, and this can produce radical changes in their configuration (Gile 1975; Gibbens *et al.* 1983). If they do survive for several seasons, they may induce a morphodynamic response in which erosion is accelerated in the intervening areas. Hesp (1988) described a sequence in which sand was first trapped by plants, but in which the acceleration of flow between the new mounds stimulated erosion, which added to the relative relief. In some parts of the world, this erosional phase in the development of vegetated sand mounds may signify land degradation (Nickling and Wolfe 1994).

Other types of dune are the result of more complicated interplay between a plant cover and dune formation. Although the presence of vegetation may indicate the inactivity of a dune landscape, this need not be so. Some authors have proposed that many types of dune can form in the presence of vegetation, including fully developed linear dunes (Tsoar and Møller 1986; Tsoar 1989; Thomas and Tsoar 1990), but the discussion here is confined to two types whose formation is widely acknowledged to be dependent on a vegetation cover, namely blowouts and parabolic dunes.

Blowouts (Fig. 5.33) are largely deflational, but unlike yardangs and pans, blowouts are eroded into loose sand, albeit bound to an extent by plant roots. They are bare hollows in otherwise vegetated dunes, and are very common indeed on coasts and in stabilized dune fields on the desert margins. Typical blowouts on Dutch coastal dunes are between about 10 m and 30 m long, with rare ones reaching 100 m (Jungerius 1984), the longer axis being commonly parallel to the direction of the prevailing wind.

Blowouts develop from patches of sand, which are laid bare by many processes, including disturbance by animals and people, localized aridity in dry spells, or extremely high winds. In some coastal dunes blowouts may be a symptom of a negative sand budget, where the dune is no longer being supplied from the beach, and is gradually degrading. The higher parts of a vegetated dune landscape are more vulnerable to desiccation and disturbance, and are hence the most common locus of blowouts. Most bare patches revegetate naturally and quickly; only a few become

Fig. 5.33 A large blowout in coastal sand dunes at Blakeney Point, north Norfolk, UK.

blowouts, and of these an even smaller number become large. In a period of nine years, about half of a population of Dutch blowouts disappeared (Jungerius and van der Meulen 1989).

When sand is laid bare, the roughness height, z_0 (Chapter 2), decreases and allows greater surface drag and sediment entrainment. Erosion may lower the surface sufficiently to permit a further increase in wind speed by a funnelling effect, and this in turn may increase the rate of erosion (Nordstrom *et al.* 1986). In theory, erosion proceeds until the blowout is so wide that funnelling is less pronounced, at which point the blowout could be said to be 'in equilibrium'. A part of this negative feedback may be that sand cannot be carried up the steepening slopes. There may then be 'spontaneous' stabilization and eventually revegetation. Erosion may also be halted if a hard or wet basement is uncovered by erosion (Fig. 5.33).

Active blowouts export sand and most of this falls among plants and cannot be easily rè-entrained (Rutin 1983). Most is deposited as a thin downwind 'plume', but if the erosion is severe, a dune may form on the downwind side and advance over nearby vegetation, enlarging the area of bare sand, a process that may eventually lead to the formation of a parabolic dune (below). In temperate coastal dune landscapes, as much sediment is moved by water in blowouts as by wind (Jungerius and van der Meulen 1989). These processes include rainsplash, rilling and soil flowage (Rutin 1983), all sometimes stimulated by the water-repellent qualities of dune soils (Witter *et al.* 1991).

Parabolic dunes (or 'hairpin' dunes) are dunes with U- or V-shaped plan forms, in which the arms point upwind (Fig. 5.34). They are characteristic of the vegetated margins of deserts, as in the Thar Desert of India, where they cover extensive areas and where they may reach many tens of metres high (Wasson *et al.* 1983), the Jafurah Sand Sea in eastern Saudi Arabia (Anton and Vincent 1986), the Kalahari (Eriksson *et al.* 1989) and Arizona (Hack 1941). They also occur in vegetated cold-climate dunes, as in Canada and the central United States (Filion and Morisset 1983; Kolm 1985) and characterize many stabilized dune areas, where they are no longer active, as in Poland (Fig. 8.9). They are also very common in coastal locations. Pye (1993b) subdivided them on the basis of their length:width ratio. Parabolic dunes with length : width ratios less than 0.4 were 'lunate'; those between 0.4 and 1.0 were 'hemicyclic'; those between 1.0 and 3.0 were 'lobate'; and those greater than 3.0 were 'elongate'.

It is widely believed that parabolic dunes develop from blowouts. The dune grows in size as it feeds on sand from the erosion of underlying sediments, but eventually the supply of sand may decrease as a firm base is exposed in the hollow, and this and its size slow the progress of the advancing apex (the relation of size and movement in dunes is explained above). The few measurements that have been made of the movement of the apex, show a wide range of celerities.

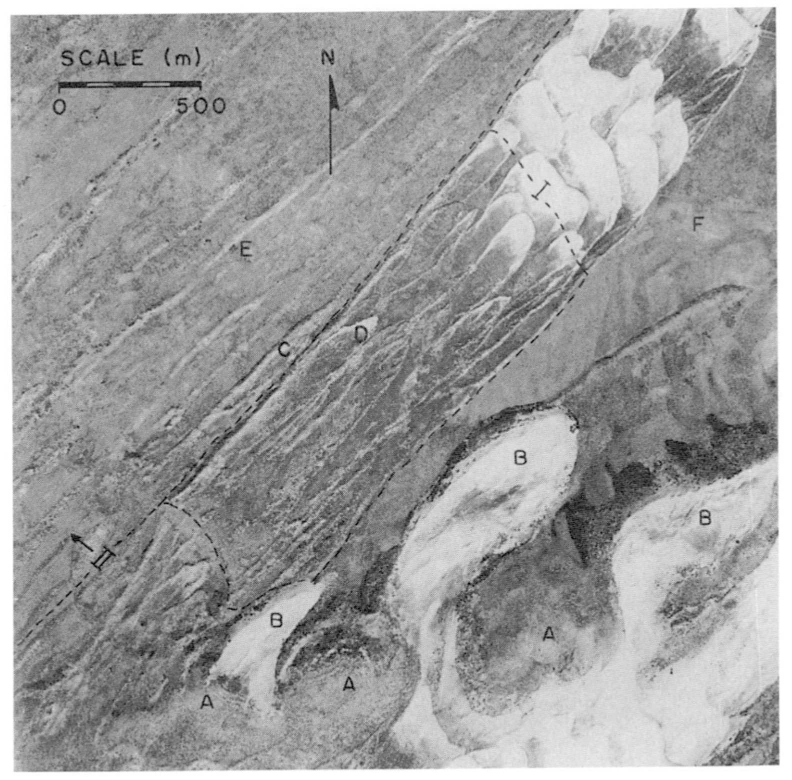

Fig. 5.34 Parabolic dunes at various stages of development in Idaho (after Chadwick *et al.* 1965).

Heavily vegetated parabolic dunes in northern Australia appear to move at only 0.05 m yr^{-1} (Story 1982), and Cooper's (1958) and Pye's (1982) direct measurements and McKee's (1979b) review also showed most to be slowly migrating at about 2 m yr^{-1}, but Hesp *et al.* (1988) measured maximum rates of 13 m yr^{-1} in coastal Cape Province, South Arica.

The role of vegetation in parabolic dune formation is said to be in protecting the less-mobile arms against wind action, and so allowing the central part to advance downwind. The association with vegetation is clear in most cases, the shape of the dune depending on the type of vegetation, be it grasses or trees, among which it develops (Filion and Morisset 1983). High wind velocities, as on the coast, and in late glacial times, may be necessary to overcome the resistance offered by vegetation. It is not always easy to relate dimensions to wind patterns (Kolm 1985), but in some cases it has been shown that higher winds generate more elongate dunes (Gaylord and Dawson 1987). Parabolic dunes may require a wind regime above a certain threshold of constancy, such as one with sea-breezes (Robertson-Rintoul 1990). In Wyoming, parabolic dunes are best developed in the divides between mountain ranges where wind directions are constant (Kolm 1985). Parabolic dunes may be a symptom of restricted sand supply, for where there are large supplies of sand in lightly vegetated landscapes, vegetated parabolic dunes give way to mobile dunes (McKee 1966).

In most parabolic dunes, there are signs of the migration of a series of dunes down the same path, building up a series of concentric, 'nested' dunes (Fig. 8.9). These testify to alternating periods of activity and inactivity. The explanation of the extraordinarily large area of vegetated parabolic dunes in the Thar may lie in its unique post-glacial climatic history (Chapter 8).

Blowouts (and to a lesser extent parabolic dunes) are usually formed on a framework of larger dunes that has been stabilized by vegetation, after either a change in climate (as on many desert margins) or a decrease in sand supply (as in coastal situations). The formation of these secondary types of dune is also often accompanied by degradation of slopes by water erosion (Chapter 8), and these two forms of disruption, especially if interrupted by renewed dune formation in small pockets (say in drier climatic phases) and yet further phases of degradation, create at first very chaotic dune topography and later very subdued relief. The process can be described as one of increasing 'dune entropy' (Warren 1988b). Complex climatic (or land use) histories are not at all unusual, and probably explain chaotic and subdued dune topography in many desert-margin and coastal situations. Examples include the calcareous, sandy 'machair' of the Hebrides and nearby coasts of Scotland (Angus and Elliott 1992; Carter and Wilson 1993) and the wide sandy sheets of parts of northern Australia (Mabbutt 1968).

Coastal dunes

'Coastal dune' is not an entirely satisfactory category. In one sense the coast can be regarded merely as a supply of sand where there is little vegetation; this viewpoint is difficult to challenge. Where, as in coastal Sinai, sand blows off a beach and into a hyper-arid desert, the dunes adopt common patterns of free dunes after a very narrow transitional zone just behind the beach. However, where vegetation can hold the sand, for a season or more, there is a complex interaction between littoral and aeolian processes, and this gives coastal dunes a special character. Even here, sand blown only a few metres inland from the zone of beach–dune interaction enters dune systems that are virtually indistinguishable from the conditions on desert margins where there is also a light covering of plants (the dunes here being mostly blowouts and parabolic dunes). The discussion here, therefore, makes some points about the broad framework of coastal dune fields and then concentrates on coastal foredunes, where beach and dune interact in such an interesting fashion. It finishes with a discussion of 'lunettes', which are coastal dunes on inland pans in semi-arid areas.

Coastal dune landscapes

Coastal dune landscapes are the result of an intimate and complex mix of aeolian processes with structural geology, and marine, estuarine, fluvial, slope, pedological, ecological and cultural processes. The broad sequence of causality is from the landscape setting, through marine and estuarine sediment supply processes, to aeolian and ecological processes, and finally to the pedological, slope and sometimes the fluvial processes that modify the dunes, once created. Real coastal dunes experience many interruptions, reversals and restarts in this sequence.

Coastal dune formation has been very considerably complicated in most cases by geologically very recent and quite rapid change in the conditions under which they form. These come from two directions. First are those connected with climatic and sea-level change. Second are those connected with interference by people. The first is discussed in Chapter 8; the second in Chapter 9.

Most coastal dune fields are found on gently shelving coastal platforms. These are created by coastal erosion, by tectonic subsidence or when rising sea levels encounter low-relief areas previously eroded by fluvial erosion. It is the bays on a coast that normally accumulate dunes. There are cases where sand is raised up cliffs from the beach and where dunes are found on top of the cliffs, but cliff-top dunes are confined to environments with strong winds and cliffs with slope angles that allow sand transport (Jennings 1967; Tsoar and Blumberg 1991; Jackson and Nevin 1992).

Within embayments, coastal dunes are most commonly found on marine sedimentary platforms, particularly on bars and spits. Slightly falling sea levels in the later Holocene stranded many of these features, preparing them as platforms for coastal dune development. Dunes may vary in character according to whether they overlie a gravel beach or spit or a series of sandy bars (Carter and Wilson 1993).

Sand supply to the beach–dune system

Given a suitable site, the extent and character of the dunes depends next on the rate of supply of sand and the rhythms of supply. In turn these are determined by two factors: the existence of a source of sediment, and of longshore or onshore drift strong enough to carry it. Sources of sand are various. Some of the largest coastal dune systems are linked to large rivers. Thus the coastal dunefields of northern Sinai and the Negev in Israel are derived ultimately from the mouth of the Nile (Issar *et al.* 1989); the Dutch dunes derive much of their sand from the mouth of the Rhine; those in the huge dunefields of Aquitaine in south-western France from the Gironde; those of

Languedoc from the Rhône; those of the Coto Doñana in southern Spain from the Guadalquivir (Klijn 1990); and the extensive dunes in Oregon largely from the Columbia River.

Another, though less prolific, source of sand in coastal dunefields is coastal erosion, which undoubtedly supplies most of the sand in the dunes of northern East Anglia in eastern England. The sand on these coasts can be traced ultimately to the erosion of the soft cliffs in glacial drift on the coast of Yorkshire from where it is taken down the Lincolnshire coast and across the Wash to East Anglia (East Anglian Coastal Research Programme 1977). Offshore supplies may also supply sand to coastal dunes in some parts of the world, but only on coasts where there are wide, relatively shallow, continental shelves. During the last glacial period these shelves were exposed by lower sea levels and accumulated fluvial and in places glacio-fluvial sands, which became the source for some of the sand on neighbouring coasts. The western and southern coasts of the North Sea and the Baltic have more dunes supplied in this way than the steeper coasts of the Mediterranean (Klijn 1990).

It is not uncommon for sand to be blown off a beach and work its way through a coastal dune system, only to end up in an estuary; the estuary 'pumps' the sand back to the beach, and the system is in a state of near-constant renewal (Carter and Wilson 1993).

Given a source of sand, the rate at which it is carried by littoral drift to the parent beach depends on the wind environment (strength and directionality), which determines the strength of littoral drift, and on competition from other 'sediment sinks' such as offshore marine canyons and sand banks. Beaches that are aligned obliquely to the waves with the greatest power are those with the strongest littoral drift. The direction of greatest wave power is a function of the annual wind pattern and of the distance or fetch over which it blows. Thus in East Anglia in eastern England, waves coming from the north have the greatest fetch, and littoral drift is therefore southerly, carrying sand derived from erosion on northerly cliffs to the south. In Oregon, the main drift is southerly from the mouth of the Columbia, driven by storm waves coming south-east across the North Pacific Ocean.

Beach–dune interaction

Beaches and their foredunes are strongly coupled systems, with very active feedbacks between, which are good examples of the morphodynamic systems mentioned above (Sherman and Bauer 1993). Dunes are usually an integral part of the total pattern of littoral drift. They act as buffers or stores of sand, which feed the beach at times when it is starved of sediment and absorb sand when it is being supplied at rates greater than the beach itself can absorb. Where the prevailing wind is nearly parallel to the coastline and the vegetation is not too vigorous, sand may be actively transported along the coast in the dunes, which take it from one beach and deliver it to another. This is very noticeable on the southern Cape coast in South Africa, where dunes carry sediment over headlands from one beach system to the other (McLachlan and Burns 1992).

Beach/dune systems are one of the most dynamic of geomorphological systems, and are constantly in a state of flux at many scales. For example, while a coastal dune system in north-western Ireland might be regarded as being in 'equilibrium' at the scale of decades (although the nature of the equilibrium might be difficult to establish), it yet may experience net short-term interchanges between beach and dune of the order of $10^4\ m^3\ km^{-1}\ yr^{-1}$, and also be slowly retreating when viewed at the scale of centuries (Carter and Wilson 1993).

At a macro-scale (of the order of several decades), an often quoted system for relating dune type to beach type is that of Short and Hesp (1982; Table 5.2). Using some simple assumptions, Sherman and Bauer (1993) showed that a dissipative beach could yield 140 per cent more sand than a reflective one. They pointed out that the Short–Hesp scheme cannot be regarded as universal, since magnitude/frequency relations are different in different environments. Tropical shores, like the ones that formed the basis of the Short–Hesp scheme, generally experience low-energy wave conditions, but may suffer extreme tropical storms that might destroy dunes once every 100 years or so. Most coasts, moreover, go through successions in which they are at one time aggrading, sometimes spending short periods in equilibrium, but at other times they are retrograding, and the dune landscapes behind the beaches respond in kind, for example degrading when a sand supply is cut off, or being rejuvenated when it reappears.

Sherman and Bauer (1993) also modified Psuty's (1992) system of classification of dune types according to sand budget on the beach and the dune (Table 5.3). They gave extensive examples of dunes in each of these conditions.

Table 5.2 The Short and Hesp (1982) morphodynamic model of dune–beach interaction as modified by Sherman and Bauer (1993)

Morphodynamic beach state	Frequency: type of dune scarping	Potential aeolian transport: foredune size	Probability of foredune destruction (per 100 years)	Nature of dominant dunes
Dissipative	Low: continuous scarp	High: large	Moderate	Large-scale transgressive dune sheets
Intermediate	Moderate: scarps in rip embayments (spaced, 1 km)	High/moderate: large/moderate	Moderate/high	Large-scale parabolics to dune sheets
Intermediate	Moderate: scarps in rip embayments (spaced 0.5–1 km)	Moderate: moderate	High	Large-scale parabolics: large blowouts
Intermediate	Moderate: scarps in rip embayments (spaced <500 m)	Moderate/low: moderate/small	Moderate/low	Discrete blowouts
Reflective	High: continuous scarps	Low: small	Low	Foredunes scarping small blowouts

At the decadal scale there are further processes that can have a marked influence on dune formation. There may, for example, be phases of dune erosion on one part of the coast that may supply sand to dunes on another, and different stages of dune development can coexist on quite short stretches of coast (Psuty 1992; Carter and Wilson 1993). Coastal changes, either natural, as when river mouths are closed, or artificial, as when sea walls cut off sand movement along the coast or when coastal marshes are reclaimed (altering the tidal prism in an estuary), lead to the replacement of one type of dune formation by another (Carter *et al.* 1992a). There may also be great rhythmic pulses of sand, in dunes and on the beach, that move down the coast at rates of a few hundred metres a year, as on Long Island (Psuty 1989). There may also be cyclic, progressive or irregular changes in off-shore processes, which may redirect currents and cause more or less sand to be delivered to the beach and then to the dunes (Armon and McCann 1979). The mysterious appearance of the dunes at Studland in Dorset after the seventeenth century or of massive dunes on the Danish coast about AD 1000 may have been associated with some such change (Diver 1933; Skarregaard 1989).

At the micro-scale, there are three primary controls on the rate at which a beach can supply sand to a dune: grain size, beach width and beach slope. However, these controls are difficult to understand. The usual aeolian transport equations, discussed in Chapter 2, are of limited application, for there are slope effects (of many kinds), the effects of a

Table 5.3 Sherman and Bauer's (1993) modification of Psuty's (1992) system of dune/beach budget classification

Beach budget	Dune budget	Morphology
Positive	Positive	Beach or dune ridges
Positive	Steady state	Indeterminate
Positive	Negative	Blowouts and deflation hollows
Steady state	Positive	*In situ* dune growth
Steady state	Steady state	Indeterminate
Steady state	Negative	Blowouts and deflation hollows
Negative	Positive	Dune growth and onshore migration
Negative	Steady state	Indeterminate
Negative	Negative	Dune erosion and washover

developing boundary layer as the wind blows off a rougher beach onto a smoother beach (or vice versa), and very variable moisture levels. There are also morphodynamic complications, as wind erosion or deposition smoothens the pre-existing profile that had been created by marine processes (Sherman and Bauer 1993).

At one time it was thought that rainfall, which in north-western Europe comes mostly in winter, would inhibit aeolian sand movement. This is probably true for very heavy rain, which saturates the surface and inhibits all movement (Arens 1995). Winter rainfall of about 1570 mm on the Oregon coast has also been shown to reduce seasonal sand transport to 36 per cent of its potential (Hunter *et al.* 1983) and in northern Sinai, with a winter rainfall of only about 97 mm, the reduction is 14 per cent (Tsoar 1974). However, experiments have now shown that driving rain, far from suppressing sand movement on a beach, may actually enhance it by splashing up grains into the path of the wind (Sarre 1988).

Most sand is moved from the beach to the dune in medium to high winds (Chapter 2), so that it is the distribution of these winds rather than of lighter winds that is important. In Europe, strong winds goes some way to explaining the development of the distribution of dunes (Fig. 5.35), though there are numerous exceptions. It may be that beaches aligned obliquely to prevailing strong winds experience the highest rates of aeolian sand movement, for it has been found that winds blowing directly on-shore carry far less sand than those blowing along the beach (Arens 1994).

Sea-breezes play an important role in many coastal dune fields. These can be strong even on temperate coasts (Hunter and Richmond 1988), but they are strongest and most regular in warmer climates, especially in summer, when land–sea temperature contrasts are greatest (Flohn 1969). On the coast of Israel, for example, they occur on 80 per cent of summer days (Halevy and Steinberger 1974). Though sea-breezes can be felt as far inland as 300 km, most are confined to a much narrower coastal zone (Illenberger and Rust 1988), so that sand transport potential declines inland and dunes move less quickly. This may explain the way in which many coastal dunefields are confined to a narrow zone only a few kilometres wide. The contrast between the smoothness of the sea and the roughness of the land means that the breeze over the dunes may be deflected by up to 35° from its angle of incidence at the coast, to the left in the northern hemisphere and to the right in the southern, and this has its effect on the orientation of dunes (Warren 1976b).

The 'cliffing' of dunes by storm waves, discussed below, is a further interactive beach/dune process at the micro-scale. In the Alexandria dune system on the eastern South African coast, it was estimated that 12 per cent of the sand delivered to the dunes from the beach each year is returned to the beach by 'cliffing' (Illenberger and Rust 1988); the figure is probably very different in other conditions.

Dunes on strongly prograding beaches in the Netherlands can receive up to 25 m^3 (m-width)$^{-1}$ yr^{-1} of sand over stretches of the order of a kilometre, and up to 75 m^3 (m-width)$^{-1}$ yr^{-1} for shorter stretches (Arens and Wiersma 1994). Arens and Wiersma developed a classification of coastal dunes that depended on the sediment budget (its strength, either positive or negative) and the amount of interference they experienced.

Ecological and pedological processes

An understanding of the ecological controls and biology of the plants that grow on coastal dunes is important to understanding the dunes themselves. A small, select set of higher plants can withstand the high salinity and the ephemeral seed beds of the upper shore, where they may be subject to high rates of burial by sand. The first plants to colonize are very hardy, *r*-selected species that trap sand during their short life-spans and allow the colonization of more-vigorous though less-hardy species that are able to trap more sand.

The most widespread and important of plants in this second group, and probably the single most important plant in dune formation, is the grass *Ammophila arenaria* (marram), about whose practical ecology there is now a vast literature, because of its role as the biological mainstay of coastal protection (Chapter 9). In other parts of the world there are indigenous plants that fill the marram niche, but marram has been introduced very widely indeed as a dune management tool. It is accompanied by a small group of other plants, but they play very minor roles in dune formation. Other dune species, such as *Hippophäe rhamnoides* (sea buckthorn), can withstand even greater rates of burial, but are not as well suited to growth on the foreshore.

Marram disperses primarily by water-borne rhizome fragments, and is hence generally restricted to the near-shore zone. It grows in clumps that are especially good at holding sand (Fig. 5.36), and grows

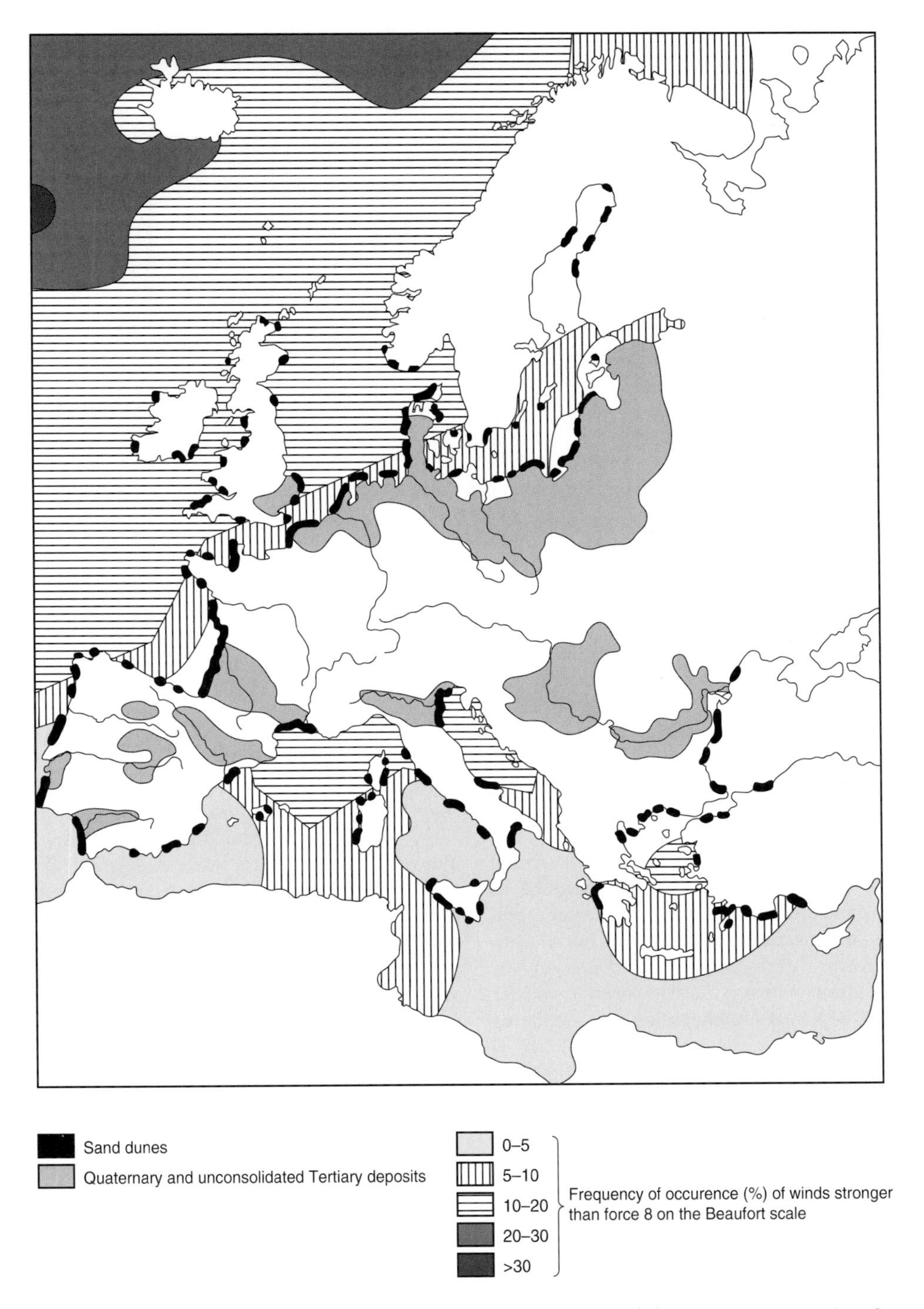

Fig. 5.35 The distribution of coastal dunes. Frequency of occurrence (%) of winds stronger than force 8 on the Beaufort scale and the pattern of January wind speeds on European coasts (after Klijn 1990).

Fig. 5.36 Marram grass growing vigorously in a coastal foredune at Braunton Burrows, south-west England.

best in accumulating sand. The reasons why this should be so are not well understood, but it has been observed that new marram roots are unusually vigorous and rapidly replace old ones as they die off (Willis 1989), and that, like its relative in north-eastern North America (*Ammophila breveligulata*), it has greater vegetative growth when being buried in sand (Seliskar 1994). *A. breveligulata*, however, seems to produce smaller foredunes, when compared on the same coast to dunes accompanying *A. arenaria* (Seablom and Wiedemann 1994).

Estimates of the rate of burial that marram can withstand were reviewed by Carter (1988) who found that most of the quoted rate ($1\,m\,yr^{-1}$) seemed to depend on little experimental evidence. Carter's best estimate was that marram could thrive at burial rates of about $0.25\,m\,yr^{-1}$, but that continued growth was possible to about $0.6\,m\,yr^{-1}$. However, van der Meulen (1990) reported Dutch studies that estimated only 0.1–$0.2\,m\,yr^{-1}$ as the tolerable limit. It is probable that the rate is variable, depending on other controls, particularly on nutrients (especially nitrogen), timing of burial in relation to growth periods and water supply. It is probable that flotsam at the high tide line is a vital source of nutrients for the growth of marram, and there are likely to be many different controls on the amounts and nutrient-content of flotsam (Gerlach 1992).

On dunes far from the foreshore, where there is less input of sand and increasing stability of the surface, marram (or its equivalents) ceases to thrive, for various reasons, but probably largely because the stability it has itself promoted, for this allows other plants to invade and compete with it. A series of plant communities, often quoted as the classic example of a plant succession (Ranwell 1972), may then follow, ending ultimately in acidophilous woodland. Accompanying the plant succession there are changes in the soil, for sandy soils are very permeable and are thus quickly leached of nutrients, especially calcium (which controls acidity and thus the whole gamut of other biological process) (Salisbury 1922). Most dune soils rapidly become podzolized (Wilson 1992). In some Australian coastal dunes, the highly leached, almost white and structureless A_2 podzol horizon may be 18 m thick (Thompson and Bowman 1984), though it is rarely more than 0.5 m thick in Europe. In New Zealand, the soils on a long series of coastal dune ridges slowly lose phosphorus over many centuries, and thus also lose their ability to support plants (Syers and Walker 1969).

Primary dune forms

Psuty (1989) made a useful distinction between 'primary' coastal dunes (whose development was linked directly to coastal processes) and 'secondary' ones, which were coastal only in the sense that the source of their sand was ultimately coastal. Primary coastal dunes include the ephemeral dunes of the back-shore and the main shore-parallel dune ridge. Secondary forms include sand sheets, blowouts and parabolics, which occur in stabilized dunes regardless of the origin of their sand, and are discussed above.

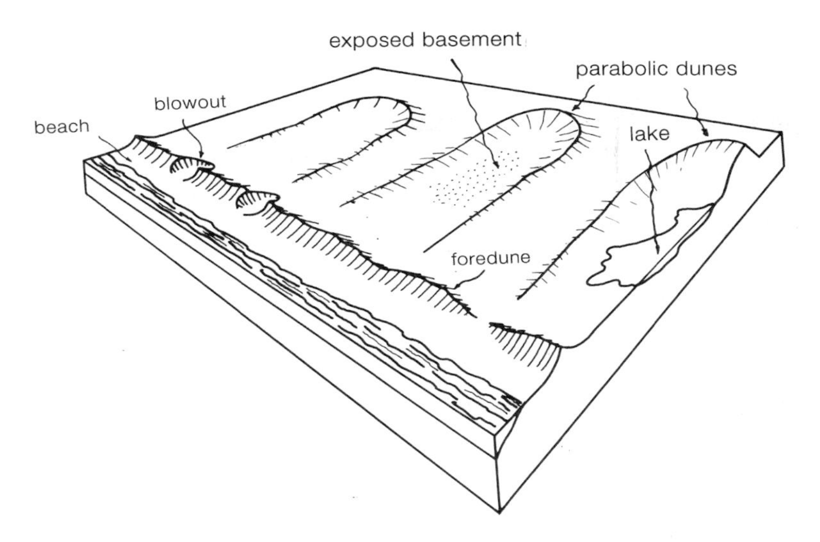

Fig. 5.37 Schematic diagram of coastal dune types.

Where the supply of sand to coastal dunes is slow, sand may be spread inland as a thin sheet of sand (Carter 1990). Where sand supply is greater, sand accumulates round the culms of marram (or an equivalent species) and is responsible for the creation of coastal foredunes, which are ridges parallel to the shore (Fig. 5.37). Where the growth of vegetation is strong, as on most coasts in northern Europe, the foredune is an almost continuous ridge parallel to the beach. But where vegetation is not as vigorous, as on hotter or more arid coasts, foredunes are more a discontinuous zone of vegetated sand mounds (see above) (Hesp 1987).

Foredunes may grow at quite noticeable rates (Fig. 5.38). Maximum rates of foredune growth in Ireland are between 3 and 4 $m^3 m^{-1}$ per week for short periods, and 0.3 and 0.7 $m yr^{-1}$ over longer periods (Carter 1990).

A primary control on the height of foredunes is the supply of sand, itself linked to the rate at which the coast is prograding. In north-western Ireland most foredunes do not exceed 3 or 4 m in height before being replaced, seaward, by a new ridge (Carter 1990). On rapidly prograding beaches with active onshore movement of sand in the wind, there is a succession of ridges, each signifying an abandoned shoreline (Bird 1990). The hollows between the ridges (the 'dune slacks') may be temporarily or permanently filled with fresh or brackish water pools. If these pools are large, wave action on their surfaces may erode the dunes,

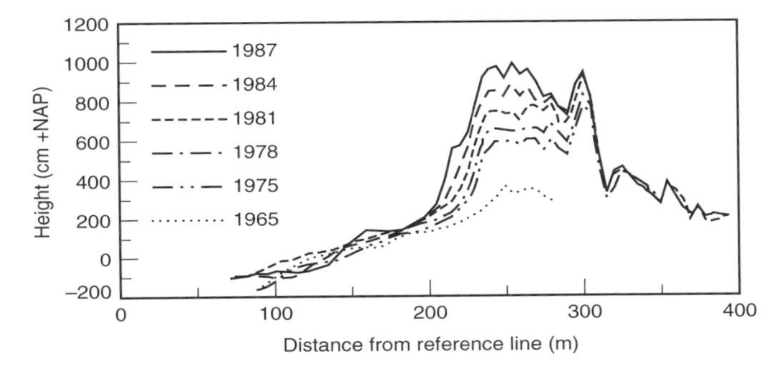

Fig. 5.38 The growth of coastal foredunes in the Netherlands (after Arens 1994).

creating cliffs and coastal platforms. Streams can achieve a similar effect by eroding the edges of coastal dunes (Langford 1989).

Another control on the height of foredunes is the balance between aerodynamic roughness and wind speed-up (described above), but these processes are poorly understood. Arens (1994) found that there was a very great increase in speed-up on dunes between 6 and 10 m high, but little between those at 10 and 23 m high. He attributed this to greater deflection and roughness in higher dunes. Deflection of the wind would limit growth by diverting sand from dunes of a certain height. Arens' finding, that much of the sand deposited on foredunes in high winds was out of suspension, further supports this argument, since this would be most readily diverted.

Yet another control on the sizes of foredunes is the frequency with which the ridge is destroyed by erosion, for on many coasts the coastal foredune is subject to frequent destruction as surges and storms 'cliff' its seaward face (Gerlach 1992). Erosion accompanying hurricanes on the eastern seaboard of the United States, can produce rapid retreat in coastal foredunes of up to 100 m (Sexton and Hays 1992). Where the foredune is massive, as on some Irish coastal dunes, cliffing may produce landslides and slumps (Carter 1990). 'Washover' is another process in which storm waves may break through a low or narrow section of dune, and destroy the dune by washing sand onto marshes or estuaries behind. Most washover sites, like dune cliffing, are merely temporary breaches and are rapidly repaired by new dune growth. As the dunes become isolated from the beach, they are transformed by the creation of blowouts and parabolic dunes, as described above.

If the supply of sand to coastal dunes is very great (by reason of strong on-shore winds, or of a rapid supply from off-shore or by littoral drift), or where the vegetation has a tenuous hold, as on arid coastlines, the sand may bury the vegetation and begin to form dunes very similar to desert dunes. These dunefields are known as 'transgressive'. The commonest forms here are transverse dunes, probably because of the influence of the fairly constant sea-breezes (Bird 1990). Transgressive dunes may also form after very severe disturbance (Chapter 9). There is a range of behaviour between those dunes in which the vegetation is dense enough to prevent almost any erosion, and those on which there is little or no interference of vegetation in the dune-forming process (Short and Hesp 1982).

Lunettes and clay dunes

Lunettes are a distinctive dune form associated with the coasts of pans (Chapter 3; Fig. 3.6). Their half-moon shape (from whence their name) is a reflection of the smooth log-spiral ('zeta-form') wave- and current-formed shore of the parent pan rather than of any aeolian process. Many, though not all, lunettes are formed of clay-sized particles; a characteristic they share with other clay dunes, also found mostly, but not exclusively, in coastal positions.

Lunettes have been described from many environments (for example, Goudie and Thomas 1986), but have been most thoroughly studied in Australia (Bowler 1983). Bowler described individual cross-sections in which there were sandy and clayey phases, and also zones round pans where sandy dunes were characteristic, and others where the dunes were more clayey. The sandy dunes behave either as free dunes or as vegetated coastal dunes, depending on local microclimates and disturbance regimes. Clay dunes, whose clay content may be as high as 77 per cent (Bowler 1973), and which often also have high salt contents, become rapidly immobilized as the clay aggregates disperse and cohere under the influence of percolating rainfall (Dare-Edwards 1983). They also adopt rather different shapes to sandy dunes, with much lower angles overall, and generally steeper windward than lee faces. Many clay- and salt-cemented dunes, like those in the Mojave, are eroded into yardangs after cementation (Blackwelder 1934) or, because their salt content discourages vegetation, become deeply gullied by runoff.

Smooth crescentic lunettes, for all their striking shape, are probably only a minority among lake-shore dunes in semi-arid areas. Most have much more irregular plan forms (for example, Coque and Jauzein 1967).

The question of how clay, which would normally travel in suspension, has come to form dunes was asked many years ago (Coffey 1909), and the answer was quickly discovered, although it has been elaborated since. Clay dunes, it transpired, were created when pellets of aggregated clay travelled by creep or saltation, in other words as if they were sand grains (Skidmore 1986a). Because these pellets cannot travel far before they are disaggregated by mechanical bombardment, they either form a dune close to their source, or are dispersed to form a loess-like sheet (Chapter 4). Clay pellets travelling as sand have been observed to come from three sources: alluvial deposits, especially shortly after flooding; bare soil surfaces; and saline depressions. This last source is probably the most prolific, and hence the one with which clay dunes are most commonly associated. Clay dunes associated with alluvial plains have been described by Wasson (1983a,b) in the Strzelecki Desert of Australia.

In salt lakes, pellets are created by clay bonding, by salt-cementation, or by algal aggregation. All these processes are encouraged by seasonal wetting and drying (Bowler 1980, 1983). The necessary environment is quite precisely defined. There must be seasonal or tidal flooding. A strong seasonal pattern is helpful, particularly one in which a marked dry season is accompanied by a wind that is strong enough to entrain the pellets and constant enough to carry them in one direction, and is followed by a wet season in which the lake refills, bringing new sediment, and in which the dune is bodily cemented together (Bowler 1968). Clay dunes are therefore the products of semi-arid rather than arid climates. Many clay dunes, however, are clearly the result of multiple phases of formation, and their origin dates well back in the Holocene or late Pleistocene (Bowler 1983; Goudie and Thomas 1986). Bowler has suggested that they are the consequence of particular climatic histories, for example the transition of a wet to a dry climate.

Conclusion

This chapter has shown that there have indeed been immense advances in the study of dunes in the last decade or so. They have been due to three main factors: better technology of all sorts, an injection of research funds from several sources, and the willingness of dune geomorphologists to broaden their horizons to meteorology and other related disciplines. Their problems, none the less, are far from solved. They may be closer to understanding some very basic dune processes, as at the toes of transverse dunes, but these discoveries have opened up further horizons of uncertainty. Careful field studies of the basic process on larger, three-dimensional features like linear and star dunes have brought huge advances, but the number of these studies is still very few, and there are major unsolved issues. The perception that primary coastal dunes are complex morphodynamic systems is still quite new, and will undoubtedly open up a great and fruitful series of research questions.

Further reading

There is a huge literature on dunes which has been greatly augmented in recent years, notably by the books by Pye and Tsoar (1990) and Lancaster (1995). Review articles on particular dune types were provided for star dunes by Lancaster (1989b) and for linear dunes by Tsoar (1989). Other reviews of desert dune types were provided by Cooke *et al.* (1993) and Lancaster (1994). The role of vegetation in desert dune development was reviewed by Thomas and Tsoar (1990). There has been a surge in publications about coastal dunes, largely in response to fears about their role in sea-level rise. Recent collections with some excellent review papers include those by Carter *et al.* (1992b), Bakker *et al.* (1990), Nordstrom *et al.* (1990) and Psuty (1988).

CHAPTER 6

Sand seas, dunefields and sand sheets

Terminology and definitions

Few sand dunes exist in isolation. Most cluster, sometimes in very large bodies in which there are dune patterns of notable regularity. These collections of dunes are called *sand seas* or *dunefields*. Sand seas are also called *ergs* in the northern Sahara, *edeyen* in Libya, *qoz* in the Sahara, *koum (kum)* and *peski* in central Asia, and *nafud (nefud)* in Arabia. Other large bodies of aeolian sand, *sand sheets* or *streaks*, have no recognizable dune forms, and are of greater total extent than the accumulations with dunes (Table 6.1). Altogether these accumulations – sand seas, dunefields, sand sheets and streaks – account for about a quarter of desert landscapes. However, few sand seas are continuously covered with aeolian sand for they usually include interdune outcrops of the underlying rock, soil cover, or lacustrine and fluvial deposits.

Cooke *et al.* (1993) set the lower size limit for a sand sea at 30 000 km^2, this being an inflexion point on the distribution curve of sand-sea size given by Wilson (1973), and also being the size of the smallest water body called a 'sea'. For them, smaller accumulations were dunefields (Fig. 6.1).

Distribution

Wilson (1973) found 58 sand bodies with an area greater than 12 000 km^2 (Table 6.2). In the northern hemisphere there are major sand seas in the Sahara (Fig. 6.2a), Arabia (Fig. 6.2b), and Asia (Fig. 6.2c). In the southern hemisphere there are large sand seas in Australia (Fig. 6.2d), and southern Africa (Fig. 6.2e), but very much smaller sand bodies on the west coast of South America.

North African sand seas cover over 2.5×10^6 km^2. Central Asian sand seas cover at least 750 000 km^2. Australian sand seas are now estimated to cover 3.07×10^6 km^2 (Wasson *et al.* 1988), much of it stabilized. Thomas and Shaw (1991) reported that the Mega-Kalahari, which stretches from South Africa to the Zaïre River, covered 2.5×10^6 km^2. It is therefore the largest contiguous sand sea, although mostly stabilized. Wilson (1973) showed that the largest active sand sea is Rub' al Khāli in Saudi Arabia covering 560 000 km^2, rising to 770 000 km^2 of continuous active sand cover when all the adjoining sand seas (Jafura, Dahana and Nefud) are included. The proportion of each desert covered by aeolian sand varies greatly from 2 per cent in North America to over 50 per cent in Australia (Mabbutt 1977). In the north polar region of Mars there are four sand seas totalling an area of 680 000 km^2 (Lancaster and Greeley 1990).

Mechanisms of accumulation

To form, a sand sea or a dunefield needs a supply of sand. The sand is usually delivered by the wind from nearby non-aeolian sources: the majority of large sand bodies are close to the deposits of rivers, seas or lakes. Sometimes, however, there is good evidence that sand has travelled great distances in aeolian transport before coming to rest in a sand sea or a dunefield (Table 6.3; Fryberger and Ahlbrandt 1979). Wilson (1973) also showed that sand could and probably did travel great distances across the Sahara in the wind, suggesting that many of the distal sand seas of the Sahara (those towards the end of the aeolian sand paths and therefore in the southern or western Sahara) were formed from this far-travelled

Table 6.1 Fryberger and Goudie's (1981) table of relative dune extent

	Thar	Takla Makan	Namib	Kalahari	Saudi Arabian	Ala Shan	South Sahara	North Sahara	North East Sahara	West Sahara	Average
A Linear dunes (total)	13.96	22.12	32.55	85.55	49.81	1.44	24.08	22.84	17.01	35.49	30.54
Simple and compound	13.96	18.91	18.50	85.85	26.24	1.44	24.08	5.74	2.41	35.49	23.26
Feathered	—	—	—	—	4.36	—	—	3.56	1.13	—	0.91
With crescentic superimposed	—	3.21	—	—	—	—	—	4.02	7.32	—	1.46
With stars superimposed	—	—	14.34	—	19.21	—	—	9.52	6.15	—	4.92
B Crescentic (total)	54.29	36.91	11.80	0.59	14.91	27.01	28.37	33.34	14.53	19.17	24.09
Single barchanoid ridges	8.96	3.21	11.80	—	0.59	8.62	4.08	0.06	—	0.65	3.80
Megabarchans	—	—	—	—	—	—	—	7.18	1.98	—	0.92
Complex barchanoid ridges	16.65	33.70	—	—	14.32	18.39	24.29	26.10	12.55	18.52	16.45
Parabolics	26.68	—	—	0.59	—	—	—	—	—	—	2.93
C Star dunes	—	—	9.92	—	5.34	2.87	—	7.92	23.92	—	5.00
D Dome dunes	—	7.40	—	—	—	0.86	—	—	0.80	—	0.90
E Sheets and streaks	31.75	33.56	45.44	13.56	23.24	67.82	47.54	35.92	39.25	45.34	38.34
F Undifferentiated	—	—	—	—	6.71	—	—	—	4.50	—	1.12

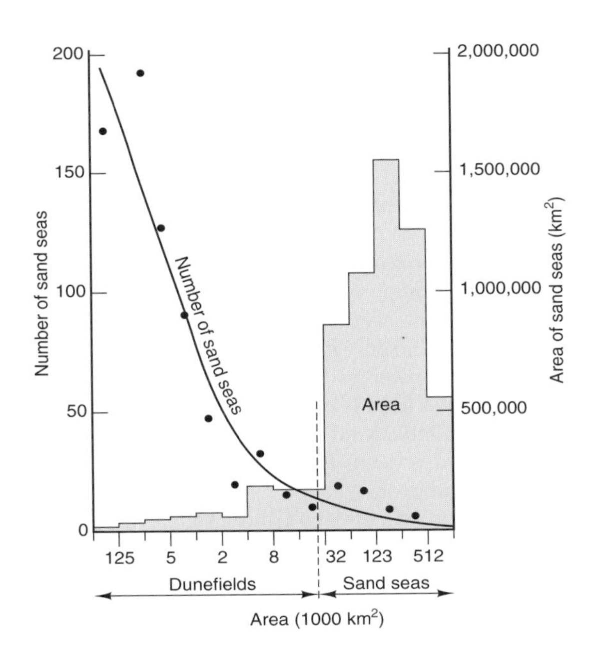

Fig. 6.1 The size distribution of major sand seas (after Wilson 1973).

sand. Fryberger and Ahlbrandt (1979) used sand transport equations (Chapter 2) to demonstrate that present-day wind environments had the potential to move 'geologically significant' amounts of sand. Sand seas therefore occur either close to their source where significant quantities of sand are injected into the aeolian system or where a climatic gradient or topographic barrier arrests the progress of far-travelled sand.

Supply controls

The most frequent control on the accumulation of large sand bodies is proximity to a sand supply. Non-aeolian sources may be fluvial, marine or lacustrine, the sand bodies occurring along river banks or the shores of lakes and seas. Sand seas close to their sources include most of the dunefields in North and South America, and the sand seas of the western Sahara and of central Asia. In the Simpson/Strzelecki sand sea in Australia, aeolian sand is interdigitated with alluvial and lacustrine source material. Here, Wasson *et al.* (1988) argued that sand was not far-travelled, but had been deflated out of local non-aeolian deposits. In the north-western

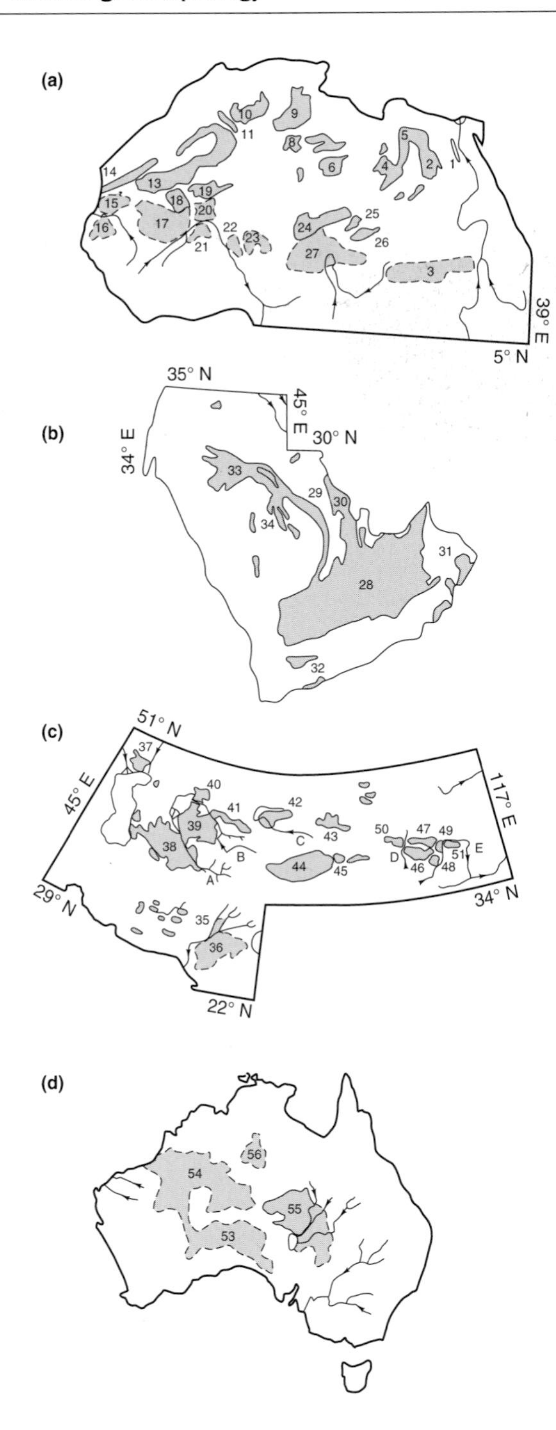

Fig. 6.2 Maps of major sand seas (based largely on Wilson 1973): (a) the Sahara; (b) Arabia; (c) Asia; (d) Australia; (e) southern Africa.

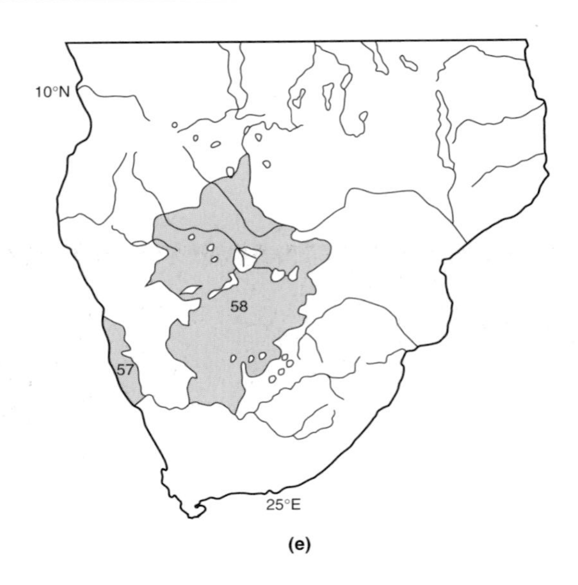

Sahara, Erg Chech in Algeria derives sand from the alluvial deposits of the Atlas or central Saharan mountains (Wilson 1971, 1973; Mainguet and Chemin 1983). The Wahiba Sands of Oman derive sands from the deposits of a marine shelf (Warren 1988c), and the dunefields in the southern deserts of Iraq have a local alluvial source (Al-Janabi *et al.* 1988). In some cases, such as the Kalahari Desert in southern Africa (Thomas and Shaw 1993), sand seas have developed by reworking underlying deposits, some of which are the remnants of earlier sand seas. The most obvious examples of ties between local sources and sand bodies are in coastal dunes (Chapter 5).

Climatic controls

Some sand seas have moved away from their source area or are composed of far-travelled sand. Their location is controlled by zones 'where a sufficient reduction of wind energy exists along the direction of sand drift' (Fryberger and Ahlbrandt 1979: 454). Dunes move or sand is transported as long as there is sufficient available wind energy, and dunes come to a halt and sand is deposited once the wind's ability to carry the sand drops.

For sand to be removed from an area there must be a high resultant sand drift potential as well as a high total sand drift potential (these terms are introduced in Chapter 2). On the other hand, sand generally accumulates where the overall resultant movement is low. This accumulation may be in regions of low total wind energy, but may also be in

Table 6.2 World sand seas (larger than 12 000 km^2) (after Wilson 1973)

No.	Name	Area (km^2)	A*	SC†	BO‡
	North Africa				
1.	Abu Moharik		L		
2.	Great Sand Sea	105 000	L		m
3.	Sudanese Qoz	240 000	F		m
4.	Erg Rebiana	65 000	L		
5.	Erg Calanscio	62 000	L		
6.	Edeyen Murzuq	61 000	L	Q	m
7.	Edeyen Ubari	62 000	L		m
8.	Issaouane-N-Irarraren	38 500	L	O	m
9.	Erg Oriental	192 000	L	O	m
10.	Erg Occidental	103 000	L	O	m
11.	Erg er Raoui	11 000	L	O	m
12.	Erg Iguidi	68 000	L	O	m
13.	Erg Chech-Adrar	319 000	L	O	m
14.	North Mauretanian Erg	85 000	L	O	m
15.	South Mauretanian Erg	65 000	F	Q	m
16.	Trarza and Cayor Erg	57 000	F		m
17.	Ouarane, Aouker, Aklé, etc.	206 000	L/F	O	m
18.	El Mréyé	63 000	L	Q	md
19.	Erg Tombuctou	66 000	L	O	m
20.	Erg Azouad	69 000	F	O	m
21.	Erg Gourma	43 000	F	O	m
22.	West Azouak	35 000	F	O	m
23.	East Azouak	34 000	L/F	O	m
24.	Erg Bilma/Ténéré	155 000	L	Q	d
25.	Erg Foch	13 000	L	O	d
26.	Erg Djourab	45 000	L	O	d
27.	Erg Kanem	294 000	F		m
	Arabia				
28.	Rub al Khali	560 000	L	Q	m
29.	Al Dahana	51 000	L	O	m
30.	Al Jafura	57 000	L	O	m
31.	Ramlat Wahibah	16 000	L		m
32.	Ramlat Sabatayn	14 000	L		
33.	Al Nefud	72 000	L	Q	m
34.	'Nafud complex'	25 000	L	O	m
	Asia				
35.	Thal Desert	18 000	F		
36.	Thar Desert	214 000	F		
37.	Ryn Peski	24 000	L		
38.	Peski Kara-Kum	380 000	L		m
39.	Peski Kyzyl-Kum	276 000	L		m
40.	Peski Priaralskye	56 000	L		m
41.	Peski Muyunkum	38 000	L		m
42.	Peski Sary Isnikotrav	65 000	L		m
43.	Peski Dzosotin	47 000	L		
44.	Takla Makan	247 000	L		
45.	East Takla Makan	14 000	L		
46.	South Ala Shan	65 000	L		
47.	North Ala Shan	44 000	L		
48.	South-east Ala Shan	14 000	L		
49.	East Ala Shan	12 000	L		
50.	West Ala Shan	27 000	L		
51.	Ordos	17 000	L		
52.	'Peski Lop Nor'	18 000	L		
	Australia				
53.	Victoria Desert	300 000	F		d
54.	Great Sandy—Gibson Desert	630 000	F		d
55.	Simpson Desert	300 000	L/F		md
56.	'Northern Desert'	81 000	F		d
	South Africa				
57.	Namib Desert	32 000	L		m
58.	Kalahari Desert		F		

Gaps in the table are due to absence of data.
* Activity: L = active erg; F = fixed erg.
† Sand-cover: O = open 20–80%; Q = quasi-closed 80–100%.
‡ Bedform order: m = mega-dunes predominant; d = dunes predominant; md = dunes predominant, some mega-dunes.

Table 6.3 Indicators of long-distance sand transport (after Fryberger and Ahlbrandt 1979)

- Sand sheets extending hundreds of kilometres, and aligned with present-day drift resultants
- Borders of sand seas that are straight or gently curving, and aligned with present-day drift resultant
- Occurrence of large sand bodies in areas where it is unlikely that concentration was by other processes, i.e. fluvial
- Conversely, lack of sand bodies in windy sand source areas, i.e. alluvial plains
- Extended aeolian sand bodies crossing drainages and hills
- General shapes of sand seas that follow trends of present-day resultant drift

areas of high total energy but low resultant energy. It is possible to have a complex, high-energy regime in which total amounts of sand transport are great, but there is no overall, dominating direction of movement. An example is the Grand Erg Oriental near Ghudamis in Libya where a complex, high total energy, low resultant energy wind environment creates star dunes and where sand accumulates (Breed *et al.* 1979b). Another approach that calculates potential sand flux from meteorological records was used by Wilson (1971) to map sand flows across the Sahara. He estimated rates of deflation of sand from different types of desert surface and related these to calculated sand flow rates to provide a map

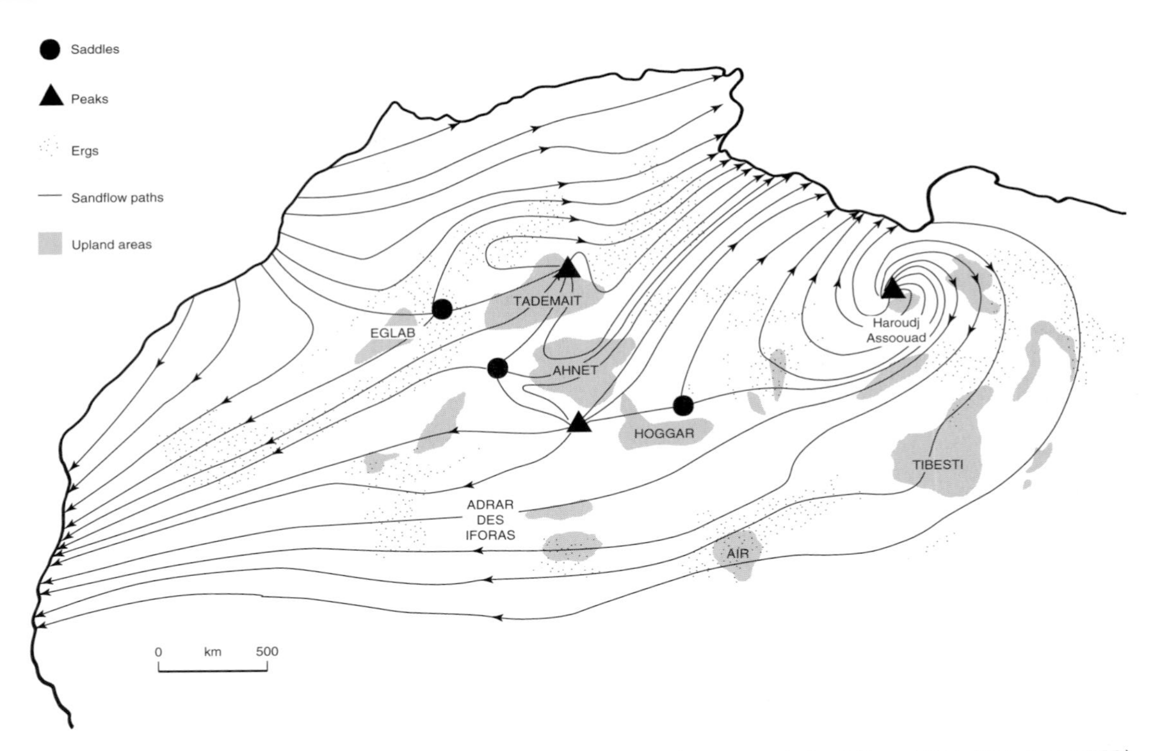

Fig. 6.3 Sand flow across the Sahara showing major source and deposition areas (after Wilson 1971; Mainguet 1978).

which showed source and deposition areas in the Saharan sand seas (Fig. 6.3).

In the sub-tropical deserts, high-energy wind environments are often found in the trade wind belts fringing the sub-tropical anticyclones such as the Arabian peninsula in Saudi Arabia, while the low-energy environments are found near the centre of high-pressure systems and in the equatorial doldrums (Chapter 1). Thus the Sahel zone on the southern margin of the Sahara Desert is an area of low energy subject to the accumulation of drift from higher-energy Sahara in north-easterly trade winds (Mainguet and Chemin 1983). Coasts are also often high energy environments (Eldridge 1980) so that material is deflated from the coastline of coastal deserts such as the Namib (see below) and Atacama and carried inland to areas of lower wind energy where sand accumulates.

Topographic controls

Regional climatic patterns are not the only cause of reduced competence of the wind to carry sand, and the development of sand seas is often related to the patterns of regional topography. Wilson (1973) noted that most sand seas lay in topographic basins, and terminated where there was a pronounced break-of-slope, but sand seas also often develop close to upland areas. The topographic control of accumulation is therefore either a basin or an obstacle.

Some of the more striking examples of sand seas that have developed in topographic basins include the Simpson/Strzelecki in Australia (Wasson *et al.* 1988), the Erg Issaouane, Algeria (Fig. 6.4; Wilson 1973), and the Taklimakan in China (Zhu Zhenda 1984).

Where sand seas have accumulated close to upland areas, it is often because the upland area extends across the regional trend of sand drift, as for example in the Algerian sand seas where a south-easterly sand flow has brought the sand up against the highlands of Tinhrert and Tademait. There sand accumulations occur in much the same way as with the *anchored dunes* described in Chapter 5. Another effect of the intrusion of an upland into a sand stream is to force the flow to diverge and accelerate, so that sand may be rapidly transported over or round the highland area, and accumulation of sand is thus discouraged on the upland itself (Fig. 6.4; Wilson 1973).

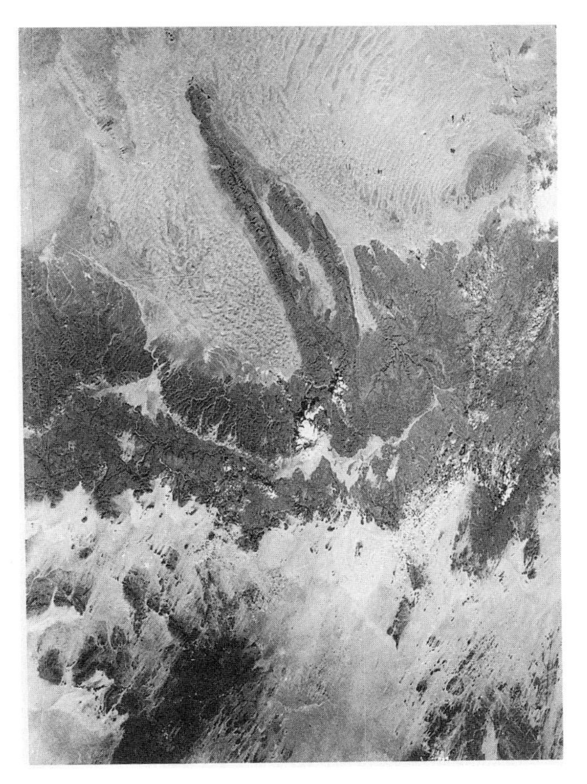

Fig. 6.4 Shuttle image of Erg Issaouane, Algeria (top left), a sand sea in a topographic basin.

Episodic accumulation

The temporal pattern of input of sand can very considerably affect the character of a sand sea or dunefield. The input is frequently pulsed rather than continuous, so that many sand seas display complex, multicyclic origins. These episodes of sand input are sometimes a direct response to periods of greater aeolian activity as a result of climate change (Chapter 8), but may also be a more indirect response to the activity of a non-aeolian source. Periodic activity of aeolian sands is well documented in the rock record (Loope 1985; Kocurek 1988).

The Algodones dunefield is one of the most-studied of sand bodies and has been shown conclusively to have resulted from an episodic input of sand. The dunefield is elongated, being 75 km long and 8 km wide at its widest point, stretching from the north-west to south-east along the south-east border of Imperial Valley in southern California (Fig. 6.5). Beaches of former Lake Cahuilla are said to be the source of sand (for example, Norris and Norris 1961; McCoy *et al.* 1967). McCoy *et al.* (1967) calculated the volume of sand, and estimated that the time required to produce the whole dunefield was 160 000 years. They argued for a single input of sand in the middle Pleistocene. Subsequent work showed, however, that Lake Cahuilla had expanded to the shoreline from which the sand was derived as recently as 750 years ago, and consequently the input was not a unique, one-off event. None the less, supply of sand does seem to have stopped at some time in the past, and Sweet *et al.* (1988) estimated a maximum age for the dunefield of 37 000 years and a minimum of 750 years. They were also able to calculate an average rate of advance of the dunefield of 13.5 m $(1000\ \text{yr})^{-1}$.

Elsewhere, Blount and Lancaster (1990) reported that sand sources in the Gran Desierto of north-west Mexico included ancient and contemporary fluvial sands from the Colorado River (Fig. 6.6), littoral sands from the Gulf of California and local alluvial sources, and that there had been at least three periods of sand input from these various sources, each producing sands with distinctive textural, mineralogical and spectral characteristics.

Development and dune patterns

Because some sand seas are tied to their source while others have moved long distances, the distinction can be made between *static* and *migrating* sand seas and dunefields. Migrating sand seas were termed 'dynamic' by Pye and Tsoar (1990).

Often the centres of sand seas contain greater volumes of sand, expressed as EST (estimated spreadout thickness; Chapter 5), than the margins. This can be manifested in a number of ways. In some sand seas, dunes at the margin are smaller and less complex than those at the centre. Examples of sand seas and dunefields where this is the case include White Sands in New Mexico (McKee 1966), the Great Sand Dunes in Colorado (Fryberger *et al.* 1979), and the Namib Sand Sea (Lancaster 1983a). There may also be sand sheets and zibars rather than differentiated dune forms at the margin (for example, Algodones; Nielsen and Kocurek 1986). Sometimes the thinning of sand at the margins is effected by more widely spaced dunes or wider interdune corridors, as in the Saharan sand seas (Mainguet and Chemin 1983), but in others, such as the Namib (Lancaster

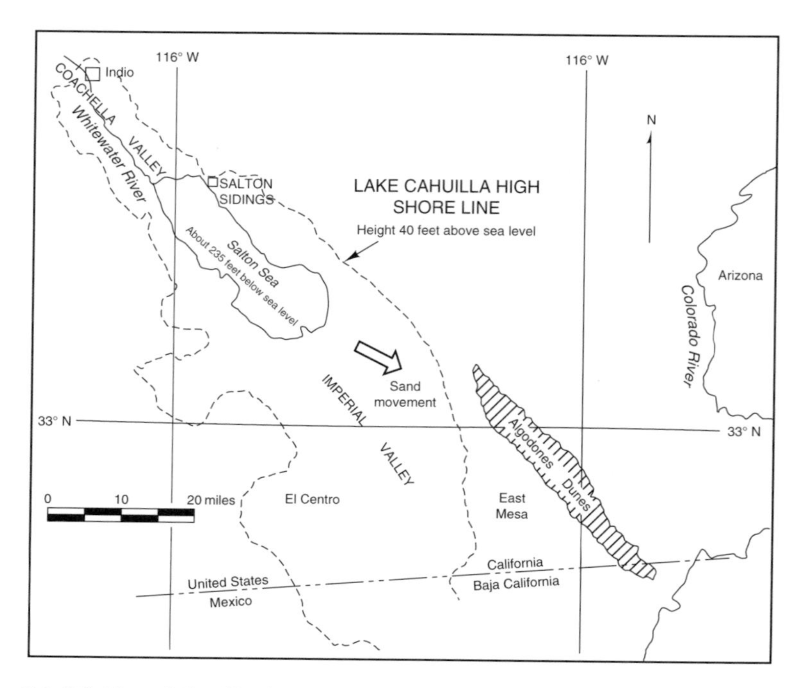

Fig. 6.5 Map of the Algodones dunefield, southern California (after McCoy *et al.* 1967).

1983a), the smaller dunes are more closely spaced than at the centre of the sand sea even though the total volume of sand is less at the margin. Recent work by Bullard *et al.* (1995) showed that even in the Kalahari, which is supposedly dominated by linear dunes, there is in fact a very complex pattern of dune forms).

Great thicknesses of sand are most likely to accumulate at the centre of sand seas with complex wind regimes. High-energy, unimodal and acute bimodal wind regimes, on the other hand, encourage rapid throughput of sand and thin sand cover. It may be that the greater ESTs are in those areas where the local wind regime encourages accumulation while thinner spreads of sand are in areas of rapid throughput of sand. Sand seas commonly have well-defined boundaries. The sharpness of boundary, particularly at their downwind edge, may imply that they are migrating.

Porter (1986, 1987) attempted to develop a general model of dune associations within a sand sea. The model showed small barchanoid or transverse dunes and sand sheets at the downwind and lateral margins, zibars on the trailing (upwind) edge where slower-moving coarse sands were left, and interiors with large, complex dunes. Although the model may apply to simple sand seas in unidirectional wind regimes, local environmental conditions, particularly of sand supply and wind regime, are generally much more important in determining patterns of dune type (Chapter 5). In addition, Fryberger and Goudie (1981), for example, showed that while linear dunes were the most common dune type in sand seas, great tracts of aeolian sand deposits were undifferentiated sheets (Table 6.1).

One striking regional-scale dune pattern that has caused considerable comment has been the so-called 'wheelrounds' or 'whorls', demonstrated particularly by the linear dunes of the southern African and Australian dunefields (Fig. 6.7), but also to some lesser extent others in the dunes of the Sahara. In both southern Africa and Australia the sand seas are dominated by linear dunes almost to the exclusion of other types, and in both cases the pattern has been taken to indicate an anticlockwise movement of sand, supposedly around the continental anticyclone, although there has been some debate about the timing of the movement and the location of the centre of the anticyclone (Brookfield 1970; Goudie 1970; Lancaster 1981; Thomas and Goudie 1984).

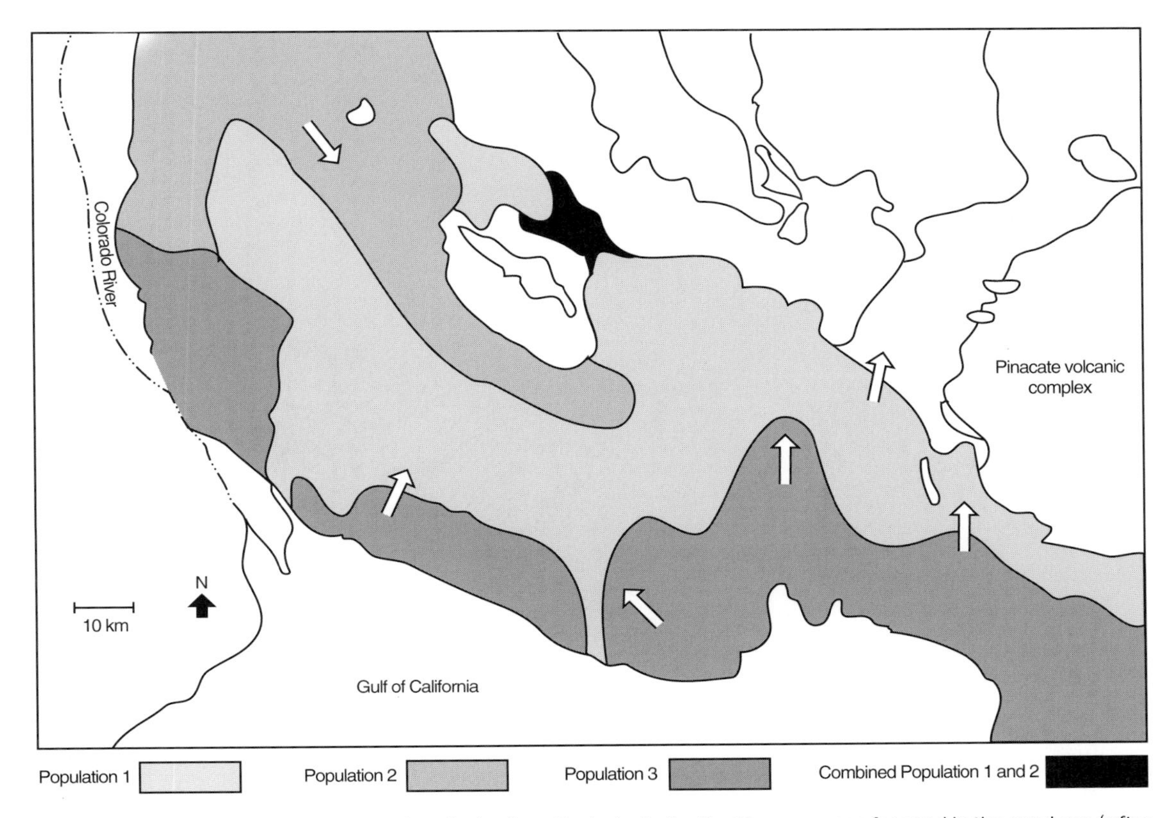

Fig. 6.6 Mixing of different sand populations in the Gran Desierto, indicating three sources for sand in the sand sea (after Blount and Lancaster 1990).

Complex origins: the Namib Sand Sea

The development of most sand seas is controlled by some combination of the three major controls on accumulation (sand supply, climate and topography). The Namib Sand Sea, which covers 34 000 km^2 on the south-west coast of Africa, is a good example of the interaction of these factors. It is bounded to the south by the perennial Orange River and to the north by the ephemeral Kuiseb River. Its eastern margin is the Great Escarpment of the southern African plateau. The core of the sand sea is dominated by large, widely spaced, complex linear dunes. On the western, coastal margin there are barchan and transverse dunes. On the eastern edge of the sand sea there are star dunes and star dune chains (Fig. 6.8).

There is a good body of wind data for the Namib, particularly for its northern margin (Lancaster *et al.* 1984). Lancaster (1985a) used it to calculate potential sand movement (using the expression $Q \propto V^3$; Chapter 2), which gave him patterns of potential sand flow. This demonstrated that the southern and central parts of the sand sea experienced high-energy, unimodal wind regimes, while further inland and to the north, regimes were lower-energy and exhibited greater directional variability (Fig. 6.9).

Lancaster (1985a) therefore postulated that sand moved from higher-energy coastal and southern source areas NNW and NE towards lower-energy northern and central parts of the sand sea. Unimodal, high-energy winds at the coast created the barchan and transverse dunes; bimodal, lower energy regimes in the centre of the sand sea created the large, complex linear dunes; and complex wind regimes at the eastern margin and low-energy wind regimes led to the formation of star dunes.

The northern margin of the Namib sand sea is marked by the Kuiseb River, which runs in a deep canyon for much of its length (Fig. 6.10). This suggests that the northward progress of the dunes is checked by the canyon from which inblown sand is flushed by ephemeral floods (although it may also be

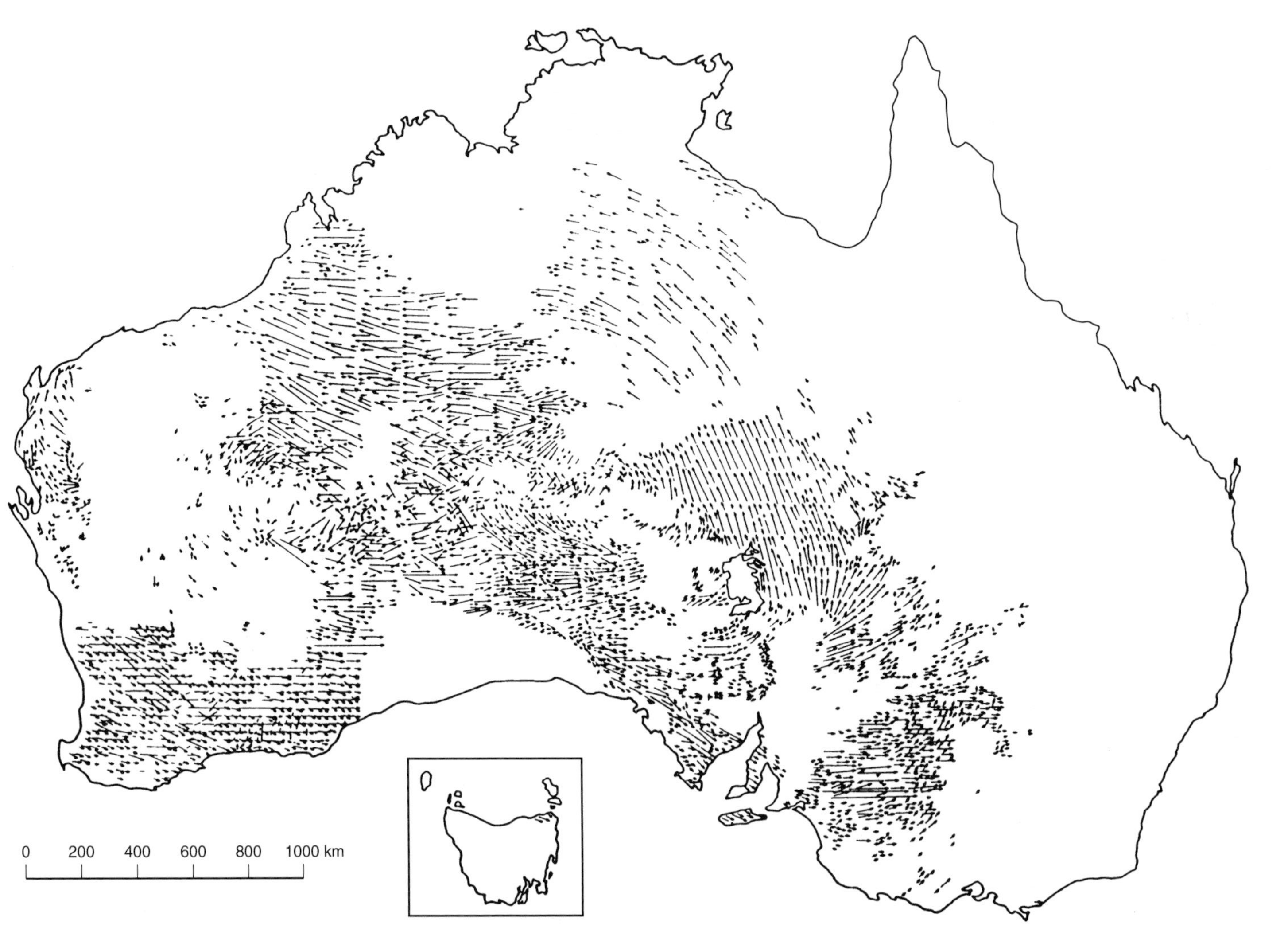

Fig. 6.7 Map of dune orientations (inferred sand transport directions) in Australia. On the map if adjoining quadrates have an orientation within 5° of each other then an arrow has been drawn through them so that the length of the arrow is a guide to the directional continuity of dunes (after Wasson *et al.* 1988).

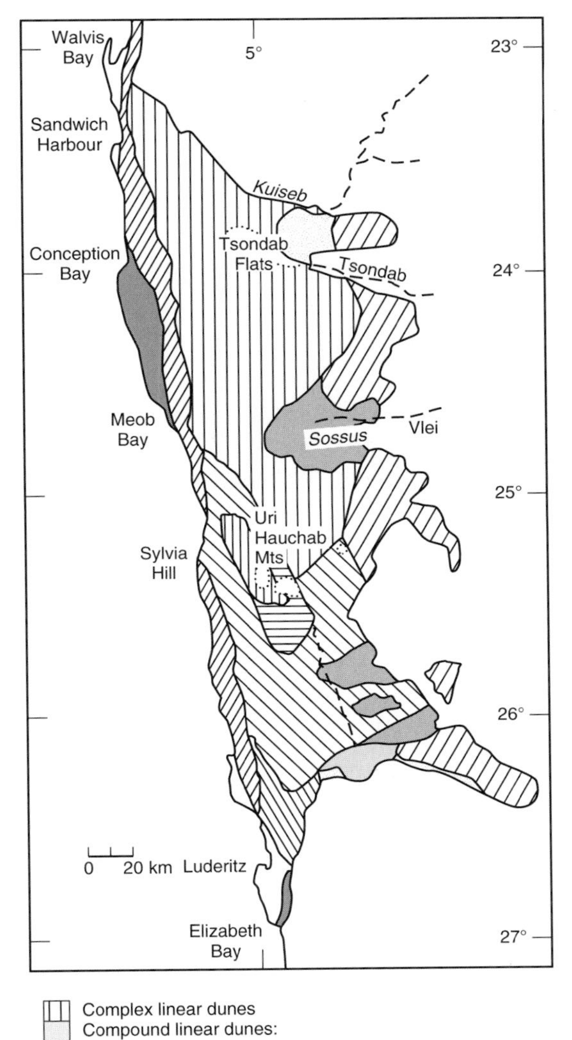

Fig. 6.8 Map of dune types in the Namib Sand Sea (after Lancaster 1983a).

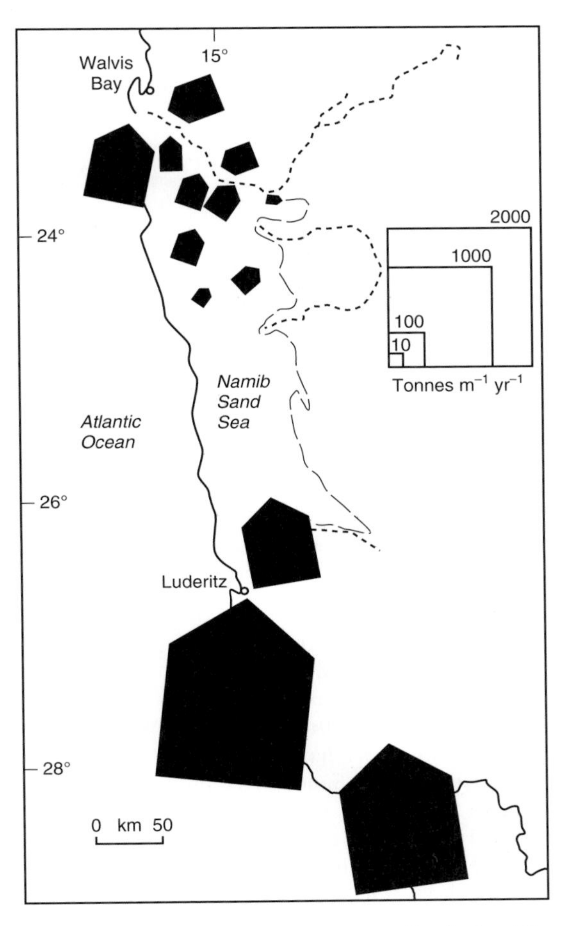

Fig. 6.9 Map of potential sand flow in the Namib Sand Sea based on wind data (from Lancaster 1985a).

slowly filling with sand (Goudie 1972)). However, the wind data also suggest the increasing importance of northerly and easterly winds in the northern Namib (Harmse 1982), so that it may be that there is a climatic control on northward movement of sand as well as the topographic control provided by the Kuiseb. Thus the Namib is associated with its coastal source; the greatest accumulation is at the centre and towards the eastern margin where the wind regime becomes lower in energy and more complex; and there is a topographic limit on its northward progress.

Sand sheets

Table 6.1 demonstrates that many accumulations of sand have no differentiated dune forms and slip faces are largely absent. These are 'sheets', 'streaks' or 'stringers'. Occasionally, this lack of differentiation characterizes a whole sand body, but more usually the

Fig. 6.10 The northward advance of sand into the bed of the ephemeral Kuiseb River at the limit of the Namib Sand Sea.

sand sheet is part of a larger sand sea, and most usually it is marginal to the dunes which dominate the central part of the sand sea.

Fryberger and Goudie (1981) showed that these undifferentiated sand bodies made up 39.46 per cent of the extent of aeolian depositional deserts. The largest, the Selima Sand Sheet on the borders of southern Egypt, Libya and northern Sudan, covers around 100 000 km^2 (Breed *et al.* 1987). Others occur in the Ténéré Desert in Niger (Warren 1971), in Saudi Arabia (Holm 1960), and in the southern Namib Desert (Lancaster 1989a; Corbett 1993). Sand sheets also occur in cold-climate settings where they are often known as *coversands* (Chapter 8).

For the Selima Sand Sheet, Breed *et al.* (1987) found a coarse surface layer of sand and gravel which acted as an armour apparently preventing sand from being mobilized into dunes. The importance of coarse sand has also been noted for sand sheets in Arctic Canada (McKenna-Neuman and Gilbert 1986), and deserts in Australia (Mabbutt 1980), Arabia (Khalaf *et al.* 1984; Khalaf 1989) and California (Kocurek and Nielson 1986). Kocurek and Nielson (1986) identified four further factors which seemed to be conducive to the formation of smaller sand sheets: an evenly spaced vegetation cover; a high groundwater table; periodic or seasonal flooding; and the development of surface crusts. All of these act to inhibit the development of dunes and slip faces. It is also possible that some sand sheets represent the erosional remnants of previous sand seas (Fryberger *et al.* 1984).

Conclusion

Careful mapping of sand seas, dunefields and sand sheets has become possible because of satellite imagery, and recent work shows that some are close to the sources of their sand, while others are composed of sand that has travelled in the wind. There remain remarkably few studies of whole sand seas, and the development and controls of dune patterns within sand seas is a major area for future research.

Further reading

Some general discussion of the distribution of sand seas is to be found in McKee (1979a), and the chapter by Breed *et al.* (1979a) maps and describes the surface features and climatic regimes of eight major sand seas. Key papers on the development of sand seas are those by Fryberger and Ahlbrandt (1979), Porter (1986, 1987) and Wilson (especially 1973). There are also some studies of individual sand seas of which the best include Wasson *et al.* (1988) on mapping sand sea attributes in Australia, Lancaster (1985a) on controls of sand sea formation in the Namib Desert, and Bullard *et al.* (1995) on variation of dune form in the south-western Kalahari.

CHAPTER 7 Dune sediments

Introduction

Geomorphologists can learn much about dune processes from dune sediments, and much about past environments from dune sediments and from dust accumulated in loess, in ice-caps and on the ocean floor. Sadly, little is known about the sedimentary characteristics of contemporary or recently active dunes, and most glaringly, although understandably, there is only a handful of studies of the internal structure of active dunes. This chapter introduces some key concepts about aeolian sand deposits, gives some pointers to the kind of information that they can give, and provides some sources of further information. Dust and loess are dealt with in Chapter 4.

Characteristics of aeolian sand grains: mineralogy, shape, and colour

Nineteenth-century geologists believed that desert dune sands were golden yellow, very well rounded, 'millet seed' quartz grains. The caricature did have some truth, for the vast majority of dune sands are composed of quartz, as is a high proportion of coarse dust (including the material that makes up most loess). This is largely because of the commonness of quartz in rocks and its chemical stability. It is not soluble in the Eh and pH environment of most of the terrestrial surface, has a crystal lattice that is very difficult for hydrogen ions to penetrate, and is hard enough to resist rapid wearing down by physical processes. Other rock-forming minerals are more susceptible to chemical weathering and only make up dune sands close to sources. For example, some coastal dunes, close to offshore deposits derived from shelly marine fauna, have high proportions of calcite and aragonite (calcium carbonate), and some desert dunes, close to dry lakes in which the salts are deposited, are composed of gypsum (hydrated calcium sulphate). Occasionally deposits of volcanic sand act as further sources of aeolian sand (Edgett and Lancaster 1993). Silicate minerals other than quartz more readily break down to clay-sized particles, which, when entrained by the wind, are carried as fine dust, much of which ends up very far from its sources in the oceans or in the soils of humid areas (Chapter 4). Some of these clays are built into dunes, especially as pan-fringing lunettes (Chapter 5), but only when they travel in the wind as aggregates of sand size.

The early caricature of dune sands may have worked for mineralogy, but it was much less reliable about the shape of aeolian sand grains. The first problem in this area is the dangerous ambiguity in the word 'shape', and here it is useful to adopt Barrett's (1980) clarification: *sphericity* is overall shape at a grain scale; *roundness* is shape at the scale of the edge of a grain, i.e. the sharpness of edges; and *surface texture* is shape at a microscopic scale.

Thinking about the shape of aeolian grains is based primarily on theoretical models of how they might be affected by three processes: entrainment, transport and attrition (though, as we shall see, theory has not yet begun to approach the complexity of real aeolian environments). In entrainment, the dominant effect is said to be the angle through which a grain must be rotated before it can be moved. A long, thin particle, if upright, needs to be rotated through a smaller angle before it is dislodged than a spherical particle, and through a smaller angle than if the long particle were recumbent (Li and Komar 1986). On the other hand, a long thin particle could wedge more firmly between others, and a more spherical (or rounder) grain would roll more easily, while a long grain would have to be dragged along. Irregularly shaped particles (less round or with rougher surface textures) would have more contact points with others than would smooth ones, and so would be harder to dislodge. Theory also suggests that shape affects the value of aerodynamic drag of larger sands (Willetts and Rice 1983), but, following Nickling's (1988) study, it appears that shape differences have to be very great – as between the platy, shelly sands and the rounded, quartzose sands in the experiments of Willetts and Rice (1983) – before shape has an appreciable effect.

Theory further suggests that there should be important selection processes in transport. This might be determined by the 'rollability' of a grain, as defined by Winkelmolen (1971), although this method of measuring rollability is debatable. In any case, rollability is likely to affect only the coarser grains in the *creep* load. Little work has been done on the theory of how grains behave in saltation or reptation, although a crude prediction might be that more spherical, more rounded and smoother grains would travel more quickly. The theory (and evidence) for attrition in transport is discussed in Chapter 2. The balance of opinion is that grains are made more rounded and spherical in transport (as they collide), so that far-travelled sands should be more rounded, while loess is not so rounded because of the lack of grain-to-grain contact in suspension.

The empirical evidence shows a much more complex pattern than theory predicts (Fig. 7.1). Taking *sphericity* first, the evidence is of two kinds: from experiments and from field observations. As theory predicts, low-sphericity particles do indeed have lower transport rates in experiments (all other things being equal), especially at high wind speeds. Shelly or shaley grains are less effective at dislodging others, are less effectively maintained in saltation, and have lower, longer saltation trajectories (Willetts and Rice 1986). However, Willetts *et al.* (1982) found some complications in their experiments: less spherical sands moved at marginally higher rates than more rounded sands in low shear, but at lower rates in high shear. Thus the experimental evidence suggests that the sphericity of an aeolian sand may depend on the wind environment which deposited it, but that, at a first approximation, more spherical sands do move more quickly, and so would be expected to characterize far-travelled aeolian sands.

Comparisons of natural aeolian sands in the field with their reputed parent non-aeolian sands do not clarify the issue: when aeolian sands are compared to non-aeolian ones, there are cases in which less spherical grains have apparently been selected

Fig. 7.1 A scanning electron microscope image of sand grains from a linear dune in the Namib Desert. Largest grains are approximately 200 μm across. (Photo: Liz McClain.)

(Stapor *et al.* 1983), and others in which more spherical grains seem to have been taken (Shepard and Young 1961). It is probable, following the experimental evidence, that selection depends on the range of shapes available and on ambient wind conditions.

As for *roundness*, the field evidence does not show that all aeolian sand grains are rounded. A study of 21 600 grains from a variety of dune types led Goudie and Watson (1981) to the conclusion that

> 'the roundness of desert quartz dune sand has been greatly exaggerated in the past. Most is sub-rounded, and sub-angularity is not uncommon. Smaller grains are more angular than larger grains and position on a dune seems to have little influence on grain shape' (Goudie and Watson 1981: 190).

This conclusion confirmed many earlier studies, but all the observations are of sands from a great diversity of aeolian environments, in some of which the grains had probably travelled only a short distance by aeolian transport, and others where they were likely to have been blown about a great deal. Some may have been more rounded than others by non-aeolian processes before they entered aeolian transport. Thus, although they clearly show that roundness is a poor diagnostic of the aeolian environment, they do not resolve the question about whether aeolian transport increases roundness (by whatever process). Moreover, rounding probably occurs more in some aeolian environments than others. Thus Thomas (1987) pointed out that sand grains were likely to be much less rounded on linear dunes, where they were blown about infrequently before being incorporated into a fairly stable deposit, than on transverse dunes which migrated downwind, continually exhuming and reworking the sand.

Finally, there is the *surface texture* of aeolian sands. When seen under a microscope, most dune sands have a 'frosted' appearance, a feature that appears to be the result of two processes: chemical solution and re-precipitation of silica; and mechanical abrasion (Wilson 1979). Not only do most grains experience both processes, at one time or another, but they are probably also mutually reinforcing, for solution is more effective on newly abraded grain surfaces and on the small particles that are dislodged in abrasion, and abrasion is more effective on grains that have suffered some solution and reprecipitation (Krinsley and Smalley 1972). The solubility of silica is considerably raised at high pH, and pH is generally high in dry environments. Frosting can indeed be produced experimentally by subjecting quartz grains to saline solutions (McGee *et al.* 1988).

Attrition in transport produces characteristic features that can be seen in the electron microscope. 'Upturned plates' in particular are said to be a good diagnostic feature (Kaldi *et al.* 1978). It may even be possible to relate the spacing of these plates to wind velocity, though the velocity on record would only be the last before deposition, and the technique has only been tested on grains abraded in the laboratory (Wellendorf and Krinsley 1980). Le Ribault (1978) also identified 'crescent-shaped impact features' with sharp edges as diagnostic of wind-transported particles.

The last element of the nineteenth-century caricature of dune sands was their colour. To say that most dune sands are red or deep yellow is not far from the truth, though it is certainly not true to say that all dune sands have these colours. The red colour comes from staining with iron oxide, giving redness of varying intensity (Gardner and Pye 1981). The colour seldom extends through a whole column of sand, being most commonly in the upper layers of a profile. Wasson (1983a) reported that the red coating was composed of iron with mixtures of the clay minerals kaolinite, illite and montmorillonite. These pigments attach themselves more readily to quartz, and this is part of the explanation for the commonness of red sands, since quartz is the commonest mineral in dune sands.

There has long been debate about the origin of the iron oxide (Walker 1979), some seeing it as having a source within the grains themselves (Norris 1969) and others believing it to have been added in dust (Bowman 1982). It is likely that it is sometimes endogenous and sometimes exogenous, but considering the widespread existence of desert varnishes on rock surfaces (Dorn and Oberlander 1982), now thought to be derived largely from iron in dust, exogenous sources are probably the more important.

Some authorities have claimed that dunes become redder as distance from the source of their sand increases, but this has been disputed with empirical data by Anton and Ince (1986) who argued that transport would abrade a red coating. None the less, in the Namib Sand Sea, sand farther from the source has been relatively stable for longer, so that star dunes of the eastern margin are indeed redder than the transverse dunes of the west coast (Chapter 6; Lancaster 1989a). In this environment, redness has developed once the sand is transported away from its marine or alluvial source, for even if some redness is knocked off in transport, more accumulates. Others claim that redness increases with the age of the sand, especially if it has been only slightly mobile, although older sands could well also be those furthest from their source.

Weathering to redness requires high Eh (oxidizing) and high pH (alkaline) conditions, both of which are common in aeolian sands. Local concentrations of ions, such as calcium or magnesium, probably accelerate reddening (Gardner and Pye 1981). Where these conditions are satisfied, deep redness may develop within 5000 years, but because of variations in the various controls, the rate of reddening must vary widely from place to place (Pye 1983). Nevertheless, this means that only fairly stable land surfaces can redden thoroughly. The world's reddest dunes are indeed found in sand seas dominated by more stable dune types. This stability may either be because levels of activity have dropped and dunes have become relatively less mobile as in the Kalahari and Australian deserts, or because the dynamic nature of the dunes is such that sand is not moved long distances in short times: active transverse dunes are rarely composed of red sands because grains are transported too frequently for iron oxide coatings to develop, whereas the sand on much less mobile linear or star dunes is frequently red.

Although redness is very common, it is not the only colour of dune sands. Pale, buff or even white sands are found in some areas, these being composed mostly of halite, gypsum or calcite, but occasionally even of very fresh quartz. Some dune sands contain variable proportions of dark grains, which are mostly heavy minerals such as magnetite, and some dune sands, especially on coastal dunes, may be darker as a result of the incorporation of organic matter.

Size

The wind is highly size-selective, picking a few sizes of grains in preference to others, and picking them according to controls such as velocity and turbulence.

It is comparatively easy, if somewhat laborious, to collect sand samples, and laser granulometers now make analysis much quicker than the older method of sieving. Consequently, there is a surfeit of reports of grain size data. Often, however, these studies have been marred by poor experimental methods, tenuous relationship to theory, and inadequate sampling frameworks, and the results are given for disparate and poorly located samples. Despite these reservations, a very large body of information about grain size variations on dunes is now available.

The underlying principle of grain size analysis is that different geomorphological processes manufacture deposits with different grain size distributions: in other words, each sedimentary environment has its own grain size 'fingerprint'. Udden (1914) and Wentworth (1922) were among the first to note that grain size distributions appeared to be log-normal, although, as Visher (1969) pointed out, no satisfactory explanation has been forwarded for this phenomenon. Krumbein (1938) developed the 'phi scale' to accommodate the log-normal distributions, such that:

$$\text{grain size in phi } (\phi) = -\log_2 d$$

where d is the grain size in mm. Characteristically, when grain size distribution in phi (ϕ) units is plotted on cumulative arithmetic probability paper the distribution approximates a straight line. Mean, median, standard deviation (sorting), skewness and kurtosis, often calculated using formulae from Folk and Ward (1957), are the descriptive statistics most frequently reported for dune sands (Table 7.1). Note that finer samples have higher phi values of mean grain size, and that skew towards coarser grains gives positive values.

It is usual for a combination of different processes to prevail in a sedimentary environment so that most samples represent the mixing of several populations. Mixing is demonstrated by a series of straight lines on a cumulative probability plot, each with its own gradient (Fig. 7.2; Visher 1969). The mixing of

Table 7.1 Grain size parameters (after Folk and Ward 1957)

Graphic Mean (M_z)

$$M_Z = (\phi16 + \phi50 + \phi84)/3$$

Inclusive Graphic Standard Deviation (σ_I)

$$\sigma_I = (\phi84 - \phi16)/4 + (\phi95 - \phi5)/6.6$$

Inclusive Graphic Skewness (Sk_I)

$$Sk_I = \frac{\phi16 + \phi84 + 2\phi50}{2(\phi84 - \phi16)} + \frac{\phi5 + \phi95 - 2\phi50}{2(\phi95 - \phi5)}$$

Graphic Kurtosis (K_G)

$$K_G = (\phi95 - \phi5)/2.44(\phi75 - \phi25)$$

where ϕn is the grain size in ϕ (phi) units of particles of the nth centile on the cumulative frequency distribution.

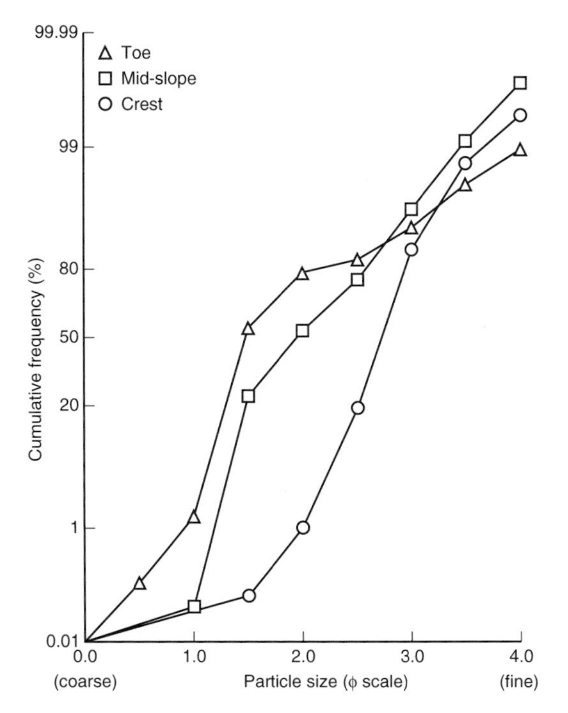

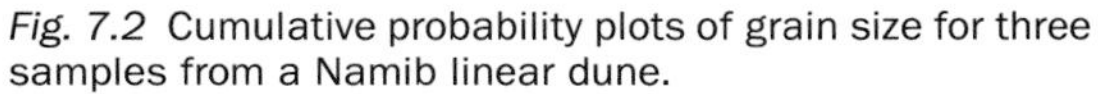

Fig. 7.2 Cumulative probability plots of grain size for three samples from a Namib linear dune.

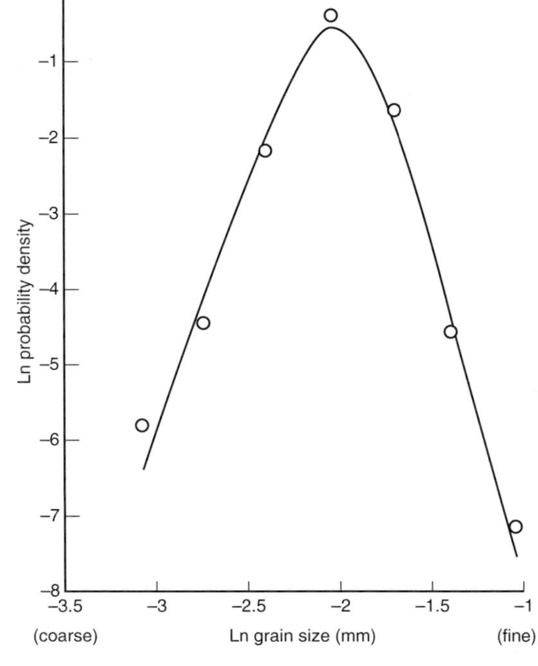

Fig. 7.3 Log-hyperbolic plot of grain size for the crest sample plotted in Fig. 7.2.

grain size populations is most strikingly seen in bimodal distributions in dune sands, about which there is a considerable literature (Taira and Scholle 1979). 'Lag' deposits are one of the types of aeolian sand that are commonly bimodal. Crocker (1946) reported bimodal sediments in the deflationary corridors between linear dunes in the Simpson Desert, Australia, and Warren (1972) described bimodal zibar sands in the Ténéré Desert. In the Ténéré, it could be hypothesized that larger grains shield the smaller, the two modes being around 350 μm (1.5 phi) and 125 μm (3.0 phi), while the intermediate grains are removed from and blown downwind to accumulate in dunes. Similar processes may operate on transverse dunes (Vincent 1984) and on linear dunes (Folk 1971b), and evidence has been found in relict Kalahari Sands (Binda and Hildred 1973).

Objections to the use of grain size to discriminate depositional environments have frequently been voiced. One group of objections has been to the use of log-normal transformations. The argument here is that grain size distributions are not log-normal at all (Friedman 1961), so that in recent years there has been a move in some quarters away from the cumulative frequency plot of log-transformed grain size parameters expressed in phi units towards the use of log-hyperbolic plots (Bagnold 1937; Bagnold and Barndorff-Nielsen 1980; Barndorff-Nielsen *et al.* 1982; Hartmann and Christiansen 1988) or skew log-Laplace distributions (Fieller *et al.* 1984; Flenley *et al.* 1987). Log-hyperbolic plots require both grain size and grain frequency scales to be transformed logarithmically. When the transformed data are plotted a hyperbolic curve is produced, and the curve can then be described by using a number of parameters. The skew log-Laplace approach also uses log-log transforms but its supporters claim that it is a simplified method. The relative merits of the approaches have been widely debated, for example by Wyrwoll and Smyth (1985, 1988) and Christiansen and Hartmann (1988). Wyrwoll and Smyth (1985) claimed that log-normal transforms were still the most effective tool. Whatever their shortcomings they will remain widely used so as to maintain comparability with existing data.

The second objection is even more radical. Some authorities believe that it is simply not possible to discriminate sedimentary environments using grain size distributions, whatever the statistical method for transforming the data (Shepard and Young 1961; Moiola and Weiser 1968; McLaren 1981). McLaren (1981) believed that grain size distributions were controlled far more by the nature of the source of the sediment than by the particular mode of transport or the environment of deposition, and his conclusion was that 'a grain size distribution cannot, by itself, identify the environment of deposition with any certainty' (McLaren 1981: 623).

Notwithstanding this pessimism, it does appear that aeolian sands do have distinctive characteristics. Size, though not diagnostic, is at least characteristic. Aeolian sands are frequently finer, better sorted and less positively (coarse) skewed than glacial, fluvial or marine deposits. Goudie *et al.* (1987) reported an average mean grain size for 1289 desert dune samples of 2.32 phi (200 μm) with average sorting of 0.53 phi. Lancaster's (1987b) samples from the crests of zibar, star and crescentic dunes in the Gran Desierto were all in the range 2.07–2.90 phi (134–238 μm). Ahlbrandt (1979) reported data for 291 coastal dune samples and 175 inland dune samples which showed the inland dunes to have a greater range of mean grain size and sorting values, although all samples were still within a relatively restricted range of 2.0–3.0 phi (125–250 μm). Furthermore, it is commonly found that where sand of this size fraction is available, it is this which is used for dune building, and that because such a restricted size range is used, aeolian deposits are always moderately to well sorted. They are also often slightly positively (coarse) skewed. Kurtosis is the least significant of the Folk and Ward parameters for discriminating aeolian sediments from others and from each other.

Patterns of grain size

Despite the objections of workers like McLaren, there are still many geomorphologists and sedimentologists who believe that there are recurring patterns of grain size in aeolian environments. Because saltation seems preferentially to carry grains around 2.0 phi (125 μm), these are removed from source sediments and carried onto dunes. Dune sediments are indeed usually finer than the surrounding interdunes, and sands on crests are generally finer than those lower on windward slopes. Coarse grains are transported down the lee slope by slip-face avalanching, and basal slip-face sands are consequently frequently coarser than crest or mid slip-face sands (Bagnold 1941).

The pattern in which sands on crests are finer than those lower on a dune is especially well-developed on transverse and barchan dunes, where it has been attributed to the transport of coarser material around the dune rather than over it (e.g. Hastenrath 1967; Warren 1976a; Lancaster 1982). But Watson (1986) found a rather less clear pattern on a Saudi Arabian barchan, even though the finest sample was still at the crest. Rather more distinctive was the description of Barndorff-Nielsen *et al.* (1982) of the increase in grain size from 208 μm (2.27 phi) at the upwind base to 225 μm (2.15 phi) at the crest of a dune and then a fall in grain size again on the lee slope.

Three different patterns have been described on linear dunes: finer crest, coarser crest and no difference. The *no difference* pattern is rare, although Sneh and Weissbrod reported that 'top to bottom changes in the mean, skewness and kurtosis are not significant enough to be able to distinguish between them ... The only tool which does enable distinction between the two slope types seems to be the sorting of the grain population' (Sneh and Weissbrod 1983: 723). Buckley (1989) also found that data aggregated from several Australian sand seas showed only very slight variations in grain size, and Warren (1971) reported data from a seif east of Adrar Madet in Niger which showed no change in grain size over most of the dune.

Finer-crest linear dunes are typical of the Namib Desert (Watson 1986; Livingstone 1987; Lancaster 1989a). Although the range of grain sizes on these linear dunes is small, the crests are composed of sand which is generally finer, better sorted and less coarse-skewed than base sands (Fig. 7.4). Just as with transverse and barchan dunes, the explanation may well be that coarser grains are too heavy to be carried in saltation up the windward slope. By contrast, work in the Kalahari Desert (Thomas 1988) and in the Australian deserts (Wasson 1983b) suggested that crest sands are coarser although better sorted than related dune-slope and interdune sands. These differences may well be attributable to the nature of source material, for if the source is finer than the preferred size of the saltation load, a *coarser-crest* pattern could be expected.

Several workers have argued that the variations found across dune cross-profiles represent discrete grain size populations. Bagnold (1941), for instance, believed that dunes could be divided into three components; crest, plinth and interdune. However, work by Watson (1986) on a Namib linear dune showed that grain size changes progressively across

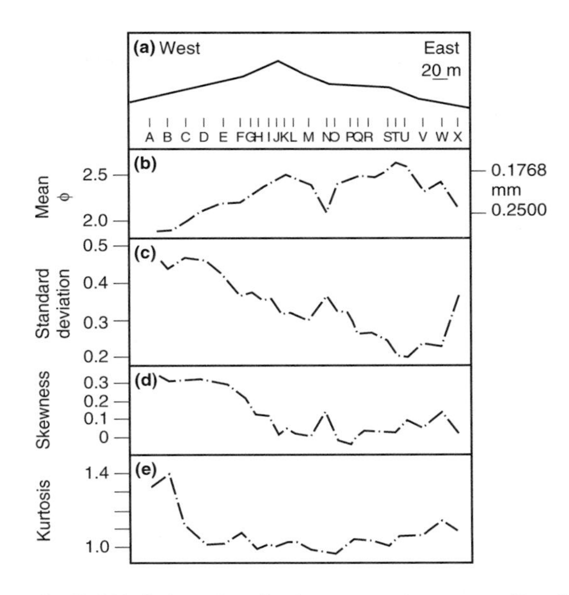

Fig. 7.4 Variation of grain size parameters over a Namib linear dune (after Livingstone 1987).

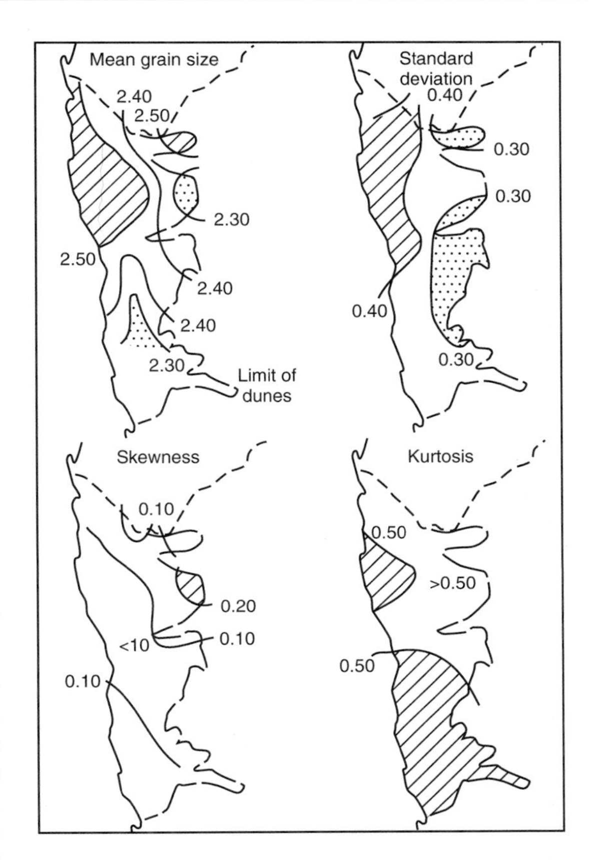

Fig. 7.5 Spatial variability of grain size (phi scale) in the Namib Sand Sea (after Lancaster 1982).

the dune, and his findings were confirmed by Livingstone (1987) who described statistically significant changes of grain size every 20–30 m on the dune cross-profile. Variations of this kind across dune profiles, as well as temporal change related to wind regime changes (Livingstone 1989b), mean that considerable caution must be exercised in comparing supposedly representative results, and show clearly that single samples can rarely be a satisfactory indication of grain size on dunes.

Variations in grain size at the scale of sand seas and dunefields have also been reported. While most dunes are made of fine sand and are well sorted, a pattern at the scale of a dunefield has sometimes been detected. This is of increased fining, better sorting and decreasing skewness in the direction of sand transport. In the Namib, for example, there is a general south-to-north fining and improvement of sorting, but overlaid on this pattern there is a body of relatively coarse, poorly sorted sand at the coast where marine sediment is being brought onshore to be incorporated into the dunes (Fig. 7.5; Besler 1980; Lancaster and Ollier 1983; Lancaster 1989a). In the south-west Kalahari a pattern of progressively finer, better sorted and less skewed sand in the overall direction of transport has also been confirmed (Lancaster 1986). Goudie *et al.* (1987) fitted trend surfaces to grain size data for the Wahiba Sands in Oman, and found a complex pattern, although they believed that there was a tendency for sands to become finer and better sorted with distance away from the source. However, grain size patterns are not always clear. Sharp's (1966) work in the Kelso dunefield in California demonstrated that a distinctive size distribution became established after the sand had been transported about 15 km, a relatively short distance when compared to transport across sand seas, and Buckley (1989) suggested that regional variation in the central Australian dunefields was slight.

Internal structure of aeolian dunes

For the geomorphologist, knowledge about the internal structure of contemporary, active dunes would be of considerable value in elucidating modes of dune formation. The information would also provide analogues for sedimentologists studying dunes in the ancient sedimentary record. Regrettably the logistical problems of cutting through active dunes and of supporting sections in dry sand have proved overwhelming; even wetted sand tends to collapse. There are consequently very few published sections showing the internal structure of an entire contemporary, desert dune cross- or long-profile, although recent advances in the use of ground-penetrating radar are promising, at present providing information about structures to a depth of about 5 m (Schenk *et al.* 1993). Notable exceptions are in the systematic work of McKee (1979c) in his study of a range of dune types. Most studies have relied on a few observations near the surface. Often it has been left to sedimentologists, either working on ancient aeolian deposits in the geological record or developing mathematical and software models, to conjecture upon the internal structure of contemporary dunes. Much useful information, summarized by Kocurek (1991), has come from these sources.

Sedimentary structures are the result of distinct sedimentary events. With each sand-moving event a single layer of sand, or lamina, is created, usually only a few millimetres thick. As the wind speed rises beyond the threshold for sand movement it picks up the finest sand grains first, so that each lamina is finest at the bottom and becomes coarser towards the new top surface.

There are three modes in which aeolian sand is deposited and hence three types of aeolian depositional surface (Bagnold 1941; Hunter 1977a). On rippled surfaces bedload grains moved by creep come to rest in sheltered positions such as the lee slopes of ripples or between adjacent grains. Bagnold (1941) termed this mode of deposition *accretion* and Hunter (1977a) called it *tractional*. These deposits tend to be packed very closely, and hence form very firm surfaces. The coarsest grains often concentrate at the upper part of ripple lee slopes so that there is clear size grading of the deposits, sometimes so striking that it is called 'pin-stripe' bedding (Fryberger and Shenck 1988). This process has now been modelled by Anderson and Bunas (1993). Accretion or tractional deposits generally characterize the lower slopes or plinths in dune landscapes and have low angles (usually less than 15°), but some are also found on the crests of transverse dunes.

Most other aeolian deposition is in the zone of flow separation in the lee of dune crests. Deposition here due to the rapid decrease of the transporting capacity of the wind was termed *sedimentation* by Bagnold and *grainfall deposition* by Hunter. Grainfall deposits show rather indistinct lamination and intermediate packing. The third type of deposition is due to the avalanching of slip faces when they are built to an angle greater than the angle of repose for sand (Chapter 5). These were termed *encroachment deposits* by Bagnold; a single lamina of this kind of deposit was called a *sandflow cross-stratum* by Hunter. These lee-slope processes generally produce high-angle beds with dips around 30°–34° (Chapter 5). Individual sandflow laminae tend to be wedge-shaped, tapering upwards; they are loosely packed, and frequently show coarser grains towards the base of the slip face, because the coarse grains outpace the finer grains in the downslope flow.

As an active transverse dune migrates downwind, sand is eroded from the windward slope and deposited in grainfall and sandflow deposits on the lee slope. These high-angle deposits facing downwind are called *foresets*, and this form of stratification is termed *cross-bedding* (or *cross-strata*). The great majority of sand dune deposits consist of this kind of bedding, and indeed these structures are considered a diagnostic feature of dune sands in the sedimentary record (Fig. 7.6). The windward slopes of dunes, especially but not solely of transverse dunes, are usually capped with a thin layer of *topset* beds, which are tractional deposits, with characteristically thin, closely packed laminae. These deposits sometimes include ripple-scale cross-bedding.

This ideal pattern can be compared with sections described from field sites. The evidence from a number of sections from contemporary, active dunes was reviewed by McKee (1979c). The review included the results of his own section from a transverse dune at White Sands National Monument, New Mexico (McKee 1966). The diagram (Fig. 7.7) shows the high-angle foreset beds of the former slip faces and the much lower-angle topset beds with ripple cross-bedding created by ripple migration on the windward slope. McKee (1979c) also reported sections for barchan, parabolic, dome, linear and star dunes.

For anything to remain of dunes in the sedimentary record, there must be net deposition. Most dunes move on and are replaced by another one and leave no deposits. Occasionally, however, erosion cuts across the deposits of an earlier dune, and the next dune migrates over the eroded remnant of the one before (Fig. 7.8). The surface between the older and newer dune is called a *bounding surface*. The surface is

Fig. 7.6 Dune structures revealed in a late Pleistocene aeolianite, the cemented remnant of a climbing dune, in Saurashtra, India. (Photo: Andrew Goudie.)

seldom horizontal, in which case each subsequent dune is forced to climb over the one before. This is known as *bedform climbing* (Rubin and Hunter 1982). If the angle of climb is great enough, it equals the angle of the dune's windward slope, and entire dunes are then preserved in the sedimentary record. More usually the distance between these bounding surfaces is small, and only a small fraction of each dune is preserved.

Based on work on dunes in the geological record, several authors have argued that there is a hierarchy of bounding surfaces (Brookfield 1977; Kocurek 1981; Rubin and Hunter 1982), but there has been some debate about the origin and classification of these. Stokes (1968) believed that bounding surfaces represented deflation surfaces whose level was determined by the local water-table. His argument was that sand was removed to this level, further removal being inhibited by moisture. Others have argued more recently that these bounding surfaces could be developed by more straightforward processes of dune migration and sediment accumulation.

Based on earlier work, Kocurek (1991) reviewed recent developments, and outlined a three-order hierarchy of bounding surfaces. In this scheme, third-order surfaces result from some change in external control, such as the direction of migration, during the deposition of a cross-strata set. Second-order bounding surfaces are the result of the migration and climb of smaller dunes on a larger dune, and first-order surfaces are created by the passage of dunes of the same size over each other as described above. Kocurek (1988, 1991) further argued that another type of surface might exist in the sedimentary record, and termed it a *super-surface*. This is a break in sedimentation brought about when conditions cease to be favourable for dune development.

The internal structure of linear dunes has proved the focus of considerable attention, not least because

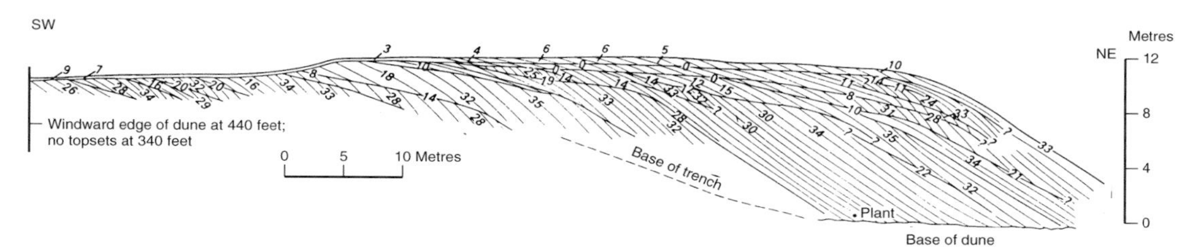

Fig. 7.7 Internal structure of an active transverse dune at White Sands National Monument, New Mexico (after McKee 1966).

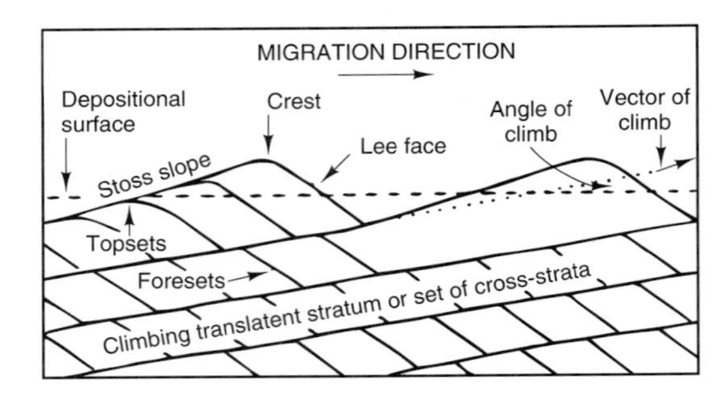

Fig. 7.8 Cross-strata resulting from bedform climbing by migrating transverse dunes (after Kocurek 1991).

of conflicting views of linear dune development (Chapter 5). The discussion usually begins with Bagnold's (1941) deduction of internal structure from his understanding of the dynamics of linear dunes, which produced a model of high angle cross-beds dipping symmetrically away from the dune crest. McKee and Tibbitts (1964) seemed to confirm Bagnold's model when they described the pits that they had dug in a dune near Sebhah Oasis in south-west Libya. Their dune, which was aligned roughly east–west, was subjected to a diurnally bi-directional wind regime such that winds blew from the south-east in the morning and north-east or north-west in the afternoon. To study the structure, they pumped water from a truck onto the dune. The wetness held the sand together and allowed them to dig pits. They found that all the beds dipped away from the crest, although they believed that the bedding also showed some evidence of lateral shift of the dune towards the north.

Two studies have challenged this rather simplistic model of symmetrical structures. In the first, Tsoar (1982) reported structures which fortuitously became visible in wet sand at the surface of a linear dune after heavy rain and subsequent wind erosion (Fig. 7.9a). Because his study dune meandered (Fig. 5.24a), deposition took place obliquely to the long axis of the dune at an angle of 20°–25°. Tsoar's description of internal structure therefore included these moderate-angle beds as well as slip-face and topset beds.

The second challenge comes from those who believe that linear dune structures are not in fact symmetrical as Bagnold deduced, because the dunes often exist in bimodal wind regimes in which the modes are of

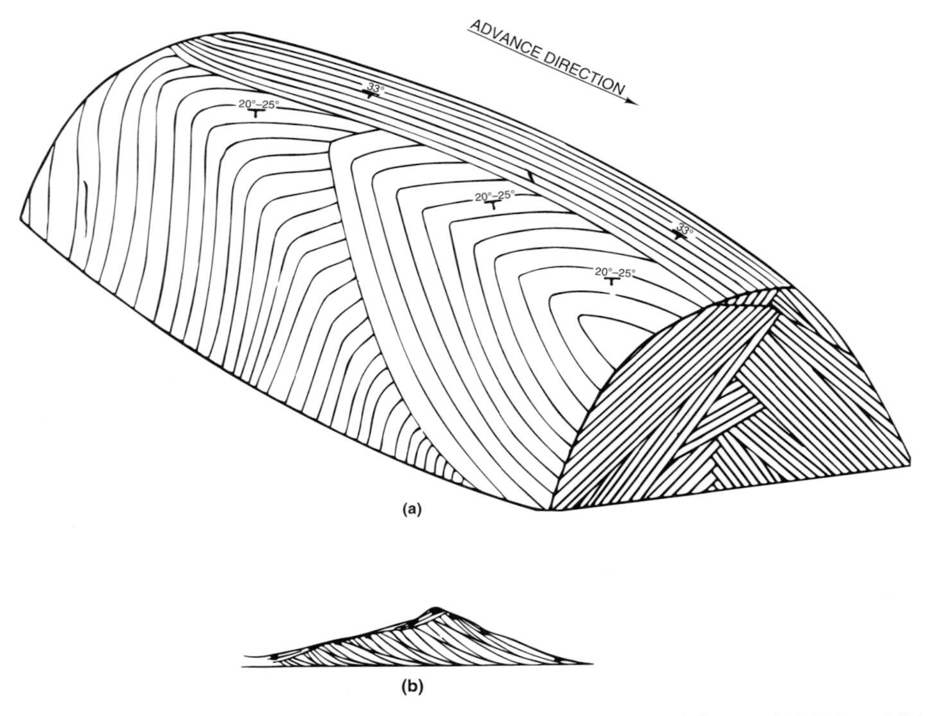

Fig. 7.9 Internal structure of linear dunes as inferred from a range of evidence by: (a) Tsoar (1982); and (b) Rubin and Hunter's (1985) model of McKee and Tibbitts' (1964) data.

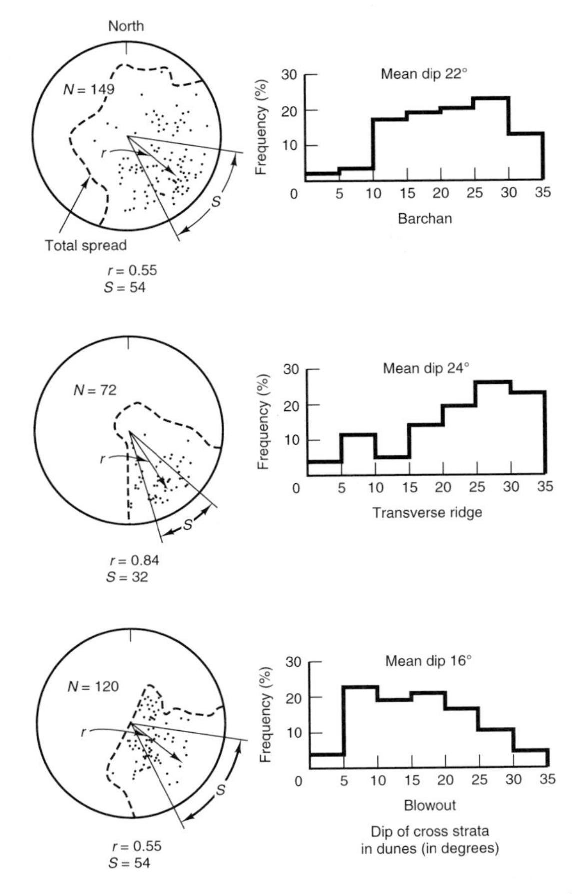

Fig. 7.10 Diagrammatic representation of dip angle and direction. The circles show compass direction of dip and angle of dip as distance from the centre (outer circle is a dip of 40°). The histograms show the frequency of occurrence of dip angles (after Ahlbrandt and Fryberger 1980).

different magnitude. In these regimes the dunes migrate laterally, albeit at a much slower rate than that at which they extend (Chapter 5). Rubin and Hunter (1985) noted that linear dunes were rarely recognized in the geological record. They hypothesized that linear dunes in asymmetrically bimodal wind regimes produced cross-beds which dipped predominantly in one dip direction, and they reinterpreted the information which McKee and Tibbitts (1964) had gleaned from their pits (Fig. 7.9b). Rubin and Hunter's belief was that 'many of the aeolian sandstones that have been attributed to transverse dunes were deposited by dunes that would be called longitudinal or linear dunes in modern deserts' (Rubin and Hunter 1985: 156).

The internal structures of dune deposits may be further affected by syn- or post-depositional changes brought about by the presence of water, ice, vegetation, roots or burrowing animals or by pressures of overburden or seismic shock. Some of the resulting structures in aeolian sands were described from the White Sands National Monument by McKee *et al.* (1971).

Although internal structures are rarely simple, a number of studies have reconstructed sand-flow patterns in the past on the premise that the high-angle beds represent former slip faces and are therefore foresets dipping in the direction of sand transport. Ahlbrandt and Fryberger (1980) used about 1000 dip angle and direction measurements for beds within known dune morphologies and deep sections in formerly active dunes in the Nebraska Sand Hills to determine past drift directions (Fig. 7.10). Glennie (1970, 1972, 1983, 1985) also used plots of dip angle and direction to distinguish transverse from linear dunes in sandstones, although his interpretation of linear dunes as forming in unimodal parallel wind regimes was subsequently challenged by Steele (1985). Further, in the light of work by Rubin and Hunter (1985, see above), linear dunes should not be relied on invariably to produce symmetrically bimodal dipmeter plots.

Conclusion

Much of the nineteenth-century caricature of aeolian sands has been overturned by subsequent studies. In recent years, alongside a general increase in the body of information about aeolian sands, two themes have become particularly important. The first, which is not unique to aeolian geomorphology, is the necessity to agree a method for representing the pattern of grain sizes present in a sample. The second concerns the internal structure of dunes. Much of what is known is based on studies of ancient dune sediments, so that links between sediments and processes are often little more than informed conjecture. Sedimentologists studying those ancient deposits rely on contemporary, geomorphological investigations to explain the processes responsible for the bedding patterns. Despite these reservations, much has been discovered about the characteristics of aeolian sand in recent years, both as a consequence of single-dune studies (Chapter 5) and as a result of the endeavours of sedimentologists.

Further reading

There is a plethora of studies of grain size. Livingstone (1987) and Watson (1986) discussed variation on single dunes, while Lancaster's studies of the Namib (Lancaster 1989a) and south-west Kalahari (Lancaster 1986) covered variations at the scale of a sand sea. Kocurek (1991) reviewed knowledge about the internal structure of desert dunes, and Goldsmith (1985) discussed coastal dunes. Other key papers are those by Hunter (1977a) and Rubin and Hunter (1982). Gardner and Pye (1981) discussed the red colour in dune sands.

CHAPTER 8

Palaeoenvironments

Introduction

The Earth's most imposing aeolian landforms are inherited rather than products of contemporary processes. This chapter discusses these features, concentrating on the types of environment that enabled their development and how they can be dated. Detailed chronologies and the nature of deposits that have lost their surface form and been incorporated into sedimentary rock are mentioned only in passing.

An appraisal of the aeolian inheritance, however, cannot avoid a discussion of the difficulty of distinguishing it from currently active landforms. Although there is no doubt about the great age and present inactivity of many aeolian features, there is a substantial minority where it is not easy to separate inert legacy from actively accruing capital. The contemporaneity of very large dunes, which is discussed below, is an example of this kind of debate. As in many other environments, geomorphologists must always be alert to facile assumptions, either that a feature is the work only of currently active processes, or that it is an inert inheritance.

Loess is a very extensive and distinctive palaeo-aeolian deposit, which because of its origin as dust is discussed in Chapter 4.

Windiness, dryness and aeolian sediment movement in the past

Graphic evidence of the intensity of aeolian activity during the Pleistocene is found in carefully dated ice cores from the crest of the Greenland ice-cap (Fig. 8.1). They show that the atmosphere in the Late Glacial Maximum, 18 000 years ago, was 40 times dustier than it is today, but also that there were abrupt alternations, before and since that time, between extremely dusty and relatively dust-free periods, as at the present (Taylor *et al.* 1993).

At least some of this inordinate dustiness must have been the result of much greater windiness, if we are to believe models of ice-age climates (on Global Climatic Models, or GCMs), using inferred climatic parameters for that period. In one model, ice-age wind velocities are 24 per cent greater in the northern hemisphere than at present in winter, and 124 per cent greater in summer; in the southern hemisphere they are 17 per cent greater in both seasons (Newell *et al.* 1981). The escalation of wind action must also have been strongly stimulated by greater aridity, for there is ample evidence that there were also changes in rainfall and evaporation in the Pleistocene (Williams 1994). Even changes in the periodicity of dry and wet spells must have had an effect (Wasson 1984). But although a change in one factor alone (windiness, aridity or variability in aridity), if great enough, could itself have initiated a period of intense sediment movement by wind, it is more likely that complex mixes of smaller changes initiated most of them. A further complication is that the dated field evidence shows that few climate changes affected the whole globe at once in the same way, and that changes in one climatic parameter were seldom in step with others.

Some of the mist is clearing, for there is now good evidence that the major climatic shifts followed the same kinds of 'Milankovitch' rhythms that triggered changes in other geomorphological processes. These rhythmic changes are produced by variations in the orbital eccentricity, precession and axial obliquity of the Earth, which combine to vary the overall solar input and its distribution between hemispheres, and thus to produce quasi-regular climate cycles (Fig. 8.2;

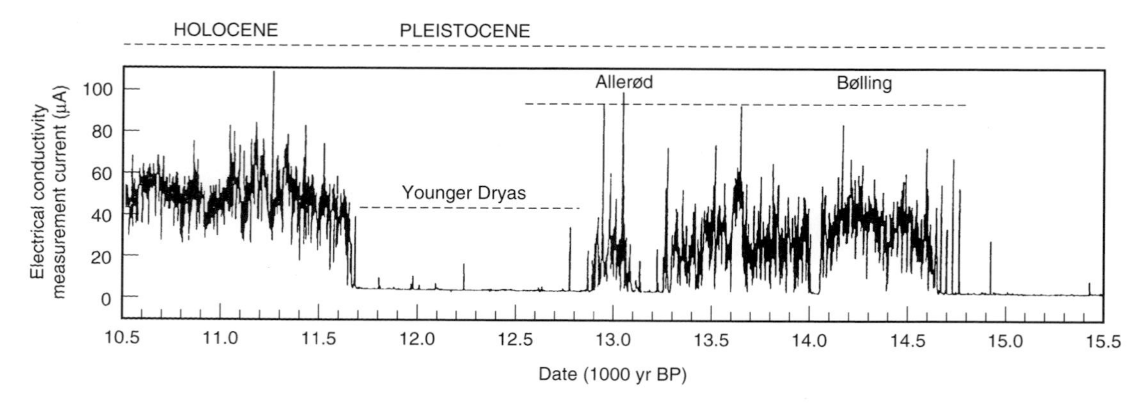

Fig. 8.1 Yearly averages of electrical conductivity (related to dust content) of a core at the summit of the Greenland ice-cap (after Taylor *et al.* 1993). The core shows remarkable (and sudden) alternations between the dustiness of cold periods and the relatively clear skies of interglacial periods.

Dawson 1992). A great range of evidence, particularly from deep-sea cores, in which the presence or absence of aeolian sediments is a crucial element, shows that there was a major climate change about 2.4 million years BP, when precession cycles began to dominate, and Earth began to cool. High-frequency, low-amplitude changes dominated from 1.8 to 0.9 million years BP, after which lower-frequency but higher-amplitude changes were much more marked (Williams 1994). Superimposed on these global patterns, interactions of the various rhythms produced different types of change at different times and in different parts of the globe.

At 125 000 years BP and again at 9000 years BP, for example, the northern hemisphere was evidently warmer in summer and cooler in winter than it is now, because of changes both in the perihelion and the tilt of Earth's axis. This may have strengthened the northern tropical monsoon circulation, and thus, by increasing rainfall in the monsoon zones, could have decreased the movement of dunes and the production of dust from places like the West African Sahel. But, by desiccating the northern continental deserts, as in China, the same set of changes probably increased sand movement and dustiness there (Kutzbach 1989). Changes like these may be part of the explanation of why, in general, the dustiest periods of the Upper Pleistocene in the Pacific did not coincide with the glacial maxima (Leinen 1989), while those in the Atlantic off West Africa did (Hooghiemstra 1989).

The complexity produced by the Milankovitch rhythms does not exhaust the intricacies of the changes that can be expected from the aeolian record. Quite apart from the climatic interference caused by tectonic uplift (as in the intensification of the Asian monsoons and of the aridity of central Asia

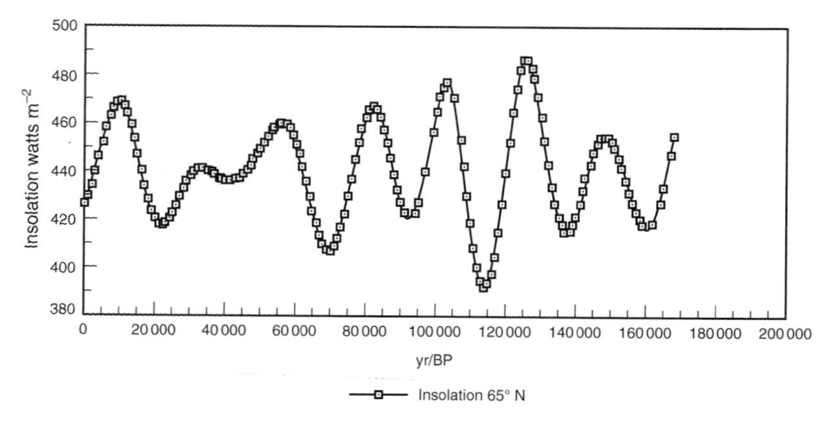

Fig. 8.2 Variations in solar input at 65° N due to changes in the 'Milankovitch' forcing factors (data from Berger and Loutre 1991).

and even of the Sahara induced by the uplifting of the Tibetan Plateau), a number of other circumstances had to coincide to allow more sand or dust to blow about at some periods than in others. The most important additional control is through the supply of sandy or dusty sediment, as the record of individual sites shows.

In the south-eastern Mediterranean, dusty and sandy periods are said to have alternated, apparently because wet periods discouraged dust production (and loess formation), but by increasing flow in the Nile, increased the supply of sand to the coast, thus producing an influx of coastal sand dunes (Issar *et al.* 1989). The controls on loess production are said to be quite different a few thousand kilometres to the west, in southern Tunisia, where it is believed that loess accumulated in moister rather than drier periods. The hypothesis is that loess accumulation required both intense production of dust-sized sediments, as in the alluvium of frequently flowing rivers, and a vegetation cover dense enough to trap the blowing dust. Because both these conditions need humidity, loess is said to have accumulated in humid periods as in the Upper Pleistocene and even Holocene (Coudé-Gaussen 1987). Further complications may have occurred in cold climates. Recent research in the Taklimakan Desert in China shows that the wind was inactive during the glacial maxima, perhaps because of frozen and snow-covered conditions, but that as sediment was released by glacial meltwaters and as the snow cover declined in the Late Glacial, the wind again began to move large quantities of sediment (Wang Yue and Gong Guangrun 1993). On coasts the controls on sediment supply were different again, for here increased supply of sand was probably the result of the coastal erosion induced by rising sea levels, as the discussion below shows.

During the 'aeolian' periods, whenever they were, dust accumulated in many places. It was absorbed into the ice, as shown above, but, more significantly, it built up as loess, which reaches hundreds of metres thick in parts of China and central Asia (Chapter 4). It also accumulated as sediment in the oceans, with notable thicknesses in the Atlantic, off West Africa (Chapter 4). Furthermore, the windier and drier conditions increased the activity of sand dunes in the present deserts, and extended the sphere of active sand movement, in some cases very considerably. The sudden changes detected in the ice core shown in Fig. 8.1 (sometimes within as little as a decade) may have allowed the preservation of the details of many of the depositional features. The deposits of aeolian dust and sand affirm that there must also have been increased erosivity by the wind, though evidence of the timing of the erosion, apart from the sedimentary outcome, is less good.

These aeolian periods in the Late Pleistocene gave us most of our heritage of aeolian landforms, for few earlier aeolian landforms have survived. This is probably because most aeolian sediments are vulnerable to erosion either by renewed wind activity, by water (when the climate becomes wetter, for whatever reason), or by waves and fluctuating sea levels, if they are at the coast. Only in basin situations (especially the oceans), where they are protected by water or later deposits, can they survive for long.

Notwithstanding this loss of primary material, the interests of aeolian geomorphologists extend back well beyond the Late Pleistocene in the geological record. The recent periods of extreme aeolian activity were only the climax of a gradual increase since the Miocene, with rapid intensification from the end of the Pliocene (Hovan *et al.* 1989; Rea 1989). The increase may have been gradual, but, by analogy with the Pleistocene, it was more probably punctuated by smaller-scale fluctuations, of the order of tens of thousands of years, between quieter and more active episodes. Even the Miocene is not the limit of interest, for the wind has been active through most of geological time, though the evidence becomes increasingly weaker down the geological column. It is mostly in the form of sand dunes, now lithified. 'Loessites' (lithified loesses) (Edwards 1979) are rarer, and more controversial, mainly because they are difficult to distinguish from other silty rocks. Wind-erosion features become harder and harder to recognize in older and older rocks.

Discriminating relic from active aeolian features

Activity and inactivity in dunes

There are two approaches to deciding whether a dune is active. The indirect way is to relate dune activity to climatic indices, in the hope that a map of climatic factors will show a map of active dunes. Goudie (1992) collected information from a number of sources to indicate that when average rainfall exceeded between 100 and 300 mm yr^{-1}, the vegetation cover became too dense for aeolian activity. Cooke *et al.* (1993: 242–3) reviewed the climatic indices of dune activity developed by, for example, Ash and Wasson (1983), Lancaster (1988b) and Wasson and Nanninga (1986).

Climatic indices are unavoidably imprecise for two reasons, one spatial and one temporal. Spatial imprecision arises from the scarcity of weather stations, which means that the division of active from inactive dunes must necessarily be very imprecise. Temporal imprecision arises from the notorious interannual variability of semi-arid climates. For example, recent climatic records for the south-west Kalahari, where mean annual precipitation is around 150 mm and where dunes are partially vegetated, show that there are many runs of years when Lancaster's (1988b) mobility index either considerably exceeds or falls well short of the threshold for sand movement (Fig. 8.3; Bullard 1994).

A second approach is to look for signs of activity or inactivity on the ground, although this method also has its risks. A surface is undoubtedly 'active' if it experiences frequent sand movement, as indicated by the presence of extensive rippled surfaces, although the exact extent of rippling that distinguishes active from inactive surfaces needs to be debated. The relationship between an active surface and an active dune is a further problem. One sign of active dunes may be widespread, active slip faces.

Indicators of inactivity, on the other hand, include evidence of fluvial or colluvial activity on dune slopes, weathering and soil formation, and the presence of vegetation. The clearest signs of stabilization are found in the external form of the dunes. Stabilization allows a crust to form, developed from dust that has been trapped on the stable surface, and reinforced by rainsplash and by the growth of algae (Barbey and Couté 1976). The crust causes the infiltration capacity of the surface to decrease, and infiltration is further discouraged by anti-wetting (or water-repellent) properties, imparted to the soil surface by plant debris (Dekker and Jungerius 1990). Decreased infiltration encourages storm runoff and this is concentrated into rills and even gullies, which degrade the dune surface (Bridge and Ross 1983). Sediment is transferred to interdunes, where it is deposited in sandy alluvial fans (Thompson and Bowman 1984; Yair 1990). Over the years these processes combine to reduce slope angles from the 15°–30° range of the original dunes to a 3°–4° range (Talbot and Williams 1978; Pye 1983), or less (Goudie *et al.* 1993).

Many dunes have been stabilized so thoroughly, for so long, and in such humid conditions, that deep and thoroughly weathered soils have developed in their surface sediments. Fine particles, derived from dust, are incorporated deep into the surface horizons (Goudie and Thomas 1986; Orme and Tchakarian 1986; Tsoar and Møller 1986). Iron in minerals may be released by weathering and may stain the upper horizons of sandy soils deep red (Gardner and Pye 1981). If rainfall and temperature are high, even quartz grains may be weathered and split, as in humid tropical stabilized coastal dune sands (Pye 1983). In the early stages of soil formation on stabilized dunes, throughflow is concentrated in columnar zones, between which leaching is much less thorough (Dekker and Ritsema 1994). Throughflow in the columns translocates first carbonate and then iron, which may be redeposited as pisoliths or thin iron-cemented layers along the bedding planes lower down the profile (Pye 1983).

With further weathering and leaching, deeper, more thoroughly altered soil profiles develop. Depending on the prevailing climate, carbonate, silica or iron dissolved further up the soil profile are redeposited in lower soil horizons, or removed in the drainage water. In humid conditions, especially on tropical coastal dunes, percolation can produce spectacular soil formation. Some bleached, podzolic A_2 horizons in eastern Australia are 18 m thick (Thompson and Bowman 1984). Below the A_2, a tough, humus-cemented B horizon or 'alios' may accumulate (Bourcart 1928; Simonett 1949; Bowden 1983; Pye 1983). Alios became a major problem in the reclamation of the Landes in south-western France (Chapter 9). In less humid areas, ferallitic soil profiles develop on ancient dune sands, reddened down to one or two metres, some with clay-rich B horizons (Daveau 1965; Warren 1968; Gardner 1981; Felix-Henningson 1984).

When aeolianite (see page 141) is stabilized, it exhibits even clearer evidence of inactivity, for the surface is then exposed to subaerial weathering, and solution of the carbonate produces karst topography, with pipes, karren and caves, as on Bermuda. If undercut, as by waves, these slopes can maintain vertical or even overhanging faces. Exposed surfaces can develop a tough re-cemented patina or crust (Gardner 1983; McKee and Ward 1983; Pye 1983).

Fig. 8.3 Data for a number of stations in the Kalahari showing the natural year-by-year variability which occurs in the P/PE climate. The graphs plot available moisture against windiness (% $W > V_t$) and use an index of mobility developed by Lancaster (1988b). Each point represents data for one year (after Bullard 1994).

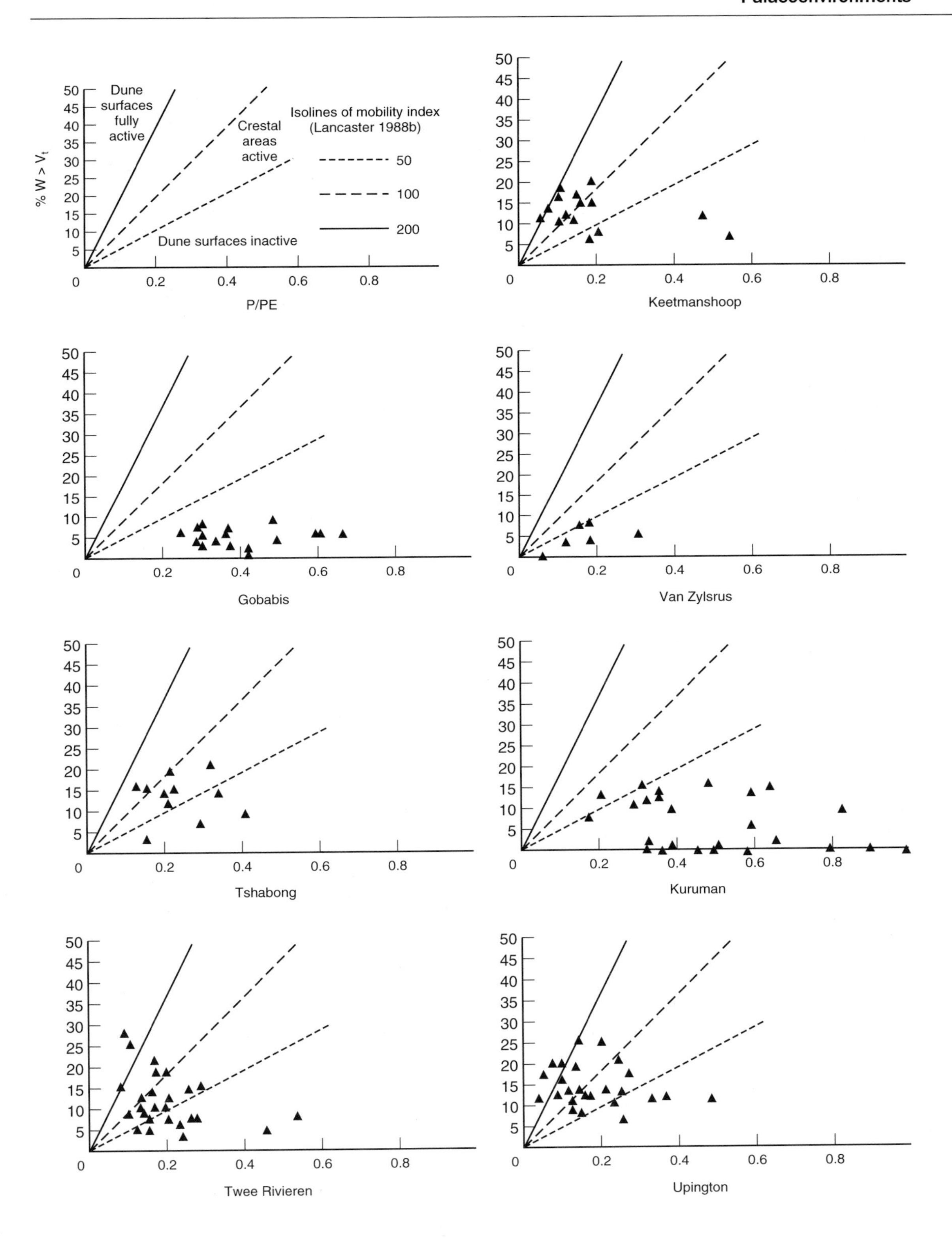
Dune surfaces fully active
Crestal areas active
Dune surfaces inactive
Isolines of mobility index (Lancaster 1988b)
50
100
200
% W > V$_t$
P/PE
Keetmanshoop
Gobabis
Van Zylsrus
Tshabong
Kuruman
Twee Rivieren
Upington

Vegetation cover is the third potential indicator of inactivity (and hence of the relic nature of the dune). As with slope and soil observations, there are situations, as in the northern Kalahari (Grove 1969) or the continental dune sands of South America (Tricart 1984), where aeolian sands are today covered by dense woodland, and where therefore they are indubitably relic.

At the drier end of the spectrum of vegetated dunes, it might be thought that vegetation cover, being directly implicated in the control of surface activity by the wind, would be potentially the most sensitive of signs of activity. However, the distinction between the activity or inactivity of a dune in these transitional situations is seldom simple. Slopes may not be very degraded, soils and weathering may not have proceeded very far, and although some authors believe that there may be a fairly precise threshold of vegetation cover above which aeolian activity cannot take place (Ash and Wasson 1983; Wasson and Nanninga 1986), recent work is showing that this boundary is also not easy to establish (Thomas and Tsoar 1990; Livingstone and Thomas 1993; Wiggs *et al.* 1994). There are large areas of transition in the semi-arid world where vegetation cover is discontinuous and where, because of great climatic variability, cover is also very variable from year to year.

The extent of this kind of intermittent activity makes it necessary to interpose a class of dune between the extremes of *active* and *relic* (Livingstone and Thomas 1993). This is the *episodically active dune*, on which sand movement may be at a low or negligible level for long periods (rippled surfaces and avalanche faces are rare or absent), but on which the indicators of lengthy inactivity are also absent. Episodically active dunes occur in transitional zones, probably equivalent to areas with rainfall in the range of 100–300 mm yr^{-1} in the tropics. In these zones, the carry-over of vegetation from wet to dry years and its slow re-establishment after dry years, means that the dunes do not suddenly flip from being active to being inactive. Only long dry periods, or periods of erratic interannual rains establish full devegetation; only a sustained run of wet years, and low variability re-establish a completely protective vegetation cover. For most of the time there may be enough sand movement to allow some form of dune formation, and the next question, as yet unanswered, is whether recognizable dune patterns can be established in these conditions, or whether full activity in a former, more arid or windier environment is necessary for the development of anything but minor, chaotic patterns (as described in Chapter 5).

There is a further reason for not taking the simple presence of vegetation to indicate dune inactivity. Significant sand movement on even the most active linear or star dune seldom extends to the lower slopes, which therefore tend to be colonized by vegetation. For instance, Jutson (1934) noticed that the downwind ends of linear dunes in Western Australia were devoid of vegetation, while the more stable upwind ends were colonized. There is no need to invoke any climatic gradient to explain this phenomenon. The relative inactivity of sand on the lower slopes, combined with the moisture-holding capacity of the sand, predisposes the slopes to invasion by vegetation. In this case, sand surface inactivity is the *cause* of vegetation cover, not an *effect* of it.

The contemporaneity of large dunes

The very size of the biggest dunes (even in modern deserts) has produced speculation that they may be relics (Besler 1980, 1982; Glennie 1983). One hypothesis is that only high wind speeds, as in the Late Pleistocene, had the aerodynamic capacity to form such huge features. An alternative hypothesis is simpler: large dunes need long periods to form. A first approximation of the time of accumulation can be found by dividing the modern rate of sand flow (in $m^3\,(m\text{-width})^{-1}\,yr^{-1}$), derived from wind observations (Chapter 2), by the cross-sectional area of the dunes. Using this method, Wilson (1972a) suggested that 100 m high dunes in the Grand Erg Oriental were at least 10 000 years old, and Lancaster (1989b) put the age of the largest dunes in the Namib at 42 000 years. If rates of sand movement are thought to have been greater (or less) during some periods in the past, the dune is given a shorter (or longer) life-span; there is no need to invoke higher wind speeds as an aerodynamic prerequisite.

In this second hypothesis, moreover, age is not a simple, absolute control on size. The rate of accumulation of a dune would vary according to two main factors. The first would be the rate of supply of sand, which in turn would depend on its availability and the wind speed (which is therefore as much a part of the structure of this second hypothesis as of the first, though in a more concrete form). The second control would apply only to linear dunes: it is the angle between the principal wind directions (obtuse angles favouring the build-up, narrower angles favouring the extension of linear dunes, as explained in Chapter 5 (Fig. 5.26)).

If large dunes are merely old dunes, then long periods of uninterrupted sand movement would be required for their formation. Cooke *et al.* (1993) proposed that such long periods could have been supplied by the so-called Milankovitch cycles (mentioned above). The strongest of these is the 125 000 year cycle, a period which would easily allow the growth of the huge dunes of the Grand Erg Oriental or the Namib, referred to above. The fact that these large dunes have life-spans (as calculated by the method above) that are less than the period allowed to them by the major cycle may indicate either that their size was curtailed by shorter cycles, or that there is some other, perhaps aerodynamic, limit to their size.

The contemporaneity of aeolian erosional features

Recognizing ancient from modern aeolian erosion features, and separating aeolian from non-aeolian erosional features presents yet another set of problems. The large yardangs of the central Sahara and elsewhere are undoubtedly ancient, if for the only reason that they could not have formed at present rates of wind erosion in anything but many thousands of years, and even at a much increased rate they would still require millennia to form. Moreover, Holocene and Late Pleistocene deposits between the yardangs clearly point to an ancient origin (Grove 1960; Hagedorn 1968; Mainguet 1968). Most smaller yardangs are incised into quite recent deposits, such as those of Late Quaternary and Holocene dry lakes, indicating their youth (Chapter 3).

The present inactivity of many, if not most, ventifacts is not difficult to establish, for many once active erosive faces are now covered by desert varnish (Laity 1987). Many ventifacts are found in ancient periglacial situations, and these too are undoubtedly inactive today. The higher wind speeds of some periods of the Late Pleistocene are thought to have been responsible for many of these features, as explained below. But the argument about the subtleness of the gradation between inactivity and activity, mentioned in relation to dunes above, must extend to many small yardangs and ventifacts. For example, quite mature ventifacts seem to have formed quite quickly in coastal dunes in northern Ireland (Wilson 1991).

Distinguishing aeolian from non-aeolian erosional features is not always simple. Recognizing the huge 'mega-yardangs' of the central Sahara as aeolian took some years. Their sheer size made their aeolian origin a difficult proposition for many workers, and although alternative explanations like 'basement fracture trends' (Pesce 1968) were explored, they are now generally rejected in favour of an aeolian explanation. 'Giant grooves' in northern Canada, though not unlike mega-yardangs in their extent and dimensions, are generally thought to be glacial in origin (Smith 1945; Lucchitta 1982). Some smaller yardangs are not unlike roches moutonnées or the scablands of Washington, which are now thought to have been formed by a massive sudden flood (Baker 1978), but problems of confusion are small.

Dating aeolian deposits

Aeolian deposits and erosional features can be dated either relatively or absolutely. *Relative dating* here means the dating of a deposit or landform according to its stratigraphic relations with other landforms. *Absolute dating* means the use of methods, such as radiometric ones, which give ages in years.

Relative dating was all that was available to early geomorphologists. In many cases it was achieved by using overlying or underlying lacustrine deposits, for many dunes (and some loesses) obstruct drainage channels and thus create shallow lakes. The method was also used for the yardangs in the central Sahara, the corridors between which were, in places and at times, invaded by ancient Lake Chad. Lacustrine deposits contain a host of relatively (and absolutely) datable materials, such as plant and animal remains and human artifacts.

It was the relative dating of adjacent lacustrine deposits that allowed early French geomorphologists to discover that the large dunes of the north-western Sahara predated the last main glacial period (Alimen *et al.* 1969), and permitted McClure (1976), to separate the dunes of the Rub' al Khāli into ones that dated from before the Holocene wet period and others that dated from the Late Pleistocene. McClure found beautifully preserved lacustrine chara limestones in pockets between the linear dunes. In Oman, Gardner (1988) used the technique to date some of the dunes of the Wahiba Sands. She also used relations between dunes and raised beaches to find that the last main period of dune activity was in the Late Holocene. Many of the early dates for the Nebraska Sand Hills were arrived at in the same way (Watts and

Wright 1966). There are many examples of the use of the techniques for dating loess, some of which are described below.

The material used for the relative dating of stratigraphically adjacent materials has varied. Palaeontological techniques are problematical because most dunes date from periods in which the fauna and flora differ little from the present, although these methods are useful for establishing palaeoenvironmental conditions. Archaeological techniques, as used in much of the north-western Sahara, are also hazardous because of the transposability of many archaeological remains such as hand axes.

The absolute dating of aeolian features relied, until the early 1980s, largely on ^{14}C dating of organic remains. Many coastal dunes and loess sections do preserve carbon, and the method has been used widely in these contexts (Wilson 1992). The lake deposits closely associated with aeolian features are especially good sources of carbonaceous material. There are, however, problems with the technique. The first is that much of the carbon buried in these and other deposits has been contaminated by the addition of carbon of a later date. The second is that ^{14}C dates can be extended back for only 40 000 years, and many dunes and loesses are older than this. Similar restrictions apply to other techniques that depend on organic remains, like the amino acid geochronology used to date coastal *dunes anciennes* in Tunisia (Miller *et al.* 1986). A further problem with desert dunes is that organic productivity is so low in the environments in which they are found, that its survival and discovery are unlikely. The problems are illustrated by an attempt to use organic remains in northern Australia (Goudie *et al.* 1993). Here the ^{14}C dates did not correspond to the those from optical dating (see below), and the authors concluded that the organic matter might have been a late introduction to the section (as in an animal burrow or a plant root).

The great advance in the last decade or so has been in the development of optical (also called 'luminescence') dating techniques. These are ideal for dating aeolian materials, both loess and dune sand. The principle depends on the release of electrons by natural radioactivity in the elements of mineral crystals. The electrons are trapped in the crystal structure, but are released if the crystals are exposed to light (or heat). Thus, when dust or sand is exposed to sunlight, the accumulated electrons are released (the slate is, as it were, wiped clean), but they begin again to build up when the grains are buried. If the deposit is carefully sampled (in the dark), and subjected to a known amount of heat (thermoluminescence or TL) or light (optically stimulated luminescence or OSL) in the laboratory, the amount of light that is released is in proportion to the time since last burial (Wintle 1993). In the OSL technique various wavelengths of light can be used, including infrared (IRSL).

TL was first developed for the archaeological dating of pottery, where the 'cleansing' agent had been heat, but since 1979, when it was realized that light could be a zeroing mechanism, TL and (after 1985) OSL have been sweeping the Quaternary geological community as techniques for dating sediments, particularly aeolian ones, where exposure to strong light is almost certain to have occurred before burial, and where the mineral suite is also usually ideal. There are many complications, like variations in local background radiation, grain size, the wavelength of the light used, the use of different minerals such as feldspars or quartz (the two most commonly used), but in skilled hands, good dating can be achieved and the number of laboratories that engage in optical dating is increasing.

The accuracy of optical techniques is the subject of great debate. Accuracy generally decreases with age, as in carbon dating (for example, Stokes and Breed 1993), but when comparisons are made between the results of different optical dating techniques on samples of the same sediments, the range can be much greater (Rendell 1995). The maximum date obtainable from a sample depends on its natural optical saturation level, in part a function of the mineral suite. Ages of up to 500 000 years have been found in some studies. Difficulties arise, in some of the methods, with the use of the technique for very young samples (Wintle 1993), though it has been claimed that infrared stimulation can give accurate dates for samples only a few decades old (Edwards 1993). Table 8.1 gives some examples of dating studies of aeolian sediments that have used optical dating techniques.

Pre-Pleistocene sand dunes

The interest of pre-Pleistocene aeolian deposits is more sedimentological and stratigraphic than geomorphological, but ancient sand dunes do hold some clues about modern dune formation (for example, almost all we know about dune bedding comes from lithified, ancient dunes rather than from modern ones; see Chapter 7), and about ancient aeolian conditions.

Table 8.1 Selected examples of the optical dating of aeolian sediments

Location	Reference
Continental dunes	
Mojave Desert, California, USA	Edwards (1993), Clarke *et al.* (1995), Rendell (1995)
Arizona, USA	Stokes and Breed (1993)
Linear dunes, Australia	Callen and Nanson (1992), Nanson *et al.* (1992), Goudie *et al.* (1993)
Linear dunes, Israel	Rendell *et al.* (1993)
Rajasthan	Singhvi *et al.* (1982)
Coastal dunes	
Australia	Gardner *et al.* (1987)
Loess	
Matmata Plateau, Tunisia	Coudé-Gaussen *et al.* (1987)
Mississippi Valley, USA	Pye and Johnson (1988)
Nebraska, USA	Maat and Johnson (1995)
Pakistan	Rendell and Townsend (1988)

Aeolian deposits have been reported from as far back in the geological column as the Precambrian in the Northwest Territories of Canada (Ross 1983) and India (Chakraborty 1992). Since those times, continental aeolian sandstones have graced many sections of the geological record (Bigarella 1979). In Britain the oldest aeolian deposits (of any great extent) are from the Devonian, when Britain was south of the equator in latitudes that now have many arid and semi-arid areas. In the Moray Firth area of Scotland, Devonian sandstones are thought to be the remnants of large star dunes (Clemmenson 1987). Other sand seas from the Devonian have been found in Ireland (Carruthers 1987).

The best known of the aeolian sandstones in northern Europe are from the Permo-Triassic, for they are good traps for oil and gas (Glennie 1972, 1983; Weber 1987). At that time north-western Europe had drifted to latitudes north of the equator that are today generally dry. The deposits are widely distributed through the North Sea oil and gas province, extending on-land in Britain and Germany, and outliers have been taken by further continental drift into Greenland. Sections in quarries in the Durham area of north-eastern England are said to show remains of large linear mega-dunes with superimposed smaller linear features (not unlike those in many contemporary sand seas), formed in response to a bimodal regime of winds from what is now the north-east and south-south-east (Fig. 8.4; Chrintz and Clemmensen 1993). Nearby in southern Scotland at this time, small sand seas formed in intermontane basins (Brookfield 1980), although here there is some disagreement about the prevailing wind direction their bedding is said to reveal (Sneh 1988).

In the western United States there are even more extensive aeolian sandstones from this period. They are well known as the red rocks into which has been carved some of the most spectacular scenery in the National Parks. Like the European aeolian sandstones, they have attracted attention because of their oil-bearing character (Lindquist 1988). A more academic function served by these, as by the European aeolian sandstones, is the reconstruction of ancient wind-circulation patterns through the analysis of dune bedding (Parrish and Peterson 1988). The best known is the Coconino Sandstone, covering thousands of square kilometres, and reaching hundreds of metres in thickness, which outcrops on the upper walls of the Grand Canyon (McKee 1933; Hunter and Rubin 1983).

The origin of the Coconino Sandstone has raised some controversy, some believing it to be marine rather than aeolian. Part of the debate concerned the tracks of animals that evidently went up the 'dunes', but never down. To those who believed in a marine origin, the explanation was that the animals had swum off the tops of marine sand waves. McKee (1944), who believed in the aeolian origin, proposed that the animals had merely slid down slip faces, leaving no footprints (another possibility is that, like pterodactyls, they flew). Other aeolian sandstones occur throughout the western United States (Kocurek *et al.* 1991), and Triassic dune

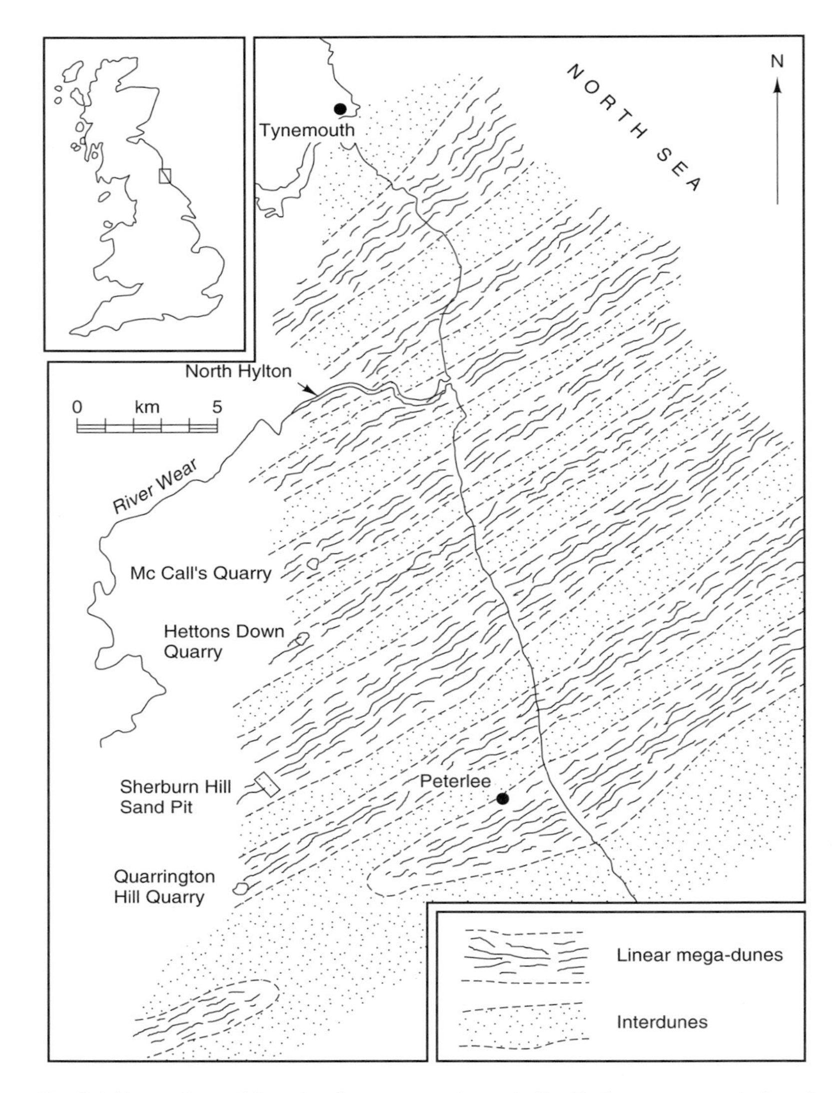

Fig. 8.4 The pattern of Permian linear mega-dunes in the Durham area, as deduced from rock outcrops (after Chrintz and Clemmensen 1993).

sandstones occur in Brazil (Bigarella and Salamuni 1961).

Pleistocene and Holocene dunes

The windier and/or drier and/or more variable conditions of the Pleistocene, described above, probably intensified dune growth in areas that remain deserts, and they certainly extended the zone of active sand dunes very considerably beyond their present extent. The area of dunes now stabilized by vegetation and soil is of the same order as, or even greater than, the area of contemporary desert covered by aeolian sand (Fig. 8.5).

The formation of the Pleistocene dunes was preceded by the creation of extraordinary quantities of potential dune sand. In the ancient cratonic deserts of Africa and Australia, a number of circumstances, especially the plate-tectonic trajectories of the continents, meant that Pleistocene aridity had been

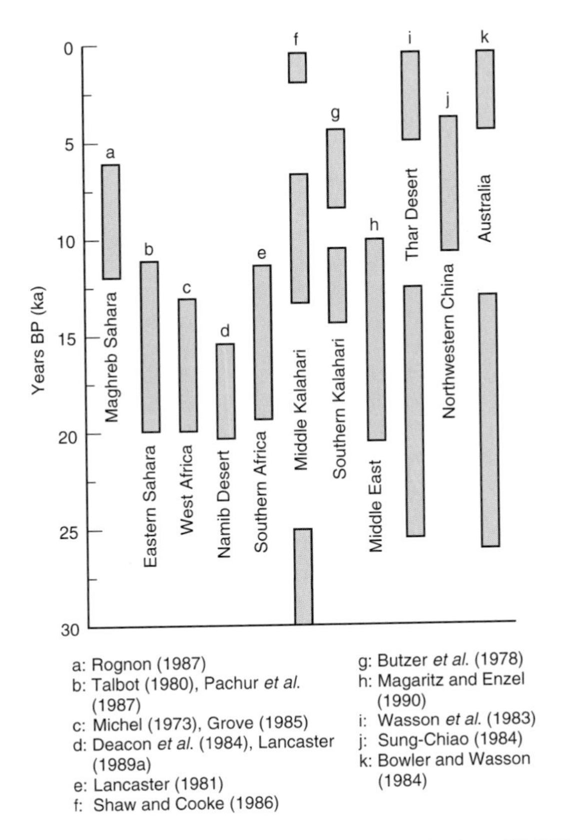

Fig. 8.5 Dry, dune-forming periods over the last 30 000 years in the world's deserts (after Tchakerian 1994).

preceded by a period of tropical deep weathering of ancient metamorphic and igneous rocks, creating large quantities of sediment which, following local tectonic warping in the Miocene, were eroded from upwarps and deposited in shallow downwarps (or 'depocentres') (Williams 1994). Some dunefields in Australia may have been formed, almost directly, from *in situ* mantles from which weathering had removed everything but quartz (Butt 1985), but most dunes in both Australia and Africa were formed from sediments eroded from these mantles, and redeposited in the basins by rivers. In the central Asian and American deserts, in contrast, sediments that subsequently went to form dunes were created by the rapid erosion of the much more strongly uplifted mountains and deposition in the accompanying basins created at this time (Williams 1994).

In the Sahara, dunes began to form from the Late Tertiary, and somewhat later in Australia. In both continents, aridity intensified after about 0.7 to 0.9 million years BP. Thereafter, the various deserts followed rather different histories of dune formation (Fig. 8.5).

Ancient dunes in present deserts

It is very probable that the intensified aeolian activity of some Pleistocene and early Holocene periods also left a legacy of dune forms in the deserts themselves. Some aspects of the contemporaneity of these dunes has been discussed above. There are also many subdued and markedly weathered dunes in areas of contemporary aridity and hyper-aridity, which are almost certainly relics of ancient dune-forming periods and of their degradation in subsequent periods wetter than the present (for example, Williams *et al.* 1987). This would be consistent with other evidence that suggests these areas have experienced alternations of aridity and humidity throughout the Pleistocene.

Ancient, now stabilized dunes in low latitudes

Figure 8.6 shows that there are extensive areas of stabilized dunes on all the continents. These are areas in which the dunes show all the signs of inactivity mentioned above. They can be divided into two groups: those that surround the present, largely tropical deserts, and those that existed in periglacial situations at higher latitudes.

The tropical and sub-tropical group includes the Kalahari Sand in southern and central Africa, which covers some 2.5×10^6 km^2, making it, by one claim, the largest sand sea on Earth (Thomas and Shaw 1991). The more fragmented, but in aggregate greater area of Sahelian dunes stretches across the middle of the continent with few breaks from the Atlantic coast of Senegal to the Nile. An example of the extent of these sands in Sudan is given in Fig. 8.7. Other large areas of tropical and sub-tropical stabilized dunes occur on the southern and eastern shores of the Mediterranean, in northern Arabia, the Thar desert in India and Pakistan, large parts of mainland Australia, Tasmania, Brazil and Venezuela.

In most of these stabilized dunefields, there are signs that there was more than one period of dune activity. The Sudanese example in Fig. 8.7 shows that there were evidently two major periods of dune formation. The first and most extensive incursion reached far to the south of the present desert edge into areas which now have up to 1000 mm mean annual rainfall. The dune forms on most of these ancient

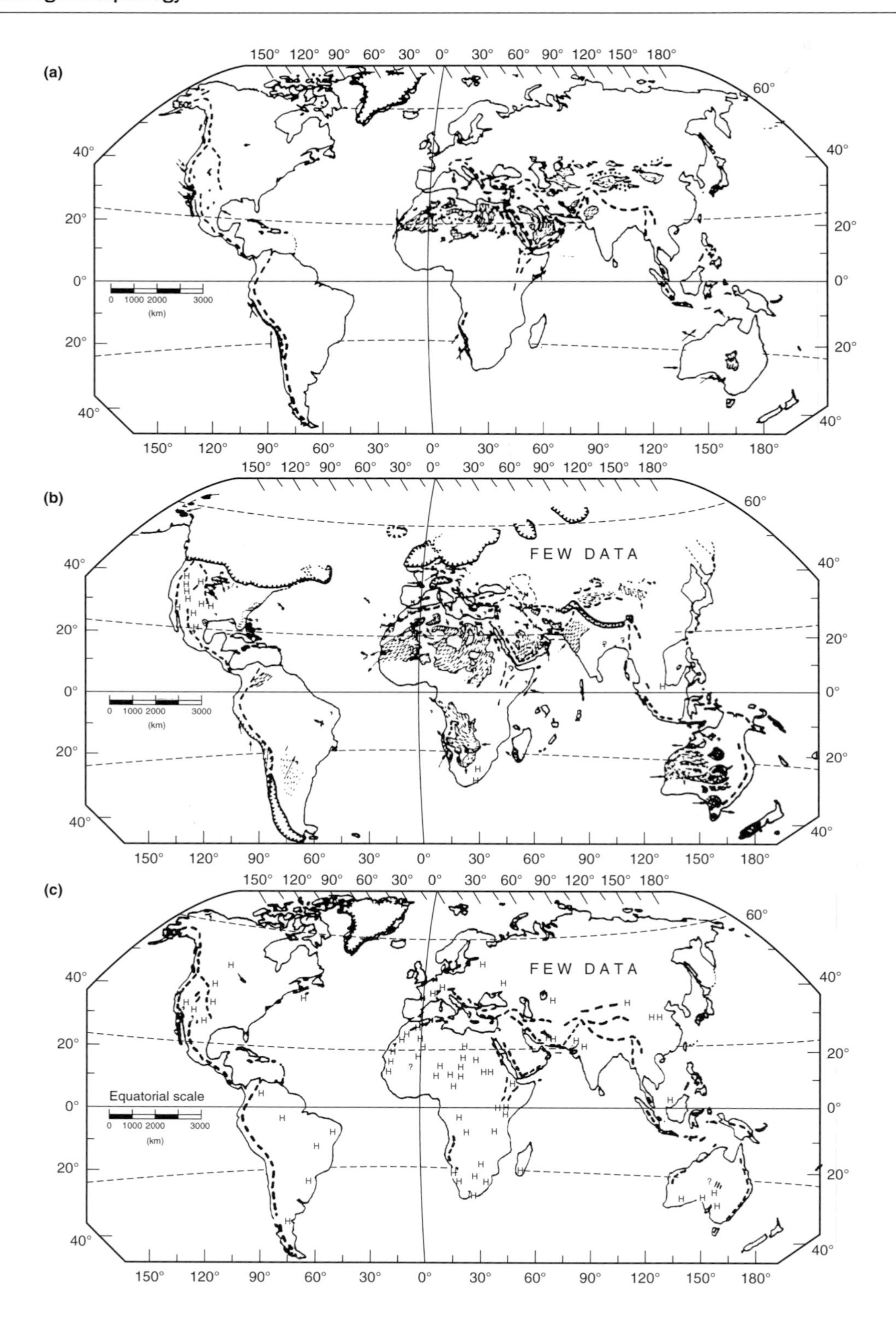
(a)
(b)
FEW DATA
(c)
FEW DATA
Equatorial scale
0 1000 2000 3000
(km)
150° 120° 90° 60° 30° 0° 30° 60° 90° 120° 150° 180°
60°
40°
20°
0°
20°
40°

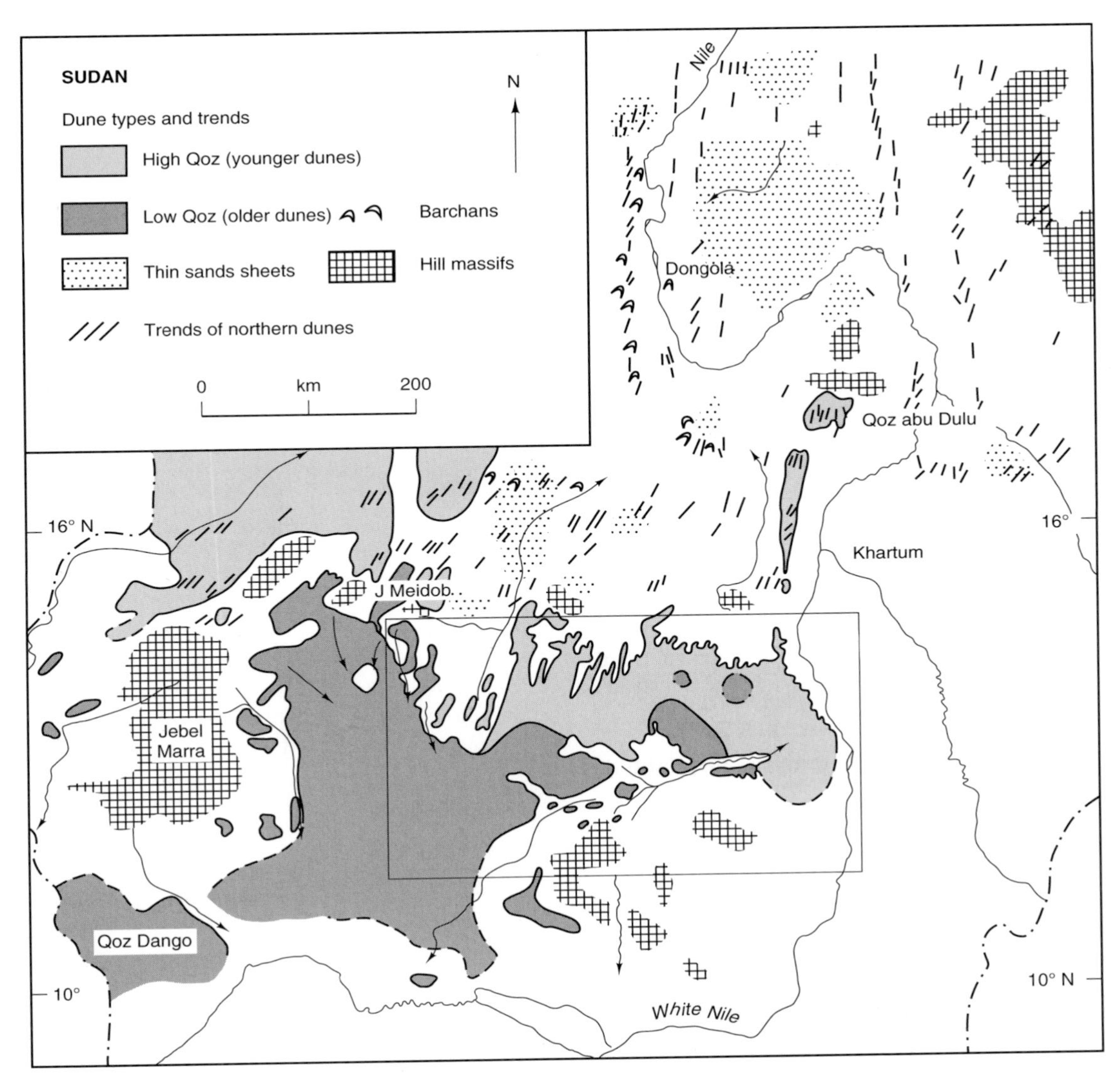

Fig. 8.7 The extent of ancient, now stabilized dunes in Sudan, showing two generations of dunes (the High and the Low Qoz) (after Warren 1970).

sands are very subdued, indeed almost impossible to detect, and the soils are deep and leached, with well-developed B horizons. The more recent dunes of the second main phase have retained their forms, in quite recognizable patterns, although their surfaces are certainly stabilized. The area is wet enough now for many of the dunes of this second phase to be cultivated (Warren 1970).

Fig. 8.6 Extent of active sand seas for (a) the present, (b) 18 000 years BP, and (c) 6000 years BP (after Sarnthein 1978). H shows humid conditions.

Where not totally subdued by water erosion, the sand in these ancient deposits is organized into dune forms with all the variety of desert dunes. In the Kalahari Sands in Botswana, Zimbabwe and South Africa the dominant dune form is linear (Thomas 1984). In the Sudan and the West African Sahel, there are a range of forms, including large and small transverse and linear dunes. There has been speculation that winds were rather different at the time of the formation of these dunes (Warren 1970), but others believe that the wind patterns were little different from those of today, though winds were perhaps stronger, and rainfall lower (Talbot 1984). The Australian stabilized dunefields are dominated by linear forms. In Australia, Wasson (1984) has speculated that dune activity might have been achieved solely by stronger winds, without any diminution of rainfall.

The stabilized dunefield with the most distinctive set of dune forms is the Thar in India and Pakistan. This may be because of the peculiar climatic history of this corner of the world (Wasson *et al.* 1983). Unlike the African Sahel, or indeed most of the now semi-arid tropics, where winds subsided as rainfall increased, the argument of Wasson *et al.* was that the Thar experienced an increase in rainfall at the same time as the re-establishment of the Asian monsoon after the retreat of the ice, bringing increasingly strong winds in towards the Asian continental low-pressure zone from the Arabian Sea. This coincidence of high winds and increasing vegetation cover provided the environment for the creation of the most extensive parabolic dunefield in the world.

Ancient, now stabilized dunes in high latitudes

In the higher latitudes, dune development (away from the coasts) seems mostly to have been a feature of early post-glacial times. It was stimulated by three circumstances: extensive areas of sandy periglacial deposits, high winds, and slow recolonization by vegetation. Sandy glacial outwash deposits are the most susceptible to reworking by the wind, and in many places, as in northern Finland, dunes are confined almost exclusively to esker trains (van Vliet-Lanoë *et al.* 1993). Where, as on the North European Plain, outwash is most extensive, so too are the dunes formed from it. The occurrence of high winds has been discussed above. As to vegetation succession, this was generally slow after the retreat of the ice (especially in northern Finland), but was complicated by the effects of fire (both natural and induced). In general, there seems to have been a major burst of dune building shortly after the retreat of the ice, and before the full recovery of forest cover, but with many smaller and more localized incidents after this (Filion *et al.* 1991).

These high-latitude stabilized dunes include a train of fragmented, stabilised dunefields that stretches across the North European Plain from the Atlantic coasts to the Ukraine, with outposts in Hungary; aeolian sands cover 20 per cent of Hungary (Borsy 1993). Even larger areas of stabilized dunes are found in central Asia and China. Few of these Old World stabilized dunefields are as large as some of those in North America, particularly the Nebraska Sand Hills, which cover 57 000 km^2, and which retain large barchan dunes (Fig. 8.8). These are merely the largest of a very widespread set of stabilized dunes in the Midwest and many other parts of the United States, as in the Mohawk Valley in New York State (Connally *et al.* 1972).

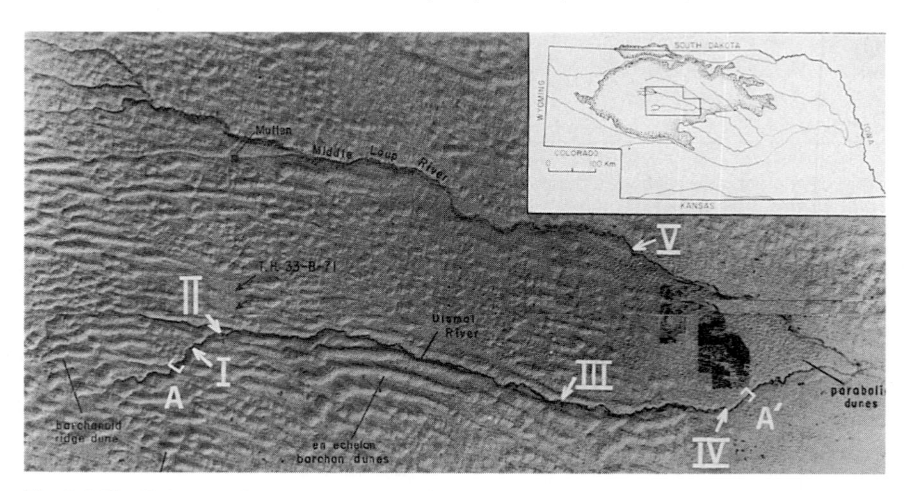

Fig. 8.8 The Nebraska Sand Hills covered in snow, on a satellite image. The snow reveals the major dune forms (after Ahlbrandt *et al.* 1983).

Fig. 8.9 Parabolic dunes in Poland (after Högbom 1923).

The high-latitude stabilized dunes, like those in the tropics and sub-tropics, also exhibit the full range of dune forms, and these have been interpreted as indicators of the wind and rainfall conditions in which they were created. The western European late-glacial and Holocene aeolian deposits, though clearly aeolian according to sedimentary evidence, exhibit few dune forms, and are known as 'coversands' (Veenstra and Winkelmolen 1971). In a few places there are remains of low parabolic dunes (as in the Breckland of eastern England (Catt 1978)), and *randwallen* (high mounds of sand evidently accumulated round the edges of woods) (Koster *et al.* 1993). The general interpretation of these pieces of evidence is that there was quite dense vegetation at the glacial maximum.

Further east in Europe the dune forms are more obvious, though often now covered by forest (Böse 1993). Some of these were clearly transverse, suggesting areas completely free of vegetation, but most are parabolic (Fig. 8.9), and therefore also suggest aeolian activity among vegetation, albeit probably less dense than it is today, and perhaps less dense than in western Europe. Since it is also probable that winds were then much stronger, as mentioned above, it may have been that even quite vigorous vegetation could have been overcome by moving sand. North of the maximum extent of the Vistulian ice in Poland, where the exposure to dune-forming conditions was for a relatively short period, the dunes are not as well developed, and have not moved as far as those beyond the maximum extent of the ice, which experienced a much longer period of exposure, some of these having moved over 100 km from their source in large river valleys (Goździk 1993). In Europe (for example in the Netherlands) there were at least three major periods of dune formation during and after the Older Dryas (Vandenberghe 1993).

In many parts of northern Europe the last phase of reactivation was well within the period of human occupation, and this and the evidence that sand blowing was irregular in time and space, suggest that disturbance by ploughing or intensive grazing could have been a factor in the reappearance of bare sand and the reformation of dunes (Chapter 9). Changes in climate were probably also involved. One consequence of the history of disturbance is a very irregular pattern of dune forms in these areas (Koster *et al.* 1993; Szczypek and Wach 1993).

Many of the North American dunefields contain much bigger dunes than those in Europe. The size of the dunes is particularly evident in the Nebraska Sand Hills, many reaching 100 m; they are both linear and barchanoid in form (Warren 1976a). Both the size of the dunes and the extent of the dunefields suggest much drier conditions than those in Europe (the western Great Plains are today much drier), and very much intensified north-westerly winds. The freshness of some of the dunes is notable (Fig. 8.10), and raises the possibility that dune formation was abruptly terminated, allowing the preservation of near-perfect dune forms.

Disturbance was probably less of an element in dune formation in the New World than in Europe, though in both Australia and Canada there are indications that activation might have been linked to fires set by early people (Conacher 1971; Filion *et al.* 1991). The dunefields on the central Great Plains, however, seem to be very sensitive to quite small changes in climate (such as the droughts that helped

Fig. 8.10 The stabilized mega-barchans of the Nebraska Sand Hills. The dunes are about 100 m high. Note the remarkable freshness of the slopes, with the slip faces only slightly degraded. This may be because the dunes were stabilized recently after a sudden change in climate.

to precipitate the Dust Bowl of the 1930s), and some dunefields in Colorado can be shown to have seen four major phases of reactivation in the last 10 000 years (Forman *et al.* 1992; Stokes and Gaylord 1993).

Ancient coastal dunes

Because of changes in sea level and because of their fragility in the face of temperate weathering and fluvial erosion, there are very few, if any coastal dunes from earlier than the late Pleistocene in high-latitude areas. However, the arid climate of the Nullarbor Plain in South Australia, and its history of falling sea levels, has allowed the preservation there of large coastal ridges dating from as early as the Pliocene (Benbow 1990), and further east on the South Australian coast, a long history of falling sea levels has helped to preserve another ancient sequence of coastal dunes rising to heights of over 100 m above present sea level (Cook *et al.* 1977). Very few other coastal dunes of this great age have been reported.

There are many more surviving coastal dunes from the late Pleistocene and early Holocene. Several events of this period stimulated coastal dune formation, all combining to favour massive incursions of marine sand. The retreating sea levels of the last ice age may have been one such event, for some authorities believe that the coastal dunefields formed in late Pleistocene times were created from sand exposed by the lower sea levels (Pye 1984). Such an exposed deposit at the mouth of the Nile may have been the origin of the dunes of northern Sinai in Egypt and the Negev in Israel (Goring-Morris and Goldberg 1990). Retreat in the form of the marine regression of the mid-Holocene also exposed marine platforms on which dunes could readily accumulate (Carter and Wilson 1993).

However, marine transgression itself may also have favoured coastal dune formation. A widely accepted model of their formation depends upon the creation of great amounts of sediment by coastal erosion as sea levels rose again after the end of the last ice age. This scenario, the so-called 'Cooper-Thom' model, appears to have strong support in many parts of the world. For example, Lees *et al.* (1993) used optical methods to date three Holocene episodes of dune formation on the Cape York Peninsula in northern Queensland and found that two of them were closely correlated with periods of marine transgression. One of the stabilization phases was correlated with a slight regression. In the Netherlands it is generally accepted that the 'Younger Dunes' (which account for by far the greater area of Dutch coastal dunes) were formed following a marine transgression between AD 800 and 1650, when the coast retreated between 1 and 2.5 km, increasing the rate of coastal transport to

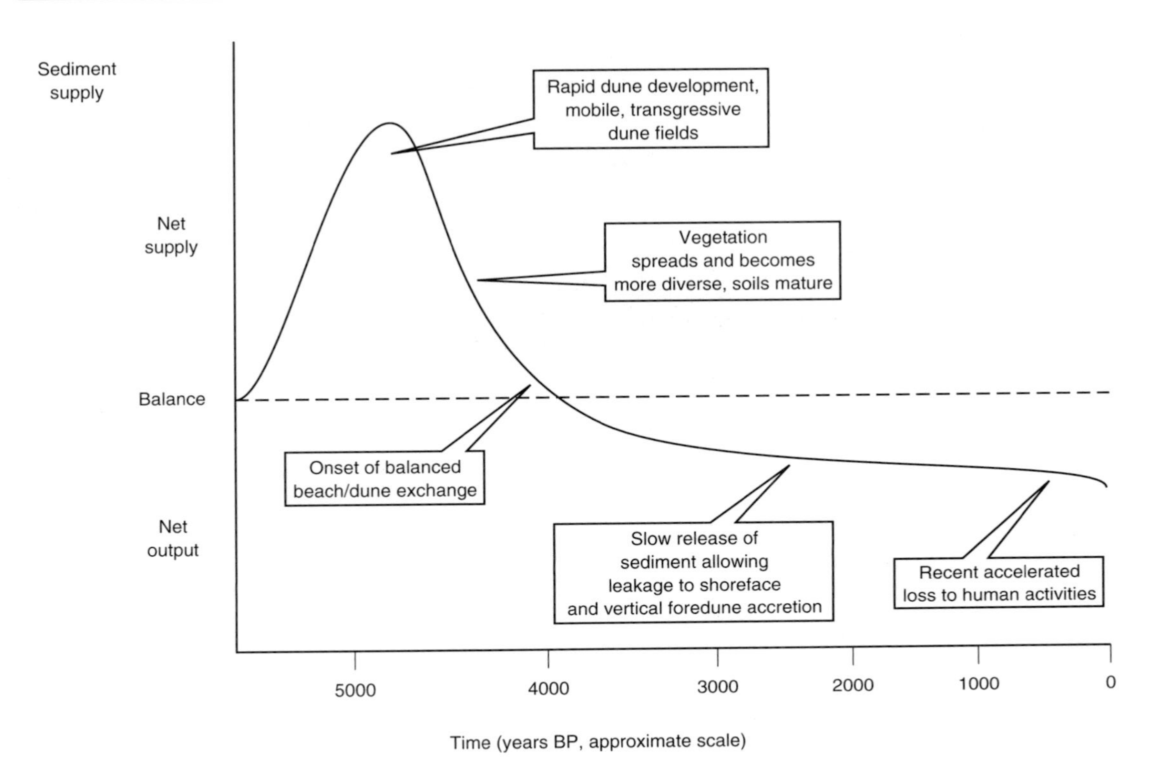

Fig. 8.11 Phases of coastal dune development in north-western Ireland (after Carter and Wilson 1993).

about $27\,m^3\,(\text{m-width})^{-1}\,yr^{-1}$ (compared with best estimates of about $3–3.5\,m^3\,(\text{m-width})^{-1}\,yr^{-1}$ today), and thus creating much of the sand that went to build the dunes (summarized by Arens and Wiersma 1994).

In some high-latitude areas, the supply of sediment was further intensified as the erosion provoked by the rising seas encountered soft, loose materials of recent glacial or fluvio-glacial origin. Most dunes on the coasts of north-west Europe, almost all of which date from the last 6000 years, are said to have been derived from this kind of sediment (Carter 1990).

Once sea levels had stabilized, most coastal dunes became inactive, and many were subject to marine erosion. In north-western Ireland, for example, coastal-dune building reached a peak about 5000 BP, and entered a phase of equilibrium or slow erosion about 4000 BP (Fig. 8.11; Carter and Wilson 1993). In places, the rising sea level invaded river valleys creating estuaries, and these trapped sediment and prevented it being delivered to dunes (Bird 1990). The practical implications of this recent history of intense coastal dune building are examined in Chapter 9.

Aeolianites

Aeolianites are cemented coastal dune sands, generally on semi-arid, tropical, high-energy shores (Fig. 8.12). Although most are highly carbonaceous, being derived from shelly marine materials, the carbonate content of the sand has wide variations and the sand itself may be mixed in varying quantities with dust, which is also often carbonate-rich, and most of which has been added since deposition and stabilization of the sand (Gardner 1983; McKee and Ward 1983). Shelly dune sands accumulate on tropical coasts for several reasons: first, tropical seas, especially those where there is marine upwelling, are more productive of shelly biogenic material than cooler seas; second, a high-energy environment (which occurs near most aeolianites) ensures that the shells are broken down and then moved towards the shore from where they can be blown onshore by the same winds that have activated the marine processes (McKee and Ward 1983); third, a semi-arid or sub-humid environment (in which most, but by no means all aeolianites occur) ensures a moisture

Fig. 8.12 An outcrop of aeolianite (cemented carbonate aeolian sands) in the Wahiba Sands of Oman. See also Fig. 7.6.

budget conducive to cementation, but not to the complete dissolution of the carbonate.

Most aeolianites are confined to a narrow coastal zone, either because of the limited effectiveness of sea breezes, or because rapid lithification prevents further movement. Carbonate sands only reach far inland where present or ancient conditions near the coast are arid, as in the Thar Desert in India, where their distribution is discontinuous, but where they reach hundreds of kilometres from the shores of the Arabian Sea (Goudie and Sperling 1977), and the Wahiba Sands of Oman, where they reach 150 km inland and continuously cover some 94 000 km^2 (Glennie 1970; Gardner 1988). When they move far inland, dunes built of carbonate material adopt the full range of dune shapes, but on more humid coasts, they adopt the forms of common coastal dunes (Chapter 5).

One of the biggest problems with aeolianites is the interpretation of their lithification. Claims that the degree and type of cementation could be interpreted chronologically, and given environmental interpretation seem to be premature (Gardner and McLaren 1993). The first problem is to account for the source of the cement, for although most of it is thought to have been derived from aragonite dissolved from the original shelly material, this cannot always account for its amount and density. Carbonate in sea spray or dust does not always seem to be able to make up the necessary difference (Gardner 1983). The second problem is the location of cementation above, below or at the water-table. If cementation occurred above the water-table, in the so-called 'vadose' zone, deposition of cement must have occurred when the solution of percolating rainwater became supersaturated for short periods. The abstraction of water by plant roots probably played an important role in this process, tight networks of lithified root channels giving some credence to this hypothesis (Amiel 1975). In other cases, complex layered zones of aeolianite are interpreted as the result of cementation at the water-table itself (Schenk and Fryberger 1988; Semeniuk and Glassford 1988).

Many aeolianites were apparently deposited in the cooler, drier climates in the Pleistocene, for many are found below present sea level, and some have been cut across by marine terraces (Gardner 1983, 1988). This, their rapid lithification, the evidence they preserve of ancient wind directions and the intercalation of soils, make aeolianites good sources of palaeo-environmental information, as on Bermuda, where they have been extensively studied (McKee and Ward 1983).

There are many situations in which coastal dune sands have been submerged by rising sea levels since the end of the last glaciation, as in Queensland (Pye 1993b). The best examples of submergence occur with aeolianites, as in Oman (Gardner 1988), perhaps because their cementation makes them resistant to reworking by marine processes. The submergence of desert dunes has been demonstrated in only a few cases, perhaps because of their vulnerability to wave action. One case is the large submerged barchans in the Arabian Gulf (Al-Hinai *et al.* 1987), which is, notably, a very low-energy marine environment. There is indirect evidence elsewhere that dunes were submerged but have not survived. In Western Australia, for example, there is good evidence that the Great Sandy Desert extended at one time into areas now submerged by rising sea levels (Jennings 1975).

Pleistocene wind-erosion features

It may be that most, and certainly the most obvious, aeolian erosional features are inherited from the Pleistocene. This is apparently true of many ventifacts, the most well developed of which are found in areas near to the ancient ice-sheets, and are now undoubtedly inactive (as in Wyoming; Sharp 1949). Even in some still very arid areas of California, a weathering rind or desert varnish testifies to the inactivity of most ventifacts. Sharp, using his observations on active ventifact formation in the Coachella Valley in southern California, speculated that ventifaction was more active in some Pleistocene periods not only because of increased windiness, but also because of an increase in the supply of abrasive sand. This may be part of the explanation for the extent of development of ventifacts in glaciated areas, for the sand there would have been brought by glacial meltwaters. There are, none the less, still some ventifacts that are clearly active (Rude 1959). Here, as elsewhere, the facile assumptions about the relic or active nature of a feature need to be carefully guarded against.

The great age of the mega-yardangs in the central Sahara (discussed in Chapter 3) is undoubted. They may have been initiated in the Miocene. They, and other major yardangs in Iran, were probably much more active than today during the windier phases of the Pleistocene. Smaller yardangs, excavated into Holocene lake deposits, and even in places into the cemented deposits accumulated round artificial sand fences are mostly much younger and many can be regarded as contemporary.

Conclusion

The importance of aeolian sediments and landforms in recent Earth history may have been underestimated. They have been a neglected field of study because of their general inaccessibility, because, until recently, few of them could be accurately dated, and because many accumulate slowly and create unobtrusive features. These problems are being rapidly and radically overcome: deserts, where most aeolian sediments occur, are now much more accessible; and more important, aeolian sediments, far from being, to all intents and purposes undatable, may become some of the more easily datable materials. Detection is also becoming easier as the characteristic size and mineral signatures of aeolian deposits become more established. Furthermore, the astonishing intensity of aeolian activity in some recent geological periods is being exposed by recent ice-core analysis.

Loess and deep-ocean aeolian sediments are revealing very large amounts of information about the Pleistocene and earlier periods (Chapter 4). Moreover, because coastal dunes are now seen to play a critical role in land–sea interaction, and because they will become even more critical if sea levels are to rise (Chapter 9), their recent history has acquired a new significance as a possible analogue of what is now to become of them. Stabilized and presently active dunes are also disclosing a fascinating past.

The deployment of the new techniques is revealing surprisingly complex patterns of Quaternary aeolian activity, and is overturning many older ideas. With the consolidation and further development of these methods, the next decade is likely to see more rapid advance in the understanding of Earth's aeolian past, and, in consequence, radical reappraisals of most of what has been described in this chapter.

Further reading

Most of the material on ancient aeolian sandy environments is rather dispersed. There have, however, been some recent collections of useful papers, such as those by Pye (1993a) and Pye and Lancaster (1993). The best introduction to thermoluminescence dating is to be found in Wintle (1993).

CHAPTER 9 Applied aeolian geomorphology

Introduction

There has been and continues to be massive investment across the world in the control of aeolian geomorphological processes. It has happened in Saharan and Arabian oases for thousands of years; on the Dutch coast since the fourteenth century; on the Danish sandlands particularly in the eighteenth and nineteenth centuries; in the Landes of south-western France from the nineteenth century; in the United States since the Dust Bowl of the 1930s; on the Israeli coast since shortly after the creation of the State in the late 1940s; on the Russian and central Asian steppes since the Stalinist period; since the 1950s in the oil-rich desert countries of the Middle East; since the early 1970s in the Sahel, North Africa, India and China; and less intensively but significantly in many other places. In most of these situations, applied aeolian geomorphology won huge resources and prestige.

These experiences teach two crucial lessons. First, investment is seldom motivated by financial return alone, for there has almost always also been a strong symbolic component. For the Dutch, aeolian geomorphology was applied for little short of national survival. In most other countries the problem is not as acute, but in many it has become a potent symbol that nature can be overcome. In the United States, for example, it at one time expressed the triumph of national destiny over the mythical Great American Desert. In the early days of the state of Israel, it signified the conquest of Levantine deserts and active coastal dunes created by the ignorance of others; and in the Soviet Union, it was to be a victory of scientific socialism over the barbarous steppes. In the less-developed world today, much applied aeolian geomorphology is little more than an emblem of the fight against desertification. In the affluent Arabian peninsula, millions of dollars are spent to bedeck highways with trees and flowers to fend off the windy wastes, when a simple fence alone would have served the practical need.

Dust, moving sand and dunes, by this evidence, generate deep-seated anxiety, which is seldom allayed with scientific facts or economic arguments alone. Their symbolic threats transform the assessment of the hazards into battles between exaggerated fear and counterbalancing complacency, both grounded in ignorance. Careful, applied aeolian geomorphology can narrow the zone of ignorance, but cannot itself avoid being judged not only in economic terms, but also in terms of the peace of mind it may bring.

The second lesson from the history of application of aeolian geomorphology is that success or failure, of whatever kind, is due not only to the application of good science (vital as that is), but also to the degree to which the science can be adapted to the prevailing culture.

This chapter is in two parts. The first covers the technical procedures, based on scientific research, used to manage aeolian activity in different situations. The second part considers some of the social, economic, administrative and political issues which arise in managing land affected by the various problems.

Although some techniques are common to all applied aeolian geomorphology – notably structures to control sediment movement and the stabilization of mobile surfaces – it has operated in four distinct natural and social circumstances: wind-erosion control on agricultural fields; the control of dust; the management of coastal dunes and dunes in semi-arid areas; and the control of sand dunes and drifting sand in deserts.

Wind erosion on agricultural fields

The extent and nature of the problem

Wind erosion can damage fields in places as moist as the West Midlands in England (where mean annual rainfall is over 1000 mm), but the main threat is to drier parts of the world such as the North American Great Plains or the Russian and central Asian Steppes. Damage comes in many forms. Above all, because erosion removes soil, it reduces water-holding capacity, and this can be critical to crop production in semi-arid climates. The vulnerable top-soil also contains a high proportion of the total clay, most of the organic matter (which together hold nutrients against leaching as well as moisture), most of the nutrients themselves, both natural and added, and almost all the seeds (Zobeck and Fryrear 1986b). Wind erosion also exposes roots. These losses mean that each 0.02 m of soil eroded is estimated to reduce crop yield by 6 per cent (Lyles 1975).

There are yet further dangers. Loose sediment can be hurled against crops, severely damaging or burying them. The locally deposited material, having had most of its fertility winnowed away, is much less valuable than the original soil, and can block roads and ditches, whose clearance can be costly, as in Lincolnshire in England after the massive 'blow' of 1968 (Robinson 1969). In the United States it was estimated in 1989 that the annual cost of wind erosion to agriculture alone (from on- and off-site processes) was $188 million (Piper 1989). Finally, wind erosion creates dust, which can be a nuisance well beyond its origin, and which brings costs that dwarf the on-farm expenses (see below). Thus there are large gains to be had from an understanding of the process.

The nature of scientific research into wind erosion

Scientific research into wind erosion on agricultural fields is, as it is with water erosion, a difficult logistic problem. There are two main issues. First is the huge variability in the rate of erosion on both temporal and spatial scales, for the greater part of a year's erosion can take place in one or two virtually unpredictable, catastrophic hours. Second is its insidiousness, for, unlike water erosion, it does not leave obvious features like rills and gullies. It may move large amounts of soil, lowering the surface by many centimetres in one place and raising it by an equivalent amount a short distance away, but because of large and seemingly random spatial variability, these changes are revealed only by careful monitoring over many years (Hennessey *et al.* 1986). Thus much less work has been done on wind than on water erosion, and there are few good estimates of rates and effects. This lack of knowledge contributes to the general tendency for inappropriate reaction.

These complexities dictate three types of scientific approach. The first, observations on field plots, is the most vulnerable to variability, and the most difficult. The second, the study of the minutiae of the process in a wind tunnel, allows the control of wind speed, soil texture, surface roughness and moisture content (among other things), but at the risk of vast oversimplification, and at considerable expense. Finally, and by far the most unreliable, is the statistical comparison of field estimates of soil loss with environmental parameters, such as rainfall and wind speed. This tactic, though widely used and improving greatly with new modelling techniques, has led to some dubious results (and policy formulations), as will be shown. For all their faults, these methods, in combination, have brought many effective answers to the technical problems faced by dryland farmers. Much of what is written in this book stems from these kinds of research, undertaken after the great North American Dust Bowl of the 1930s. What they did not do, and what they may have suppressed by the faith in technical fixes that they have encouraged, was to tackle the social forces that were the main cause of accelerated erosion, which are discussed below.

Some of the best scientific work on wind erosion was done at the United States Department of Agriculture (USDA) Wind Erosion Research Unit at Manhattan in Kansas, close to the Dust Bowl. Work began in 1935 and continues to this day (Lyles 1985). The outcome of the research can still be seen on huge acreages of the Midwest and many other countries (Fig. 9.1). Some authorities maintain that the adoption of the new methods laid to rest forever the spectre of another Dust Bowl (Schwein *et al.* 1983). It is true that dust storms have never reached even a quarter of their frequency in the 1930s (Gillette and Hanson 1989), but other authorities are more cautious, for it is still estimated that about 2×10^6 ha yr^{-1} of land are damaged by wind erosion in the United States (averages for the period 1935–1985). 'Damaged', in this context, means soil loss that is visible, said to be when it is more than 33 t ha^{-1} yr^{-1} or when there has been between 0.02 and 0.05 m vertical removal (Kimberlin *et al.* 1977).

Fig. 9.1 Whole landscapes designed to counter wind erosion: strip fields covering many thousands of square kilometres in the prairies of Montana. This satellite image shows narrow strips of ploughed and unploughed land aligned at right angles to the prevailing westerly winds.

The estimated area of damaged land rises in drought years as in 1954–1955, when it reached an all time peak of about 6×10^6 ha, and 1975–1966, when it was about 3.2×10^6 ha (Lyles 1985).

The technical assessment of wind erosion

The scientists in the USDA soon realized that wind erosion on agricultural fields was a function both of climatic factors (erosivity) and surface factors (erodibility). Following the pattern of earlier work on water erosion, they eventually produced a method of calculating the amount of wind erosion from an agricultural field (Woodruff and Siddoway 1965). This was the Wind Erosion Equation or simply the WEQ:

$$E = f(C, I, L, K, V) \tag{9.1}$$

where E is the potential erosion loss in tons acre^{-1} yr^{-1}; C is a local climatic index; I is a soil erodibility index; L is a factor relating to field shape in the prevailing wind direction (or 'fetch'); K is a ridge-roughness factor; and V is a vegetation cover index. A value for each factor was computed by a tightly prescribed process, using soil, weather and crop data for the field in question, and fed in to the WEQ to give the expected loss of soil for a particular field in a particular season. The computation procedure for each factor was based on extended empirical and theoretical research.

Though a great advance at the time, the WEQ had many problems. It was firmly based on experience in the dryer Midwest (many of its factors were empirically normalized against conditions in eastern Kansas); it was only slowly adapted to deal with changes in crops and soils through the year; it could not deal with the complex interactions between crop, weather, soil and erosion; and it generalized rather too much about wind characteristics (Argabright 1991). A more rigorous model, for use in predicting dust emissions in the United States, was developed and found to be empirically sound by Gillette and Passi (1988). In response to the criticisms and developments, a new method is being evolved by the USDA, based on further empirical research and the assumption of the widespread availability of personal computers: this is the Wind Erosion Prediction System (WEPS) (Hagen 1991). WEPS is to be published late in 1995 and its full specification will be available on the Internet (URL = http://www.weru.ksu.edu/weps.html/).

The technical control of wind erosion on agricultural fields

The WEQ is used here as a framework for discussing the processes and corresponding controls of soil erosion by wind. These factors have been arranged in descending order of inevitability. Climate (C) can only be altered with very great investment, as under glass. Only some of the inherent soil characteristics (I) can economically be manipulated, but fetch (L), surface roughness (K) and vegetation cover (V) are usually well within the control of many farmers. The success of the WEQ was that it directed attention at these factors, and at simple and feasible ways to make radical reductions in erosion.

The climatic factor (*C*) was a simple combination of mean annual wind speed and a moisture index (Chepil *et al.* 1962). It quantified what was already obvious: dry, windy places are the most vulnerable to wind erosion. Though crude, the index has proved a fair measure of soil erosion by wind, even in Britain (Briggs and France 1982). A further discussion of the effectiveness of measures of drought and wind speed is given below in relation to the production of dust (a very closely related problem), where it is shown, however, that cultivation practices can strongly override wind speed as a control of dust production.

Another climatic variable, wind direction (or drift direction) especially at times in the agricultural calendar when the soil is bare, is significant to the alignment of the control structures and practices discussed below, and is therefore another valuable piece of climatic information. Drift directions can be worked out using the kinds of data and models outlined in Chapter 2. An example of their use in a desert situation is described below.

The erodibility of the soil (*I* in the WEQ) has permanent and ephemeral elements. The more or less permanent quality that is important to wind erosion is the primary grain-size distribution of the surface soil, for, as was argued in Chapter 2, the threshold of movement of a soil (the wind speed at which it begins to be eroded) is related to grain size. The curve in Fig. 2.6 shows that it is fine sands and silts that are the most liable to be eroded, being susceptible to quite light winds. With low-density particles, as in soils derived from peat, the threshold is even lower.

However, although the size of the primary particles is important, it is for a rather different reason than implied by the curve in Fig. 2.6, for the effective grain-size distribution (the one that is important to entrainment by the wind) is seldom the same as that of the primary particles. The size of aggregates is a much more effective control, and it is through this that primary grain size operates. The USDA method took account of this by deducing the '*I*' factor from the results of a dry sieving procedure on undispersed soil. This procedure has been vindicated by recent research which found a good correlation between the results of dry sieving and the sand and silt percentages, the sand/clay ratio, the calcium carbonate content and the organic carbon content of soils (Fryrear *et al.* 1994). Sandy and silty soils are erodible largely because they do not contain the clay that is an essential element in the production and maintenance of large, stable aggregates.

It would be very expensive (though not impossible) to alter primary grain size, but aggregation can be controlled (to a degree) through another of its important controls, namely organic matter (of the right kind). Aggregation is also a function of the type of plough in use (Zobeck and Popham 1990). The use of a mouldboard plough can actually reduce erodibility by bringing large, stable aggregates to the surface (Fryrear *et al.* 1994).

In addition to increasing the effective grain size for wind erosion (in general), aggregates also provide a rough surface (see the discussion of the '*K*' factor below), and, if they are not too scattered, they also shelter finer material (Fryrear 1984). However, sand-sized aggregates are more erodible than the primary clay particles that make them up (as can be seen from Fig. 2.6), and since they rapidly break down as they collide in transport and when they return to the soil surface, their entrainment can considerably increase the rate of dust production (Gillette 1980). The ability of aggregates to resist abrasion is related to their content of clay and to the type of clay, among other things.

The erosivity : grain-size relationships discussed above are seen in the spatial pattern of wind erosion. It is soils of the critical grain sizes and low densities that erode most easily. In Britain, only sandy soils in places like the Vale of York, the West Midlands and Lincolnshire, and the loose, low-density peats of the Fens are in any real danger (Radley and Simms 1967; Pollard and Miller 1968; Fullen 1985). In northern Europe, it is also fine sands and silts, many of them already of aeolian origin from glacial times, that are the most susceptible (Fig. 9.2; Chapter 8; de Ploey 1977; Richter 1980; Møller 1986). This is especially so in dry periods, and when fields are intensively used, as was apparently the case in parts of the Netherlands and Denmark in the Middle Ages when cultivated inland dunes were extensively reactivated (Heidinga 1984; Skarregaard 1989). In the United States, it is the Florida peats, and the extensive loose loessic (Chapter 8) and fine sandy soils of the Great Plains (which were the main sufferers in the Dust Bowl) that are most at jeopardy.

The fetch over which the wind acts (*L* in the WEQ) is another easily manipulable factor. Wind speed is low behind a hedge (where *L* is also low), but is high on wide, bare plains. Research soon revealed

Fig. 9.2 Sandy soils in Denmark under active wind erosion (photo: J.-T. Møller).

relationships between the reduction of wind speeds and turbulence both upwind and downwind of barriers and their shape, porosity, resilience and seasonality (Fig. 9.3; Skidmore and Hagen 1977; Stockton and Gillette 1990). It also revealed that barriers at some angles to the wind could create eddies downwind that were even more powerful than the unobstructed wind (Seginer 1975), giving rise to the recommendation that windbreaks should be aligned as near as possible at right angles to the strongest winds. Other research showed that if windbreaks were designed solely for erosion control, they should be as high as possible, not completely impermeable to the wind and fairly close to the each other. The Soviets recommended *coulisses* (closely-spaced rows of crop stalks), as a defence against wind erosion (Zachar 1982), and there has been research on these in the United States as well.

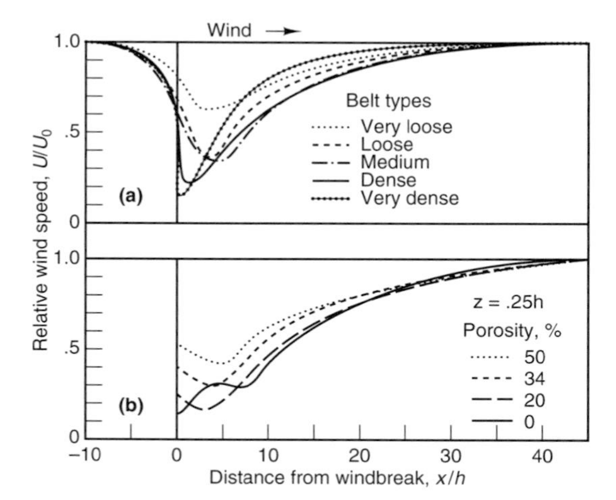

Fig. 9.3 Relative wind speeds (U/U_0) to the lee of windbreaks of different densities (a) and porosities (b) (after Heisler and Dewalle (1988) based on data from Naegeli (1946)).

Windbreaks serve so many purposes other than wind-erosion control, such as moisture control, snow-trapping, dust-trapping, wood and Christmas tree production, and amenity, that their design needs to be fitted very carefully to the local situation. The planting of windbreaks is also very symbolic (see below), but, despite these multiple purposes, windbreaks are not always economic, at least on the short-term accounting horizons of United States farmers, for they occupy space and take water that might be used by a crop (Skidmore 1986b). Fryrear (1976), for example, found that cotton yields were reduced by 10–29 per cent near windbreaks. In Niger, Smith *et al.* (1995) found that, where the water-table was beyond the roots of windbreak trees, the roots could provide serious competition with crops.

Fetch has yet other influences on wind erosion. The aeolian load increases downwind of the leading edge of a patch of loose sediment for a number of reasons. These were aggregated as the 'Fetch Effect' by Gillette *et al.* (1995), who believed it to be a combination of Chepil's 'avalanching' (here termed 'cascading' to avoid confusion with processes on the slip faces of dunes), an aerodynamic effect on the growth of the internal boundary layer, and a change over distance of the threshold velocity. Chepil's (1957) model of 'cascading' had particles being picked up at the edge of a field and dislodging others as they descended (by bombardment, see Chapter 2). These dislodged yet more particles, progressively augmenting the load of the wind. Gillette agreed that cascading is accelerated by the breakdown of aggregates, and by the smoothing of the surface as small depressions (such as furrows) filled with sediment. Chepil (1957) found that the distance to full load varied from about 60 m on highly erodible material to over 1.5 km for slightly erodible surfaces. Thus fields of exposed soil should be narrow across the wind, a pattern that is now adopted by many farmers in the erosion-prone parts of the Great Plains (Fig. 9.1). Alternating strips of stubble and bare earth not only cut down cascading; they also trap saltating sand and reduce near-ground wind speeds. They should be wider on sandier soils, and if the stubble is short.

The ridge-roughness factor (K in the WEQ) can be explained by reference to the roughness height (z_0) in Chapter 2. Experiments found that the rougher the surface, up to about 0.06 m, the lower the wind speed at the surface (Armbrust *et al.* 1964; Hagan and Armbrust 1985). The effect is greatest when the furrows are at right angles to the wind. Another way to roughen the surface is with crop residues, as is commonly done in the Sahel on loose sandy soils, but sandblast can quickly erode ridges and sediment can infill furrows (or the interstices of crop residues), so that ridging and leaving stubble are merely temporary reprieves (Skidmore 1986a). If the crop is valuable, soil stabilizers can be used (see below in connection with the fixation of desert sands), but they are seldom an economic proposition over large areas or with low-value crops.

Studies of the vegetation factor (V in the WEQ) also produced simple recommendations, and these are the most crucial of all. In short, keeping a dense vegetative cover, especially if the crop has narrow leaves and short stalks, like grasses and cereals, is the best of all the controls on wind erosion, as can be understood from the argument about the effects of vegetation on the wind velocity profile in Chapter 2. For all its critical importance, recommendations that a vegetation cover should be maintained may be difficult to meet, for the farmer, to stay in business, must respond to economic as well as to environmental stimuli. The crops that are grown on the huge fields of the North American, Australian or Russian plains demand that the soil must be tilled, and that the seedlings must go through a stage when they are all small. It is almost always in the early part of the year in these places, when crops have barely begun to grow, and offer little protection, that the worst blows occur. It is for this reason that a mix of other measures has always to be used.

The drive towards the Wind Erosion Prediction System (WEPS) is stimulating new wind erosion research, and new technology is allowing more of this to be in the field, where the processes really happen. Portable wind tunnels, some so large that they need to be transported into the field on the back of 10-t flat-loader trucks (Scott 1994), new and better anemometers and sediment traps, and new methods for estimating the erodibility of surfaces, such as vacuum cleaners (Zobeck 1989), are among the new methods.

The new research is uncovering many important facets of the wind erosion process. For example, a strong synergy has been revealed between wind and water erosion: raindrops loosen particles which can then be removed by the wind, either contemporaneously in the rainstorm, or subsequently when the soil dries out (Lyles and Schrandt 1972). The wind also attacks loose sediment that is deposited by rainwash and rills, for this is usually of just the right size to be mobilized by the wind. Soil crusts are now seen as a critical part of the erosion process in many cases (Zobeck 1991): they can increase resistance to wind erosion by 10 to 5000 times. They are promoted by heavy rainfall (Valentin 1991), high clay contents, algae and soil organic matter. Some soils are more than crusted: the whole of the upper horizons may be fused or 'hardset', and these are extremely resistant to wind erosion (but also to agricultural working) (Mullins *et al.* 1990).

For all this research, there are still very few surveys of the regional variation of wind erosion in western countries, mainly because of the absence of reliable surface indicators, the difficulty of monitoring, and the ambiguous interpretation of dust storm data (see following page). A new approach to regional survey is to model the climatic and soil controls on dust emissions, and to compare these predictions with actual emissions, on the assumption that differences reflect the rate of dust emission induced by land use activity (for example, Gillette and Passi 1988). Such an approach has allowed the isolation of the parts of eastern Australia where more dust is being produced than is predicted by environmental factors (McTainsh *et al.* 1990).

The control of dust

The extent of the problem

Chapter 4 confined itself, more or less, to dust that is naturally produced, but there are now many artificial sources and the atmospheric dust load is rising steeply. The rate at which dust is trapped in British peat bogs has increased by two to three orders of magnitude since the industrial revolution (Oldfield *et al.* 1978), and this so-called 'human volcano' may now account for a quarter to a half of atmospheric aerosols. The human volcano has many components, like industrial pollutants, besides the soil-derived particles that are the focus here, though these particles are a major ingredient.

The most notorious input of artificial dust came in the 'Dust Bowl' in the United States in the 'Dirty Thirties', referred to repeatedly in this chapter, which was an instance of the coincidence of drought and disturbance that often generates large quantities of soil-derived dust (Fig. 9.4). The problem persists in North America. Estimates of total dust production in the contiguous United States were ranging, until recently, from 528×10^6 to $1698 \times 10^6\,\mathrm{t\,yr^{-1}}$, though Gillette *et al.* (1992) have produced the much lower figure of $18.9 \times 10^6\,\mathrm{t\,yr^{-1}}$. Even so, these are large quantities, and mineral dust is the most important component of air pollution in many places. Dust control is therefore a prominent environmental issue in the United States.

Dust in other parts of the dry world is less widely reported, although it seems to be quite as serious a problem. Violent dust storms followed the ploughing up of the Russian, Ukrainian and Kazakh steppes, one notorious incident being the great dust storm of 1892 (Stebelski 1985). French (1967) reported further 'severe dust storms' in the Ukraine in 1946, 1948, 1951, 1953, 1954, 1957, 1959, 1960 and 1962. Artificially induced dust is also a major problem in parts of Australia (McTainsh *et al.* 1989). Figure 9.5 shows that there have been further problems in Africa. This particular case was apparently caused by another coincidence of drought and disturbance, probably the collection of fuelwood. In the 1970s, shortly after the start of the recent great drought in the Sahel, the quantities of dust arriving at Barbados from the Sahara rose by a factor of three over the levels of the 1960s (Prospero and Nees 1977). Disturbance was undoubtedly also a major contributor to recent increases in dustiness in central Asia and in western India (Goudie 1983). More recently still, disturbance increased dustiness in parts of Saudi Arabia by factors of about 1.5 during the military movements of the Gulf War of 1991 (El-Shobokshy and Al-Saedi 1993).

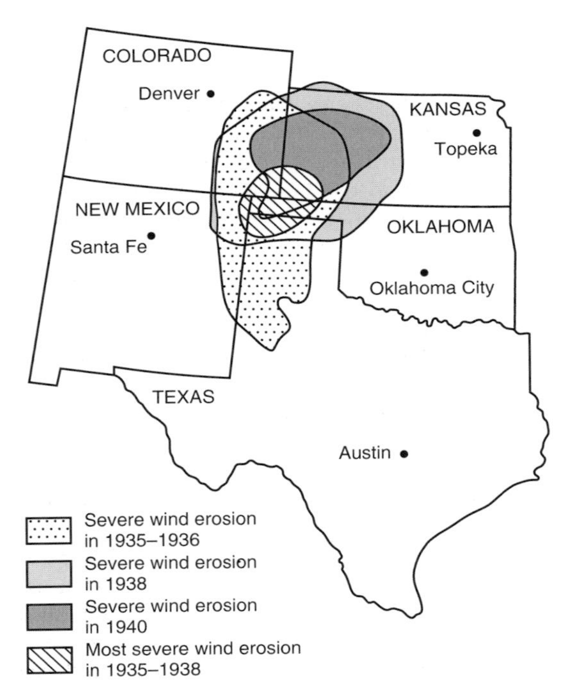

Fig. 9.4 The 'Dust Bowl' of the 1930s and 1940s as defined by Worster (1979).

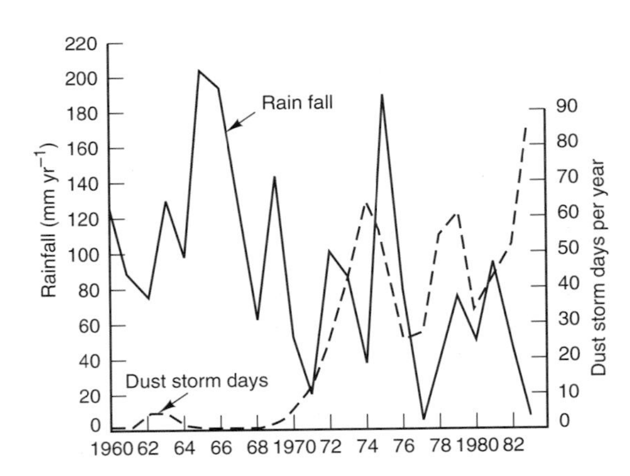

Fig. 9.5 Increased frequency of dust storms following clearance and drought at Nouakchott in Mauritania (after Middleton 1985).

Dust, whether generated naturally or artificially, could have far-reaching effects, either cooling or warming the atmosphere, depending on its particle sizes and concentrations and its combinations with other factors like cloud cover or humidity. If the dust produced large cloud nucleii, rainfall could increase; if the resulting clouds were composed of very fine droplets, rainfall might be inhibited. Most authorities believe that there are far too few data on which to base a conclusion (Hansen and Lacis 1990). The effects might be global, but are more probably regional, as within the Sahel, where dust certainly cools the ground, and may play a significant role in generating rainfall (Maley 1982). Any of these climate changes could drastically affect all kinds of geomorphological processes, but to enlarge on these effects would take this discussion far beyond its remit.

Dust also causes much more immediate problems, and it is on the alleviation of these that this short section focuses. If dense enough, clouds of dust can be a considerable hazard, as on airports and roads (Buritt and Hyers 1981), let alone to radio and television communication, and it can cause disease, both *per se* to the respiratory system (of people and animals), and when it carries salts (Wheaton 1992) or pathogens (Leathers 1981). It is, finally, an aesthetic concern.

The first task in a control strategy is to discover what creates dust. As in all applied aeolian geomorphology, there are three sets of influence: meteorological, surficial and cultural (the last being discussed below). The relative importance of the meteorological controls such as wind speed, and surface controls such as grain size and moisture are discussed in Chapter 4.

The artificial production of dust

The relationships of wind speed, drought and land use are hard to disentangle, for they work in different ways in different areas, at different times and at different scales. While some studies at the continental scale, and for runs of a few years, implicate wind speeds and perhaps drought as major controls (Gillete and Hanson 1989), others find that it is farmers ploughing particularly vulnerable soils who should bear the main responsibility. Farmers can encourage dust production in two ways: first merely by clearing the land and reducing z_0 values; and second by breaking up soil aggregates during the machine working of dry fields (Gillette *et al.* 1992). Around Lubbock, Texas, and in Arizona, for example, machine working is the main dust trigger (Brazel and Brazel 1988; Lee *et al.* 1993). The phenomenal growth of centre-pivot agriculture in the western United States, with large fields, few windbreaks and extensive use of machinery, is also thought to have been a major factor in increasing dust emissions (Breed and McCauley 1986). But even here the analysis is not simple, for Huszar's (1988) study of sod-busting on the High Plains showed a complex interaction of natural and cultural controls. The pattern of over- or under-reaction, so familiar to the history of applied aeolian geomorphology, was at work yet again.

For the United States, it is estimated that wind erosion and tillage produce about one-half of all mineral dust, another major source being traffic on unmade roads (Gillette *et al.* 1992). Vehicle-generated dust, even in peacetime, is a major nuisance in many Middle Eastern towns, where vehicles are numerous and where there is free access to the desert (Jones *et al.* 1986). Plumes of dust, emanating from vehicles travelling across the desert, can be detected on satellite images. Yet other culprits are construction sites (Nakata *et al.* 1976), mines, areas deforested for firewood, and battlegrounds – both real as in the desert war of the 1940s (Oliver 1945) and in the Gulf War, or in military training grounds (Marston 1986). The desiccation of the Aral Sea, mainly because of the abstraction of irrigation water from its feeder rivers, has produced a further and large new source of dust. This dust is highly saline, and may be producing major soil changes in the surrounding zone, let alone its effects on human health. Akiner *et al.* (1992) quoted Glazovski's (1990) estimates that on average 8200 tonnes of salt were being deflated per square kilometre from parts of the desiccated sea bed. The drying out of desert lake basins in the western United States, particularly Owen's Lake in California, largely because of water abstraction for domestic, agricultural and industrial use, has had similar dust-creating effects, though on a smaller scale (Gill 1995).

The control of accelerated dust production can be divided, like all the issues in this chapter, into two strands: technical fixes at the small scale; and social measures at the large scale. In both, zoning plays an important part, for more dust is produced, by any activity, from vulnerable soils (generally, those that are fine-textured), and these should therefore be mapped and given special attention, of whatever kind.

There has been little research into the problem of dust in desert towns, serious though it may be. Péwé *et al.* (1981) estimated that 34 000 t of dust entered Phoenix, Arizona, from the surrounding countryside in the 1972–73 season. P. Davison (pers. comm.) found that desert towns trap the dust in their interstices, so that they can clean up a dusty wind in some circumstances. He believed there to be two main issues where research might alleviate the problem in Doha, the capital of Qatar, which suffers high levels of dust in the summer. Locally generated dust is added to the stream of dust being carried from the north-west on the strong *shamal* wind. The construction work that considerably increases these inputs, mostly from newly developing areas on the periphery of town, could be controlled by spraying (see below). Second, the shape of buildings, their relations to each other and their orientation, can materially alter their propensity to pass dust onwards or to trap it. Simple changes in design, based on aerodynamic principles, could alleviate these problems (Cooke *et al.* 1982).

Technical control of dust production

Control techniques for agricultural dust production include many that are also used for wind erosion. Techniques more specifically for dust control include the spraying of various substances, also used to control desert dunes, but there are many others. Much of the research has concerned situations like dumps of mine tailings or fly ash from power stations, where large amounts of dust can be generated in places close to habitation, and where companies are worried about being sued for damages and about their image. There have been three approaches here. First is the shaping of the dump to reduce wind erosion forces (Hunt and Barrett 1989). This work is related to the research described in Chapter 5 on speed-up over sand dunes. The conclusion is that dumps should be longer in the wind-parallel direction than across the wind. Second is the manipulation of surface roughnesses with inert objects or vegetation. A third technique, applicable especially to dry lake beds as at Owen's Lake in California, is to control the saltating sand that initiates most dust entrainment (Chapter 4) with fences or strips of vegetation (Fig. 9.6; Cahill *et al.* 1995).

Managing coastal dunes and dunes in semi-arid areas

This section discusses together the management of coastal dunes and dunes in semi-arid parts of the world, for although their cultural context is very different, the technicalities and science are very similar. The constraints associated with the different cultural contexts are discussed later in this chapter.

Problems on coastal dunes

Coastal dunes can be associated with major hazards. If the dunes on the Dutch coast were ever breached,

Fig. 9.6 Fences on Owen's Lake, California, USA, designed to control the entrainment of dust, which is often initiated by the impact of saltating sand grains.

not an unlikely event, there would be catastrophe. In the Landes in south-western France, the church tower at Mimizan was at one time buried in 16 m of sand (Fenley 1948). But often, as with so many of the hazards in aeolian geomorphology, the threat, though real, is overstated. It was claimed, for example, that a great storm in 1694 buried a manor house and six farms in the Culbin Sands in Morayshire, though Edlin (1976) believed the story had been somewhat embellished. Further apocryphal tales of buried villages are found for Forvie in Aberdeenshire, Gower in South Wales, Legé in the Landes and at Tved on the south Jutland coast of Denmark, each tale suffering exaggeration somewhere in the retelling (Blanchard 1926; Lees 1982; Skarregaard 1989; Robertson-Rintoul and Ritchie 1990).

The measured judgement must be that, though seldom as cataclysmic as some of the versions of these stories, the protection that dunes provide against extreme storms is substantial (Nordstrom and Gares 1990). While coastal dunes survive, they provide protection to the whole of the province of Holland, to much of lowland Lancashire and to many other valuable areas. Coastal dunes are a dynamic component of the barrier islands that line the Atlantic and Gulf coasts of the USA (and many other coasts), and are therefore also an integral part of the management of these valuable sea-defences (Leatherman and Zaremba 1987). Moreover, the costs of the burial and exhumation of buildings and land, though small in each case, can be very large in aggregate. The potential real costs and benefits and the symbolic value of controlling the imagined terrors of coastal dunes are effective stimuli to discovering efficient means of control.

Another impetus to understanding is that many coastal dunes are at the same time a valuable resource. They cover huge areas in total, and many are in strategic positions. A quarter of the southern Australian coast is backed by dunes (Short 1988). The whole of the Belgian coast is lined by dunes, as is 80 per cent of the Dutch coast and most of the German and Polish coasts, all of the west coast of Denmark and vast stretches of coast in France and the United States. In Great Britain, which is not specially well endowed, coastal dunes cover 56 300 ha or $\frac{1}{4}$ per cent of the total land area (Fig. 9.7). Coastal dunes underlie a large proportion of cities such as The Hague, Durban and Tel Aviv. They are used as natural underground reservoirs for domestic water supply in many parts of the world. In Holland four million people draw over $40 \times 10^6\ m^3\ yr^{-1}$ of water from sand dunes, which are recharged artificially, and which act as natural filter beds (van der Meulen and Jungerius 1989).

In the densely populated European countries, coastal dunes were the last reserves of unused and accessible space in the periods of international conflict or economic expansion in the nineteenth and early twentieth centuries, and were therefore commandeered by state enterprises such as military training grounds, forestry, defences against invasion, ports and industrial estates. One of the largest forestry planting schemes ever undertaken, in the Landes of south-western France, was on coastal dune land. Napoleon appointed an engineer, Bremontier, to reclaim the Landes in 1787. The new forest covered nearly 1 million ha by 1892 and now covers about 9000 km^2 (Lowdermilk 1944). Other large plantations, though none as massive, occur in the Culbin Sands of north-east Scotland, in western Denmark and on the Baltic coast of Poland. Some coastal dunes are valuable sources of sand for concrete. In northern New South Wales and southern Queensland, beach sands, rich in zircon and rutile, have been mined from beneath coastal dunes since the 1940s; perhaps 5 per cent of a 1000 km stretch of coast will eventually be mined, leaving potentially massive devastation (Clark 1975). Coastal dunes are now being threatened by mining for titanium in southern Madagascar.

Coastal dunes are, furthermore, one of the most intensely used recreational resources the world over (and thus one of the most familiar of landforms). Many square kilometres are covered with holiday homes. Seventy per cent of Belgian coastal dunes are built up in this way (de Raeve 1989). Because of the simplicity of their ecosystems and their selection to illustrate a particular concept of successional theory in many ecology textbooks (Ranwell 1972), coastal dunes also attract thousands of students each year on field courses. They are a valuable wildlife habitat, with distinct and sometimes rare species (Doody 1989). They are attracting renewed interest because of their role in coastal protection during possible sea-level rise in the twenty-first century. All this interest, much of it conflicting, focuses on one of the most fragile of environments.

The threats to and values of investments, real or imagined, that are inherent in many coastal dunes in the developed world are potentially so great that they have created a strongly conservative attitude in managers, and this in turn has influenced the science and technology of control. But, whether restrained or innovative, scientific understanding is essential to the management of coastal dunes. It can be utilized at two

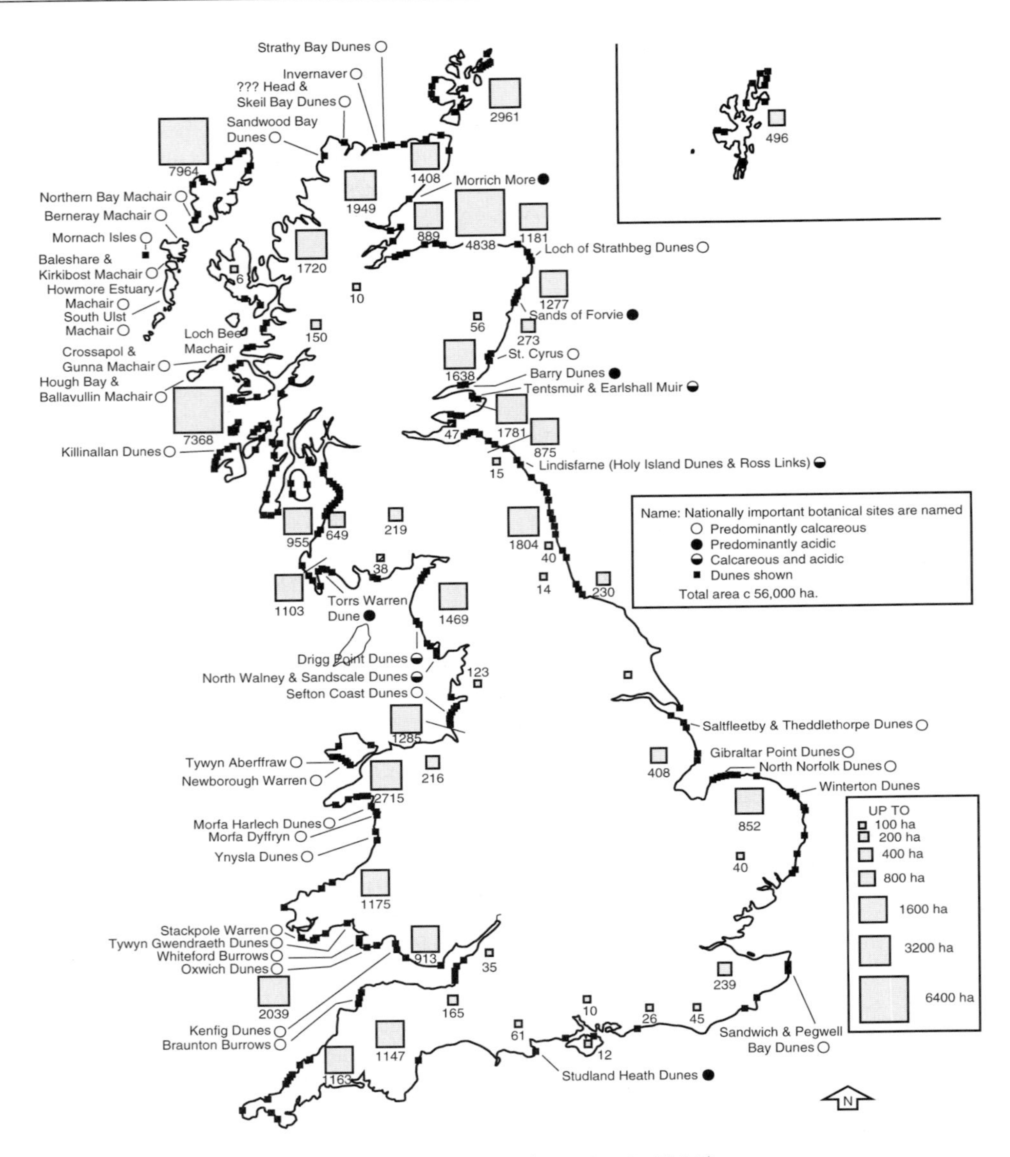

Fig. 9.7 The extent of coastal dunes in Great Britain (after Doody 1989).

different scales. At the small scale (that of a few hundred metres) management is fairly strictly a matter of aeolian geomorphology and plant ecology. At the larger scale (that of the dunefield), horizons need to expand to include the whole of a coastal sedimentary cell, centuries if not millennia of environmental change, a range of geomorphological processes and techniques, and much more profound considerations of the scientific, social, cultural and political context. A similar kind of distinction was drawn by Gares and Nordstrom (1991), who observed that small-scale management, with time-horizons of little more than 10 years, was appropriate only for heavily used coastal dunes, while the longer-term perspective applied to all types of coastal dune.

Small-scale coastal dune management

Management at the small scale relies on an understanding of the dynamics of the system as outlined in

Chapter 5. Four characteristics have practical significance. First, dunes and beach are in close connection: a change in one is rapidly passed on to the other. If artificial groynes lead to the accumulation of beach material, for example, the dunes behind will grow soon after (Nordstrom *et al.* 1986). Second, most sediment is moved only in brief, rare storms. Third, in general, it is storm waves that do the most damage to dunes, not the wind: the wind can be used to repair the damage. Fourth, in the great majority of cases, there is a marked decline in the amount of sediment in transport as one moves inland from the beach, so that sand-fixing devices are generally much less necessary inland.

This last observation is the basis of a useful division of coastal dunes into three management zones: I, the active seaward zone; II, a zone with a mosaic of very active and completely vegetated patches; and III, a densely vegetated, even forested inner zone (Jungerius and van der Meuelen 1988). This reflects Psuty's division into 'primary' dunes close to the coast and 'secondary' dunes inland, as explained in Chapter 5.

The outer zone (Zone I) is generally narrow, though in transgressive dunes it extends many hundreds of metres inland, as explained below. Parts of the second and sometimes even the third zones can be expected to be without vegetation, quite naturally, and even in quite humid climates. Blowouts appear and disappear in these zones, but need not be seen as unequivocal signs of widespread reactivation, or as in need of emergency stabilization (Jungerius 1989; Carter *et al.* 1990).

The necessary type of management activity at this small scale varies by zone. In the active, seaward zone the main objective is to trap mobile sand (usually after a storm has breached the foredune) either mechanically or with plants (if management is needed at all). The crudest method of building dunes in this zone is to create them with bulldozers, but there are more subtle and less expensive methods, of which the building of sand fences is the main one. The nearest analogue for designing fences to trap sand is the design of snow fences to protect highways and railways. The accumulation of sand (or snow) around an obstacle is a positive function of two factors: (i) the Froude number,

$$Fr = (u_* - u_{*t})/(gh)^{1/2}$$

where h is the height of the obstacle, u_* is the shear velocity, and u_{*t} the threshold shear velocity, as defined in Chapter 2; and (ii) the concentration of sand in the wind (Iversen 1986b).

Although knowledge of the design of sand-fixing fences on the European coasts is now centuries old, there are still gains to be made from research. Manohar and Bruun (1970) quoted earlier studies in which it was found that fences with a permeability of about 50 per cent were the best for sand trapping (although no sand is trapped by such a fence in winds greater than $18\,m\,s^{-1}$). In a wind-tunnel study they found, among other things, that two fairly closely spaced parallel fences were better than one, and that the optimum position of the fence in front of an existing dune depended on the ambient wind speed. Other research has shown that zig-zag fences create wider, lower, more manageable dunes than straight ones, and are more likely to trap sand brought by winds from a variety of directions (Watson 1990).

Experimental observations of accumulation round a fence with 38 per cent porosity showed that sand collects on the windward side of the fence, and soon finds its way through the interstices, creating a dune in the lee (Willetts and Phillips 1978). The crest of this dune is further from the fence in higher winds (Iversen 1986a). In many situations, 1.5 m high fences can be overtopped in one season. Because the growing dune itself begins to act as a fence, and since trapping ability varies as the height of the total obstacle, successive fences on top of the dune trap more and more sand (Watson 1990). The dune may, therefore, grow to many tens of metres in height if the fences are repeatedly replaced. The ultimate height may be controlled by the depth of rooting of plants, or the greater speed-up of the wind over higher parts of the growing dune (Rasmussen 1989).

The choice of plants as sand-accumulators in this mobile zone is restricted, for although there are other species that can survive salt spray and burial in sand, and have been used from time to time, only the grass *Ammophila arenaria* (marram grass) can be grown with confidence (Chapter 5). Even in the fourteenth century, special regulations covered the planting of marram in the Netherlands (van der Maarel 1981), and marram or '*oyat*' was Bremontier's main tool for fixing the dunes of the Landes in south-western France in the nineteenth century. The grass, a native of north-western Europe, has now been used across the world in this role, including the United States and Australia. Most authorities agree that it does the job better than its North American relative *A. breviligulata*, which is nevertheless sometimes used (Chapter 5).

In planting schemes, the spacing of the plants can be critical, as can the way in which the roots are arranged, be it vertical or horizontal. Fertilizer treatment or stabilization with sprays (see below) can speed growth. To help marram survive the few short weeks before it gets going, or to protect the sand out of the growing season, surfaces can be protected with brushwood (in many cases discarded Christmas trees, some doubtless planted first as windbreaks, as described above), or straw fences (as with inland dunes, see below). Even in the nineteenth century a kind of folk knowledge had been acquired in France about these methods. After nearly two centuries of systematic dune fixation in Europe, there is now a weighty body of knowledge on these matters, and many good handbooks (for example, Ranwell and Boar 1986; Ritchie 1989).

Dune fixation, by whatever means, is not always wholly beneficial. Dunes may trap sand that would have moved along the beach, and thus accelerate erosion at a point down-drift or downwind, by robbing it of sediment, as in New Jersey (Nordstrom and Gares 1990). On the southern South African coast, 'courageous' management plans have sanctioned the devegetation of some coastal dunes in order that a 'natural' system of sediment transfer can resume and feed sand to beaches that had apparently been denuded of sand by earlier planting programmes (Swart and Reyneke 1988).

Thus, even at this scale it is impossible to eliminate cultural influences. There are many more examples. In some cultures, for example in the Netherlands, coastal dunes are managed centrally, and uniform policies extend over long stretches of coast (Arens 1994). In parts of the United States, such as Long Island, in contrast, the dunes are managed piecemeal by the owners of the nearest property, usually a beach house. In this case there is enormous variability in the degree, type and effectiveness of defences and the dunes they generate (Nordstrom *et al.* 1986).

A final issue in this zone is the management of the 'cliffing' of dunes by storm waves (Chapter 5). Two extreme conditions can be foreseen. Cliffing has to be countered if it endangers installations behind the dunes. On the other hand, it is to be encouraged if, by feeding sand to the beach, it helps to modify wave attack, and so counter the overall erosion of the coast (Arens 1994). Judgements as to which is the most appropriate response depend on a number of other factors, both at a particular site and up- and down-drift of it. One procedure, appropriate in both conditions, is the mechanical modification of the outer dune slope so that it is better able to accommodate extreme waves. Engineering models can predict the appropriate configuration, and appear to be successful (summarized in Sherman and Bauer 1993). Over-protection of a dune, with no regard to dune–beach interaction, may create an artificially steep beach profile, which may then be very vulnerable to wave attack (Sherman and Bauer 1993). In many cases, a policy of 'managed retreat' might be more appropriate.

Inland from the mobile zone, marram ceases to thrive and is replaced by a community of other plants (whose species composition is very different in different parts of the world) underlain by successively more acid soils (Chapter 5). These communities and soils are very vulnerable to some kinds of disturbance. The A_2 horizons in the soils are usually leached and loose, and can be destroyed by deliberate physical clearance, as by ploughing. Podzolic B horizons lower down the soil profile may be so dense that they considerably impede drainage, creating saturated conditions in winter. One of the great breakthroughs in the reclamation of the Landes in south-western France in the nineteenth century was the discovery that drains cut through B_h horizons (the hard black '*alios*') could dry out the soil enough to allow trees to grow and to control malaria (Lowdermilk 1944).

These characteristics of the inland areas within coastal dune systems are usually adequately managed in intensive land use systems like horticulture or agriculture, but can render them vulnerable to less intensive, less highly capitalized land uses like grazing by domestic animals and recreation. Fire is another hazard. There has been a long, inconclusive debate about the effects of introduced rabbits (Rutin 1992), but whatever the reality, many managers control them, just in case.

The most obvious technique to control unacceptable amounts of bare sand or sand movement in these low-input management systems is to restrict usage, for example by managing the most heavily used zones intensively so that damage does not reach unacceptable proportions. Paths can be mown and their use rotated, or fertilized and sown with hard-wearing species. If usage is expected to be heavy, artificial surfaces, like boardwalks, can be laid. When all else has failed, the techniques of sand fixation that are used on the mobile edge of the dunes can be employed, though with more conventional grass species-mixtures than with marram (Ritchie 1989). In many cases, dunes have been afforested, though this has usually been more for timber than for protection, even when protection has been the excuse (Edlin 1976).

The details of dune control in semi-arid areas

On the desert margins, where there is some rainfall, albeit limited and unreliable, the technical problems of dune fixation are much the same as those in temperate coastal dunes, though perhaps with more exotic sand-fixing species. The technical solutions, however, can be different.

Where, as is usual in less developed countries, there is plentiful and cheap labour, the common practical procedure is to hold the loose sand with low brushwood or reed fences until plants can grow up enough to reduce wind speeds and hold movement (FAO 1985). If the wind is expected from one direction only, as on the Somali coast, the fences can be in rows at right angles to the wind (Fig. 9.8a), but if the winds are variable, a criss-cross pattern is advisable (Fig. 9.8b), the size of the enclosed squares being smaller on steeper, more undulating topography. Sometimes, if labour is especially plentiful, the fences can be made of clay (Zhao Songqiao 1988; Kebin and Kaigo 1989). There is now a range of techniques to help the plants to get away, such as earthenware pots with better soil to retain moisture and small sachets of water-retentive chemicals and nutrients. The new plants invariably need to be watered for quite a few years before they establish and need to be fenced against domestic stock and the poor seeking firewood, browse and hay. This needs a well organized workforce (which is not always forthcoming).

Technical aspects of coastal dune management on a large scale

Three sets of problems arise if the coastal dune manager moves beyond sand fences and planting schemes. First is the 30–50 year 'cycle' of activity between major disturbances that occurs on many coastal dunes (Gares and Nordstrom 1991; Chapter 5). Second is large-scale reactivation, which may bury valuable land. Third and last, there is the maintenance of dunes of any kind, active or fixed, as a form of coastal defence, in the light of their widespread, inevitable decline.

The management of occasional storm damage is a matter of employing small-scale methods strategically, both before and after the inevitable event, and above all, of ensuring some kind of plan for coastal development that neither endangers the dune, nor life, limb and property.

The large-scale aeolian reactivation of a coastal dunefield involves the burial of vegetation for several hundred metres and even kilometres from the coast. In these cases the dunefield has become 'transgressive' (Chapter 5). Some of the dunes on the Baltic coast of Poland, for example, were highly transgressive in the recent past and, despite forestry, some still are (Borówka 1990). Transgressive dunes may require active intervention (such as tree-planting) over large areas. There are situations in which to attempt control is futile: for example, coastal dunes on desert coasts are barely fixed at all, as on the northern coast of Sinai in Egypt (Tsoar 1990c). But there are many other situations in which control is conceivable, if problematic. The main problem of management at this scale is the range and complex interaction of possible causes, and the consequent possible extent of effective intervention. Reactivation can have natural or induced causes (or more commonly a subtle mixture of the two).

Medium-term 'natural' surges in sand supply can be created by storms which cause marine erosion of nearby foredunes, the sand then moving down the coast and being blown off onto dunes elsewhere. The changes at the Grand Dune du Pilat in Aquitaine, which at 105 m high is said to be the biggest dune in Europe, have been shown to be related to such a coastal change (Froidefond and Prud'homme 1991). Rhythmic pulses of sand and the delayed effects on a beach of offshore changes, both referred to in Chapter 5, are further causes for change, but are seldom easy to diagnose. Also at this scale, intense storms can move great volumes of sand from the beach to the dunes or can create blowouts; or droughts or hurricanes can kill plants and reduce their protective capacity (Bird 1974).

Longer-term 'natural' changes are yet more difficult to elucidate. Enhanced supply may come from intensified erosion on cliffs many hundreds of kilometres away, as in eastern England, or from a river mouth, as in the case of the dunes associated with the Rhône, Rhine, Gironde and Tagus in Europe, or from erosion following a change in sea level, nearby or remote, recently or far in the past. In few of these cases is the ultimate source of the sediment or its routing thereafter unambiguously established or even establishable.

For all this emphasis on the management of active sand movement on coastal dunes, transgressive dunes are the exception rather than the rule on temperate coasts at the present time. A much more fundamental problem is that most are geologically

(a)

(b)

Fig. 9.8 (a) Brushwood fences arranged in strips to protect planting designed to control sand movement on sand dunes in Somalia; (b) brushwood and dry grass fences arranged in a criss-cross fashion in a dune-fixing project in Rajasthan.

ephemeral, or indeed in active decay: Bird's (1985) worldwide survey showed sandy coasts in retreat almost everywhere. Thirty-eight per cent of the Netherlands coast is regressing, 39 per cent is stable and only 23 per cent is prograding (Arens and Wiersma 1994). Many coastal dunes occur on spits or barriers that are short-lived littoral features, which can grow and disappear in matters of years. Many others were created in storm conditions over the last few centuries, and are not being maintained (Goldsmith 1985), though new ones may be expected after storms in the future. Most have been built with sediment that was produced by coastal erosion as sea levels rose after the last glaciation, or in high latitudes by the marine remobilization of fluvio-glacial sand (Chapter 8). Neither source any longer yields much sand (though rising sea levels may produce more sand again, as discussed below). In many parts of the world, as in the Netherlands, beach defences (such as groynes and other measures to restrict littoral drift) can restrict the supply of sand to the dunes, and even accelerate their decline (Arens and Wiersma 1994). Unreplenished coastal dunes cannot last for ever.

In addition to erosion by the sea, the processes of decay in coastal dunes include the leaching of soils, slope processes, gradual dissipation inland and erosion by streams. Coastal dune management, in this perspective, is little more than postponing an inevitable decline. Many authorities worldwide are adopting a policy of 'managed retreat' in the face of these problems. It is argued that a constant battle with the sea must inevitably build even greater problems in the future, especially during rare storms. But as the scale of management increases, so does the intrusion of political, economic and aesthetic considerations. These are discussed on the following pages.

Coastal dunes and global warming

Coastal sand dunes could, finally, be very vulnerable if CO_2 concentrations in the atmosphere were to increase, and by causing global warming, raise sea levels and alter climatic patterns. Although coastal dunes would inevitably be affected, the nature of the change is not clear, and it is unlikely to be the same for all dunes (van der Meulen 1990; Carter 1991).

Those coastal dune systems that are isolated from sand supplies (either artificially or naturally) would be the most vulnerable. Dunes would be eroded by waves more often, and would recover more slowly in the absence of new sand. Another vulnerable set of dunes are those that have been artificially maintained against erosion. On one of these, Cape Hatteras in the eastern United States, it is estimated that, if unmanaged, there will be a loss of 22 per cent of the dune system by AD 2000, and if sea levels were to rise further, 70 per cent of the dune system would disappear (DeKimpe *et al.* 1991).

Quite another picture emerges where there is the possibility of sediment movement, onshore or alongshore, which is often the case. In these cases a more complex scenario is more likely, in which higher sea levels and greater storminess would increase erosion on some parts of the coast (particularly on vulnerable soft cliffs), and feed it to an intensified system of longshore movement. Some coastal dunes, particularly those at the distal ends of sediment transport systems, might then be engulfed in new sand. If this did not bury marram (whose growth may at the same time be stimulated by increased CO_2 and rainfall), larger dune systems, perhaps at somewhat more landward sites, would result. (The rates of burial that marram can survive are discussed in Chapter 5.) If there was a sudden increase in supply, burying even marram, then dunes might be reactivated on a large scale and themselves become a hazard (Carter 1991). This is what seems to have happened on many coasts at the end of the Pleistocene, when rising sea levels are said to have resulted in large amounts of sediment, although this sudden burial might have been in response to the very abrupt nature of the climate change discussed in Chapter 8 or to other distinctive climatic conditions of windiness or aridity.

These changes in coastal dunes in response to climate change are unlikely to be the only ones, for there may be more salt spray in stormier conditions, killing some plants, and erosion might lower water-tables creating drier conditions and also killing plants (van der Meulen 1990) (though it could also raise the water-table along with the general rise in sea level). Higher temperatures might desiccate soils and have a similar effect, but, in combination with higher levels of atmospheric CO_2, might increase the rate of plant growth (depending also on rainfall). A decrease in the cover of vegetation would release more sand to the wind. These various processes cannot be expected to act in the same way on all coastal dunes, and there will doubtless be complex and poorly anticipated interactions between 'natural' processes and the measures taken to combat them.

The management options to deal with these changes are also various, depending not only on the way the coast reacts to change, but also on the likely economic and symbolic impact. The options are 'managed retreat', selective control, full control, and even reclamation (van der Meulen *et al.* 1991).

The control of sand dunes and drifting sand in deserts

The drylands present yet another distinctive natural and socio-economic context within which applied aeolian geomorphology must operate. Two elements are familiar: the tendency for exaggeration, as in the stories of lost oases and even lost Roman legions in north-eastern Egypt (de Lancey Forth 1930); and the presence of real enough threats beneath the bluster. The dangers have some distinctive characteristics in deserts: sand and dust are much more mobile than in wetter climates and plants are much more difficult to keep alive. On the cultural side, however, there are some positive features: many deserts today share a singular characteristic in that money is not the extreme constraint that it is in most of the other spheres discussed above. The reason is oil, for many of these economies depend on maintaining oil supplies from areas suffering aeolian geomorphological problems, and oil generates the wealth with which these and other problems can be tackled.

There are two categories of aeolian problem in the drylands. One, excavation of loose soil by the wind, as around telephone poles and houses, though locally irritating, is by far the less serious, and is not discussed further. The other has to do with moving sand, either in saltation (piling against objects, blocking culverts, making driving hazardous, interfering with machinery or abrading all manner of things); or as moving dunes which bury installations. Both processes only occur where winds are strong enough and where there is abundant loose sand (in perhaps a quarter of the arid lands). In these situations, blowing sand can be a problem in almost any seasonal pattern of winds, but moving dunes are only a significant issue where winds blow from one dominant direction, in other words in areas with transverse dunes and barchans (Chapter 5). Examples are Peru, the Atlantic coast of the Sahara and the Gulf Coast of Arabia.

Folk science

For all the new technology that oil has brought, the management of sand in these environments relies first on cheap, time-tested, folk technologies. Populated oases in the Old World deserts have used techniques to withstand blowing sand and moving dunes for thousands of years. In some places, it is true, gardens can be abandoned for the few years during which they are buried by a moving dune, and brought back into cultivation when they reappear (possibly even benefiting from the fallow period) as at Faya in Chad (Capot-Rey 1957), but most oases could not have survived without routines for protection from burial.

The commonest of these ancient techniques is a fence of date-palm fronds (Fig. 9.9). Though evolved independently of the fences used on temperate coasts (described above), the successful designs share many characteristics. Most of the oases in the Sahara and Arabia are surrounded by walls of sand that accumulate round these fences, some up to 40 m high (Wehmeier 1980). Growth of $1\,m\,yr^{-1}$ has been reported (Achtnich and Homeyer 1980). Though they may hold back the sand in some years, they cannot offer permanent protection. The varying fortunes of the palm gardens are partly due to the inevitable failure of the some of the fences. In the huge El Hasa oasis in eastern Saudi Arabia, it has been estimated that $3.5\,km^2$ of palmeries were lost over 50 years, despite fences. The problem at El Hasa is at least 2000 years old (Hidore and Albokhair 1982).

Fig. 9.9 Palm frond fences protect a small oasis palm garden in Algeria.

Sand fences have by no means been superseded, but date gardens have now acquired much greater commercial value. They, their installations and dependent villages are now joined, as objects of protection, by even more valuable installations like oil wells, pipelines, factories, ports, suburban housing, railways and metalled roads. Oil companies and national governments have been investing large sums, over many years, in researching and then providing protection (Kerr and Nigra 1952).

New approaches to protection against sand drift and dune encroachment

As soon as metalled roads and railways were built in these environments, they needed protection. The first principle of protection is *avoidance*, which is a real option in the vast open spaces of the desert, and with structures like roads which are not rooted to water sources (like gardens). Routes running parallel to the drift, or ones in zones where there is little drift are clearly advisable. To avoid drift, one needs to know its volume and direction, and this can be discovered by analysing the distribution of local source areas (such as dunes) and wind direction and speed, although wind data in deserts are not always adequate to the task (Jones *et al.* 1986). In general, the control of source, as might be done for dust abatement (see above), is only possible in the rare cases where the primary source is local (for example, a river wash). Most local sources are fed from more distant ones, so that local control is only temporary, and has to be repeated ceaselessly. Most primary sources are too huge, too distant or too mysterious for control to be contemplated.

If avoidance or removal are impractical, *defence* is necessary. Some of the methods for preventing the burial of routes by sand depend on modern applications of folk understandings. A hoary piece of this kind of folk science was built upon by the celebrated 'sand engineer', Si Belkacem, in El Oued Souf in Algeria in the 1950s. He knew just how and where to protect roads with low walls of sand, covered with earth, presumably knowing local sand streams and the rudimentary aerodynamics of fences (Mason 1961). The ploy used by plate-layers on the Peruvian railway in the early twentieth century is another case of applying a visceral understanding of aeolian processes. They spread loose pebbles and grit thinly on the surface of approaching barchans. The dunes sagged and dispersed, leaving the sand to saltate harmlessly over the track, rather than blocking it as a dune (Bailey 1907). Stones at a certain critical density presumably increase the rate of erosion by encouraging turbulence, as shown experimentally, much later, by Logie (1982).

If zones of moderate sand drift have to be crossed, it was appreciated quickly that there are some general design guidelines. It is better to run a route across land that gently rises in the direction of drift, where sand movement is accelerated; it is better not to create twists or pockets, like narrow cuttings or bridge emplacements, that can accumulate sand, rather to allow it to pass quickly over the route; running the route on top of an embankment with smooth side slopes of between 1:5 and 1:6 accelerates flow and moves sand quickly over the top (Stipho 1992).

Another well-tried method, for routes or other installations, which can be extended with the use of machinery, is to reduce the erodibility of the surface. The oldest and most familiar method is spraying with water. The method is still used in many towns to 'lay' sand and dust, and it is recommended in humid climates to lay dust on dumps of fine material near mines and power-stations (Hunt and Barrett 1989), but it can be expensive in deserts, its effect is short-lived, and it cannot be used over large areas. Another way of controlling erodibility is to spread a layer of coherent soil or a coarse gravel over the eroding surface (taking care that the gravel is not, as on the Peruvian railway, too dispersed). In effect, this is manipulating the I factor (soil erodibility) in the WEQ (above). Where there is a local supply, as in El Hasa, soil spreading can be temporarily effective, but, unless very extensive, the covering is soon buried by advancing sand from places upwind which have not ben protected. Moreover, the costs of mechanically spreading the soil can be high, and yield no return other than protection (Achtnich and Homeyer 1980).

Newer forms of covering include *stabilizing substances* of many kinds. Asphalt and other oil-derived stabilizers are now popular, having come down in price (many being locally derived in oil-rich countries); they last much longer than water, but can be unsightly and even toxic. There are now many specially designed chemical sprays on the market (including latex-based ones, sodium-silicate-based ones, polyvinyls and various polymers), with varying brittleness, hardness, viscosity (and hence penetrability) and resistance to UV light or oxidation (Stipho 1992). There are also varying guidelines for application rate and pattern. Spraying in strips permits more area to be covered with the same volume of stabilizer, and, carefully done, can be almost as effective as blanket spraying. Sprays need to be carefully chosen for particular soils, depending on their texture and content of organic matter (Watson 1990).

Sprays can be used in many different ways. One tactic, advocated in an early study by Kerr and Nigra (1952), is to spray the upwind side of a dune to immobilize it, ensuring that sand is rapidly transferred up and over the immobile mound. This transferred sand creates a lee dune, for the original dune now acts as a fixed obstacle. More sand is thus trapped than would otherwise have been, and the problem is postponed until the next dune arrives. Another, more usual tactic is to spray swathes of sand alongside roads or upwind of installations, hoping to cut the sand supply. Relief depends on the extent of spraying, but even this method cannot bring permanent protection. Besides, sprays, though cheaper than they were, are as expensive as, if not more expensive than, spreading earth mechanically, and give no more permanent protection. This has not prevented well over 50 km^2 being treated in the former USSR. In Saudi Arabia, one company alone sprayed 35 km^2 between 1978 and 1983 (Watson 1990).

Research has developed many more tactics and devices, such as pits, wider than the common saltation jump length, to collect sand (though these need to be constantly emptied); solid panels that divert it away from particularly sensitive places (angles of 35°–40° to the prevailing direction of sand movement are recommended); rough surfaces to trap sand; smooth ones to encourage its movement; or aerodynamic modelling of surfaces to generate faster sand movement. Sand-trapping fences can be improved upon with various new materials (Stipho 1992). Still, none of these methods eliminates the problem; they merely alleviate and postpone it (Watson 1990; Stipho 1992).

Planting bare sand, where there is a supply of water (from sparse rain, surface irrigation water or goundwater) is a better, though still imperfect option. If the plants can be protected from sand drift as they establish, it is more permanent and provides multiple benefits (like firewood). Unlike the planting programmes in coastal dunes, grasses are seldom suitable as the primary plant in desert schemes, for they are buried quickly, root shallowly and rarely retain enough foliage through the year to act as perennial sand traps. Bushes and trees are more successful, though care must be taken to plant them at a density sufficient to reduce rather than to increase sand movement (by funnelling the wind between clumps). Some tree and bush species can tolerate only occasional watering and even quite saline irrigation water. Tamarisks, Acacias, Prosopises, Casuarinas and Eucalypts are widely used. There is a large literature to help the manager in choosing the right species (Kaul 1983; Busche *et al.* 1984).

The vast scale of some planting schemes in the oil-rich states can be gauged by the one at El Hasa in Saudi Arabia, where a belt of sand some 5 km across was planted in the early 1960s. In only the first phase 500 ha of moving sand were planted with five million trees (Achtnich and Homeyer 1980). They were irrigated for the first five years to allow roots to reach to the water-table. In the Chinese deserts, there is a major programme of similar kinds of planting alongside railways to protect them from sand encroachment, and these plantations need extensive irrigation, at least in their early years (Kebin and Kaigo 1989). Massive irrigated planting, for the same purpose, is easier to achieve along irrigation canals in the desert. This has happened along the huge Indira Gandhi Canal in Rajasthan in India and along the Kara Kum Canal in Turkmenistan.

When all else fails, and the installation is valuable enough, there are two further solutions. The first is simply the *removal* of the approaching dunes. A principle from Chapter 5 is worth recalling: small dunes travel faster than large. Small dunes, therefore, are the more immediate danger; large ones are a bigger, but a longer-term problem. This also means that to reduce the size of a dune is simply to accelerate it. Removal can be achieved with the aid of the wind itself, as when dunes are reshaped to encourage dispersal or sand trapping (Watson 1990), but the usual approach is much less subtle, namely bulk removal (Fig. 9.10). If the hazard is a set of closely packed dunes, bulldozers and trucks may be needed; if more dispersed, scrapers will do. In either case, dune removal is seldom a one-off operation: more dunes can usually be guaranteed to follow those that were first removed.

The second solution is to invest in an endless programme of *sand sweeping*, as is done on many of the super highways in Arabia. In planning a new road through a belt of moving dunes, it is necessary to ensure that the costs of either of these kinds of clearance are added to the overall costs, and that they remain on the books as long as the life of the road.

Fig. 9.10 Mechanical sand clearance off a road being covered by migrating dunes in Oman.

The cultural subtext

For all its many successes, scientific and technological research has never been able and can never hope to resolve aeolian geomorphological problems on its own. There have to be mutual adjustments of science and technology on the one hand and land use culture on the other. These adjustments are complex, usually painful, and in consequence protracted; few aeolian geomorphological problems can be solved quickly and usually require longer than a decade.

The American experience of scientific and cultural adjustment to wind erosion is the best documented and the best analysed. The lessons run counter to narrow scientific arguments that urge direct application of science regardless of culture. Four main aspects of the interaction have been found to be important. First, the acceleration of wind erosion is a product of distinctive economic, political and social environments. Second, the findings of research are absorbed in a way that is peculiar to the time and the place. Third, interaction is commonly obscured by myths about science which are also a product of the cultural milieux, and which further complicate the interaction. For example, it has been argued that a predilection for technical fixes in North America, which was a product of the mythic value attached to science at the time, obscured more important processes behind the acceleration of wind erosion, namely the social, political and economic climate (White 1986). Finally, the character of scientific and technological research into wind erosion is itself influenced by the cultural environment.

These themes recur in applied aeolian geomorphology at other times and in other places. The discussion that follows uses different spheres of applied geomorphology to illustrate different aspects of these themes.

The North American campaign against wind erosion

The culture of Dust Bowl North America had several strands. One powerful theme was modernism (the belief in the inevitability of progress and in science as its principal instrument). One consequence was the 'Fordist' dogma that fields should be vast and machines massive, with clear implications for wind erosion. A more transient consequence was the pseudo-scientific notion of 'dry-farming', which duped farmers into destroying the structure of the soil (in an attempt to conserve moisture), making it, of course, much more vulnerable to wind erosion (Riebsame 1987).

Modernism, when combined with the myth of the frontier, supported the drive to plant windbreaks or shelter-belts, but only partly as a hedge against wind erosion. Good scientific reasons could be given in many circumstances, but the rapid uptake and state support of windbreaks had other, much stronger stimuli. In one visionary and mythico-scientific doctrine of the frontier, windbreaks would even have ameliorated the hostile climate of the Great Plains. Subsidies for planting began as far back as 1865 (Griffith 1976), but the most spectacular symptom of the preoccupation was President Roosevelt's proposal for a 260 km by 3000 km shelterbelt along the 100th meridian (Zon 1935). The plan was uncannily like equally misconstrued projects for trans-Sahelian greenbelts in the 1970s and 1980s described below. Although the Roosevelt plan never materialized, there were, at the maximum in the 1960s, some 65 000 km of shelterbelt on the Great Plains (Fig. 9.11). Frontier symbolism must have been the driving force, for there has been little or no analysis of gross benefits (Riebsame 1987).

Capitalism was another strong element of the prevailing culture. It bound farmers to bankers and to salesmen of machinery, both of whom had agendas quite unrelated to the environment. It also signalled a need to maximize food production for the market, and the market had different and dissonant rhythms to those of the climate. Worster (1979) argued that the Dust Bowl was to be blamed much more on capitalism than drought, wind or soil: farmers, driven by needs for short-term gains, speculated and lost. Their capitalist culture required unfounded optimism about the weather and disregard for long-term stewardship.

North American capitalism changed only in degree, not kind in the later part of the century, for there is still much the same kind of agricultural economy, and as has been seen, wind erosion is still very damaging. The link between prices, ploughing and erosion is striking. In 1973–1974, as in the Dust Bowl, farmers were stimulated by good prices to gamble. They ploughed up 3.6 million new hectares for crops, but again they lost the gamble: there was a winter drought and 20 000 ha of the new ploughings on the Great Plains suffered losses of between 40 and 380 $t\,ha^{-1}\,yr^{-1}$ (Grant 1975). Capitalist economics may even have been winning over the frontier myth in the 1970s, for many of the early windbreaks were dying or being grubbed up (Sorensen and Marotz 1977), though the pattern of loss was very variable. Agricultural sprays, maintenance costs and the needs for bigger fields were among the reasons. Symbolism has in places withered to an indirect stimulus, as in the planting of rows of Christmas trees which was one of the few major forces behind shelter-belt planting in the late 1970s (Griffith 1976).

Science and culture in wind-erosion control

To understand the cultural relativity of North American wind-erosion science, one needs to compare it with that of another advanced scientific culture. The Soviet Union was the other major economy in which wind erosion was a serious problem, for the Russian and central Asian steppes have had a long history of accelerated wind erosion (see page 150 and the discussion about dust). It came to a head after hasty ploughing of 98.4×10^6 ha of dry virgin lands in the Trans-Volga, western Siberia and northern Kazakhstan between 1953 and 1960 (in a vain bid to outdo the North American economy). Massive losses of productive area were the consequence (French 1967). One-third of the agricultural land in the Kustanay Oblast in Kazakhstan was lost to wind erosion between 1954 and 1957, and, large as this area was, it was only one among many other seriously damaged areas.

The Soviet cultural background, however, was rather different to that of North America. Here the approach, though also suffused with modernism, existed in an otherwise rather different political culture. As in America, modernism accorded a high status to science, and demanded huge fields, mechanization and mass production. But Soviet agronomists argued, like Worster, that it was the capitalist search for profit that had brought about the American Dust Bowl; they believed that the communist experience would be different (Vilenskii 1963). Their most spectacular Soviet project (part of 'Stalin's Plan for the Transformation of Nature') would have far outdone Roosevelt's dream of a shelterbelt along the 100th meridian. It would have dissected the steppes and even the deserts with massive windbreaks, for which Lysenko promised trees that would be suitably genetically engineered. In the event the schemes covered little more than a few hundred square kilometres (Fig. 9.12).

The USA and the USSR had very different systems of agricultural extension. In Roosevelt's New Deal and its successors, 'The Great Plains Program' was a package of on-farm subsidies for things like shelterbelts, set-aside, price supports and advice (Riebsame 1987). Farmers weighed the costs and benefits of soil conservation, not always an easy equation in the erratic capitalist market, and when economic rationale conflicted with conservation, as it often did (Held and Clawson 1965), the slack was supposedly taken up by the state. As capitalism softened, more social controls were introduced, in the United States

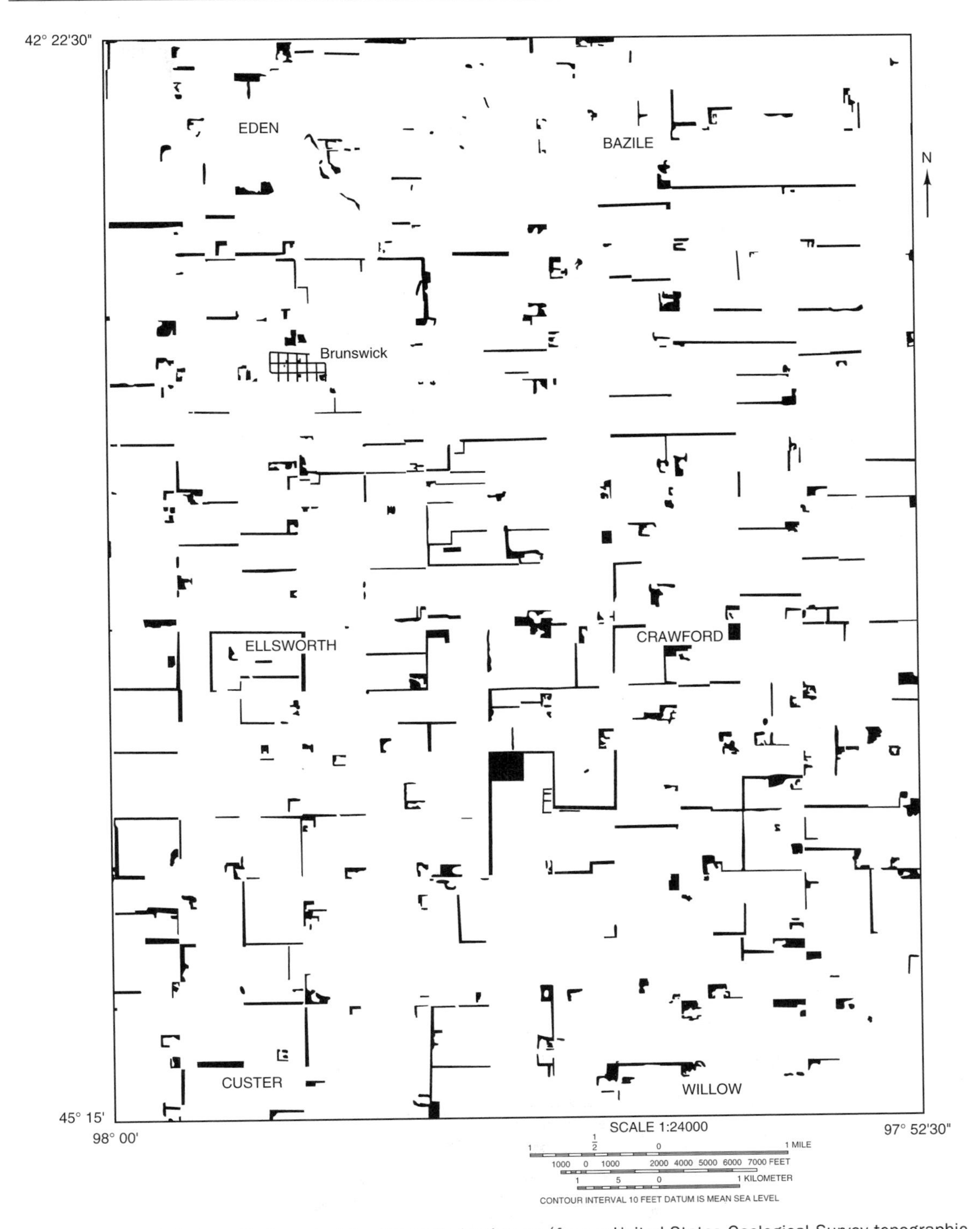

Fig. 9.11 Planted shelterbelts in part of the Nebraska landscape (from a United States Geological Survey topographic map).

Fig. 9.12 Windbreaks on a Soviet experimental farm in the Voronezh region, the prototype of a plan to cover the Union with a system of windbreaks in Stalin's Plan to Control Nature (Anon. 1955).

and in other advanced capitalist economies in dry lands, like Canada and Australia (Loopey 1991). In the Soviet Union, Stalinist thinking persisted under Khrushchev, soil conservation still being a matter of state decree from Moscow, but the ideologically indifferent Brezhnev allowed more local autonomy, more grass leys and less damaging ploughs. Under Gorbachev, there was even talk of fitting techniques to soils (Stebelski 1985).

These huge cultural and technical contrasts highlight the way these different cultures influenced the science itself. The argument that science and ideology were linked was a major part of Soviet ideology, and was forcefully restated by early Soviet soil scientists, who believed that talk of the fragility of the soil was no more than 'the humanity-hating writings of neo-malthusians' (Vilenskii 1963). For them, drought was due simply to mismanagement. Although the idea of cultural relativity in science was generally ridiculed in the West, it is not difficult to see that the WEQ was a product of North American culture. For a start it was implicitly individualist, being applied on a field-by-field basis for farmers who were free agents. The WEPS is also specifically designed to apply at the field scale (significantly termed 'the accounting region' in the WEPS). The difficulties in quantitatively estimating wind erosion rates, discussed above, explain why there have been very few regional surveys in North America. Soviet culture, depending more on centralized decision-making, put more emphasis on Union-wide surveys, despite their dubious value (Zachar 1982; Stebelski 1985).

Wind-erosion control in the context of desertification

Cultural contrasts are even greater between either of these two cultures and those in Africa (where there is actually a much wider range of cultures than in either the USA or the former USSR). In the 1970s and 1980s, the driving phobia in Africa, at least among expatriate scientists, was desertification. The expatriate campaign against desertification, such as it is, has depended heavily on sweeping overviews of damage, a style closer to that of Soviet geographers than to that of North American agronomists. The overviews purported to show huge parts of dry Africa to be suffering from wind erosion (Fig. 9.13; UNEP 1992). But the method suffers the same problems as did its Soviet predecessors, namely a paucity of real data. The African surveys were based only on the judgements of experts (as in the Soviet Union), for there are virtually no scientifically authenticated measurements of wind erosion in Africa. Moreover, the scale of the maps is such that they are of no value on a field-by-field basis, the smallest area diagnosed being about 2000 km^2.

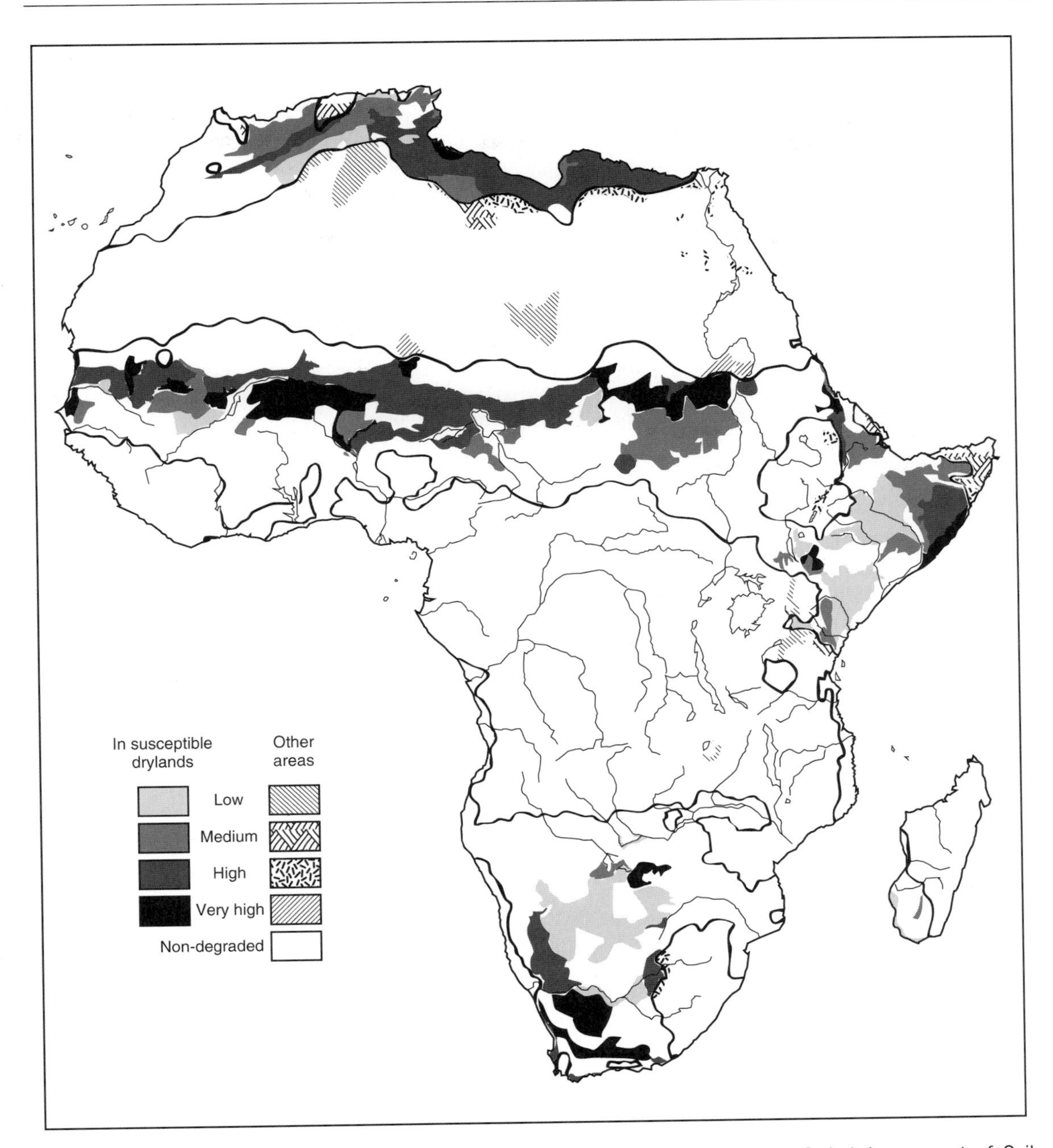

Fig. 9.13 Map from the UNEP *World Atlas of Desertification* (1993) based on FAO's Global Assessment of Soil Degradation (GLASOD), purporting to show the severity of wind erosion in drylands in Africa.

If the existence of wind erosion was being predicated without one reliable measurement, and was being plotted only at grand scales, it is difficult to escape the conclusion that the approach was based on the assumption that African farmers and pastoralists were inherently destructive, and that only extensive intervention could solve the problem. The experts assumed that most of the problem occurred on pastoral land (as can be seen in Fig. 9.13), the assessment being presumably based on the almost universal condemnation of pastoralists in the scientific community (until the last decade). But pastoralists may now have been vindicated by the scientific community, among whom a new scientific paradigm recognizes the opportunist pastoralist style as ideally suited to an erratic environment (Ellis and Swift 1988; Westoby *et al.* 1989; Warren 1995). If the estimates of wind erosion on Fig. 9.13 were adapted to the new paradigm, the map would change drastically, a change that would bring the whole approach into question.

Thus Africa, because poor and dependent on others for scientific advice and environmental aid, is being subjected to inappropriate, even dubious science, largely because it is a product of quite different cultural situations. The market-orientated WEQ/WEPS approach, described above, would be quite as inappropriate as the Soviet survey approach. It could not be applied directly to undercapitalized farmers, with small fields and intensive labour inputs (and without personal computers). Both of these styles are examples of how foreign, especially 'modernist', styles of agricultural science can harm rather than benefit Africa, a case made forcefully by Richards (1985). Misdiagnosis can at best misdirect funds and at worst encourage quite inappropriate ameliorative techniques. Richards believed that only action and understanding (including science) at the grass-roots level could bring about effective conservation.

Dust in a cultural context

Many of the North American culture-specific themes that permeated the history of wind-erosion control on agricultural fields recur in the history of the fight against dust. One that occurs in the wind erosion area, but is better illustrated for dust-control area is the cost of control in a capitalist society. The costs and the benefits of control are rarely experienced by the same people. Farmers (and other producers of dust) would almost always have to incur far greater costs in controlling dust than they could recoup in yields or other forms of return. Townspeople, as in Lubbock, Texas, which is probably the dustiest place in the USA, suffer most of the costs, such as for cleaning and wear and tear on machinery, yet only the farmers (or others) can allay them. The on-farm costs of wind erosion have been estimated to be only about 2 per cent of the off-farm costs, and these off-farm costs may be in the order of $4 billion for the western United States (Piper 1989).

One small consolation is that even dust clouds have silver linings, for most dust is quite alkaline (especially that from roads paved, as many are, with loose crushed limestone), and thus can help to neutralize emissions of acidity (another major environmental problem). Estimates show that alkaline dust is just about able to do this in the western United States, but that it is not up to the task in the eastern part of the country (Gillette *et al.* 1992).

Culture and the control of coastal dunes

Coastal (and semi-arid) dune management illustrates yet another aspect of the application of aeolian geomorphology: the increasing technical and cultural complexity of issues as the scale of concern expands.

The most obvious reason for the need to expand into cultural affairs is that the activity (or inactivity) of coastal dunes has strong cultural controls (though distinction between cultural and natural is seldom easy). The coastal dunes of Gower in South Wales are known to have been reactivated in the fourteenth century when they buried whole villages. Was this the result of a new body of sand finding its way onshore, many years after the last glaciation, from offshore fluvio-glacial deposits? Or was it because of a series of stormy years? Or was it the consequence of intense grazing of the coastal dune grasslands by local cattle (Lees 1982)? The Landes in south-western France are said by some authorities to have become transgressive when activated by medieval deforestation (Lowdermilk 1944), but the trigger might as easily have been the windy period of the fourteenth century, as on Gower. Neither case is fully resolved.

One of the more significant recent episodes in the long history of interference occurred widely on European beaches in World War II. The damage was done either in direct conflict (as in Normandy) or in training exercises (as on Gower). In Jersey, German coastal defences totally isolated the dunes from their sediment supply (Romeril 1989), but not all these works are so well recorded. Yet further major hazards are sand mining, in many other places in addition to the Australian case referred to, and from landfalls for oil or gas pipelines, as in north-east Scotland and Norfolk. With care, both of these activities can be controlled by planning regulations and any damage can be quickly repaired (Ritchie 1989). Other forms of interference may be remote, as on the Israeli coast, where it was thought that the building of the Aswan Dam on the Nile might have curtailed sediment supply to the Nile Delta and so eventually to the beaches of the Levant, many hundreds of kilometres east (and so in turn to the dunes behind them). In this case, fortunately, coastal erosion on the beaches of the delta was found to be maintaining the supply of sand to the down-drift beaches (Rohrlich and Goldsmith 1984). In southern Louisiana, likewise, the artificial restriction of sand supply from the mouth of the Mississippi River has caused erosion of sand dunes (Ritchie and Penland 1988).

Another more recent problem in which the distinction of natural from cultural causes has seldom been fully resolved, concerns recreation on dunes. Since World War II, a recreational 'wave', driven by increasing prosperity, has reached almost all the coastal dunes of Europe and North America. The indigenous sport of coastal dunes, namely golf, can be an ideal means of conservation, but not all recreation is as benign. In some cases, as in south-eastern Australia, where the impact of off-road vehicles (ORVs) has been immense, the damage can be severe and the culprits are obvious (Gilbertson 1981), but they are not always so easily identified. In Britain since the 1960s, private cars have taken the recreational wave to even remote coastal dunes of north-west Scotland, and this has set off a reflected wave of concern among countryside managers. However, when examined with care, many patches of bare sand in these areas appeared to have been denuded of vegetation more by storms than by holidaymakers (Ritchie and Mather 1971). The familiar conflict between over- and under-reaction in applied aeolian geomorphology was again in action.

In addition to the difficulty of distinguishing the cultural from the natural, the interpretation of the induced processes is itself very complex. Most coastal dunes have suffered a long history of settlement, clearance, burning and re-planting, each event adding to the complexity of interpretation. The calcareous 'machair' dune landscape of the Outer Hebrides in Scotland is one instance among many in which coastal dunes were first taken into cultivation 2000 years ago (Angus and Elliott 1992), and here as elsewhere the present landscape still owes many of its features to that event.

The changing culture of science itself has intruded into the problems of managing coastal dunes, as in the management of wind erosion on agricultural fields. Nature conservation theory has been experiencing a subtle but significant paradigm shift since the mid-twentieth century. Earlier in the century, extensive nature reserves and national parks were designated across the world's coastal dunes, and, at first, when the dominant ecological paradigm emphasized succession and stability, sand fixation was prosecuted with renewed vigour. Since the beginning of the 1980s nature conservationists across the world have changed their tune, being carried along now by a new paradigm, based on notions about chance and uncertainty (van Zoest 1992). The claim is that the new paradigm helps to explain why some characteristic dune species have been declining in frequency, for they appear to depend on there being bare sand somewhere in the system. This paradigm switch has major practical implications. If bare sand is considered necessary for the survival of some species, then disturbance needs to be encouraged. The dunes on Gower in South Wales, having survived centuries of abuse by people and nature, have now had to suffer the contrasting repercussions of these two scientific vogues. After military activity of one kind or another during World War II had laid bare great acreages of sand, conservationists built fences to keep out people. When they discovered a decline in the populations of some rare sand-dune species, they opened them up again (M. Hughes, pers. comm.).

Many coastal dunes may be almost wholly artificial, so that in trying to manage them, the manager is perpetuating an unnatural landform. On the Ainsdale coast of Lancashire there may have been few dunes of any size before local landlords began to encourage the planting of marram in the eighteenth century, when a passion for dune control swept northern Europe (Pye and Neal 1993). In Brittany some large coastal dune ridges, artificially created in the nineteenth century, have now become difficult to manage (Barrère 1992). Indeed, marram has now been so widely introduced across the world, that there may be few coastal dunes in a 'natural' state. Many others owe their existence to sand fences built at some, usually forgotten, period in the past. Dune growth can be at such a rate, in many areas, that a dune can look quite 'natural' in as little as ten years (Arens and Wiersma 1994).

Change has been produced in some areas, as on the island of Norderney on the north German coast and in Florida, by 'beach-nourishment', in which sand is fed artificially onto a beach to maintain it, some of it then inevitably being blown onto dunes (Erchinger 1992; Psuty 1993). In the Netherlands in particular, but also in some other countries, it is not only the beach that is nourished, but whole dune systems that are rebuilt artificially, by bulldozer, with sand taken from the beach, other dunes or from offshore (Adriaanse and Coosen 1991).

It is significant that the main schemes of reclamation or protection of coastal dunes in Europe and the Mediterranean, beginning with the Dutch coast and including the Landes and the coast of Palestine (Tear 1925), have all been at the instigation of centralized state institutions. Where it was not the state, it was powerful landowners (sometimes in collusion with the state), as in Lancashire and Denmark. This is because the management of coastal dunes at this scale, like that of wetlands, can only take place if there is radical legislation (as in the Netherlands from the fourteenth century and Denmark from 1539) and a large supply of capital (Jensen 1994). Continental-scale implications of coastal dune management may need even larger-scale political involvement: the European Community is already beginning to finance studies and conservation schemes on coastal dunes at this scale (Udo de Haes and Wolters 1992).

The political context of controlling dunes in semi-arid areas

Where desert and desert-margin dune-fixation occurs in developed countries, like the United States or Australia, the problem is little different from coastal dune management, but in the African, Asian and Latin American semi-arid lands the cultural context is very different. It is the same as that in which the dubiously scientific strategy for wind erosion control is being implemented in dependent, poor countries of Africa (see above). This context presents yet another aspect of the political economy of aeolian geomorphology: the globalization and bureaucratization of science.

In the West African Sahel, as in many similar climates and economies, recent droughts, coupled with locally intense usage, particularly round large settlements, have created moving dunes where none existed before, and these are undoubtedly a serious local problem (for example, Besler and Pfieffer 1992), but the seriousness of the issue in terms either of income lost or of the length of time to recovery is open to much more debate. The salient feature in the political background of dune fixation in less developed countries, as explained above, is desertification. This bandwagon provides most of the funds, nearly all of them from international aid. The planting of dunes is a major 'flagship' activity in this sphere, for it provides one of the few kinds of scheme that are big enough, quickly enough achieved and photogenic enough to give political credibility. But the projects are mostly directed by remote donors or national governments (most with vague notions about the local environment), and few of their plans bear any relation to the needs or knowledge of local people. In one scheme in the Sudan, the largest item of expense was for fences and guards to exclude local people (Tinker 1977). Like many others, this scheme was commandeering land that was already being used for other purposes. Other major programmes in this genre have taken place in Mauritania and Somalia, and there are smaller ones in many other Sahelian countries (Fagotto 1987; Goumandakoye 1990). In China, dune fixation has acquired the same mythical significance as the conquest of the steppes acquired in the Soviet Union, and is managed in much the same fashion. The title of China's case study for the United Nations Conference on Desertification in 1977 was symptomatic: *Tame the Wind, Harness the Sand and Transform the Gobi* (Peoples' Republic of China 1977).

Dune fixation in these environments is being advocated with little analysis of causes, which, as in coastal dunes, are almost always complex and difficult to diagnose. What little research there has been shows the reactivation of this kind of dune to be controlled by partly by drought, and partly by land use (Seevers *et al.* 1975; Ohmori *et al.* 1983), in seldom easily predictable mixtures. In Australia there is debate about whether ancient Aboriginal burning may or may not have been the cause (Conacher 1971). In any event, many of the denuded dunes that so concern visiting experts are probably quite normal parts of the landscape (like many blowouts in coastal dunes), being places where vegetation has failed on account of factors like the height above the water-table. Some active dunes in amongst cultivated land in semi-arid Sudan existed before the severe droughts of the 1970s (let alone the increasing population pressures) and survived unchanged through the drought and increased pressure on the land; they are not, therefore, symptoms of desertification (Ahlcrona 1988). It is true that dunes have been reactivated in many other parts of the Sahel in recent years, but much of the damage is trivial (Mortimore 1989), and in many cases, the economic or social returns from fixation are very dubious. Although here, as elsewhere in applied aeolian geomorphology, there are other kinds of return from this kind of activity, it is the often the remote donors who gain the non-financial benefit, and the peasants who have to bear the real economic costs.

Major shelterbelts, like President Roosevelt's massive scheme for the 100th meridian, are again being proposed by the major donors. They are seen, in a confused way, as bastions against advancing dunes. One such was the south Saharan 'greenbelt', proposed for the 1977 United Nations Conference on Desertification (UNCOD 1977). The idea has been revived in a Japanese scheme to plant a greenbelt across the northern Sahel from Mali to Chad (Rognon 1991). Another such project, in China, is named 'The Great Green Wall'; in the first stage, from 1978 to 1985, 5.3×10^6 ha of windbreak were planted; it is planned to be 5000 km long and cover 8×10^6 ha (Kebin and Kaigo 1989). The Algerians modelled yet another project, known as the 'Green Dam' on the Chinese one, with equally uncertain results (Ballais 1994). These megalomaniac schemes, if they come to fruition, are squandering funds on projects that have small chances of success and, even if successful, do little for most of the people who suffer environmental problems in dry lands. Most of them live far from the greenbelts, and their problems have little to do with an advancing front of dunes.

There are, notwithstanding the widespread misdirection of funds, some dune planting schemes in the less-developed world which are successful. Their common characteristic is that they involve local people and their environmental knowledge and assign them the use of the reclaimed land. These schemes, like the one in the Pushkar Valley in Rajasthan, can have multiple benefits: crop land is no longer threatened; grazing for domestic stock is conserved; fuelwood is eventually made available; and many 'forest-products' like gum and medicinal plants are provided (Fig. 9.8b; Consortium of Indian Sciences for Sustainable Development 1992). It may be that dune fixation is only viable where the rainfall is enough to allow crops on the reclaimed sandy soils.

Conclusion

Applied geomorphologists face many challenges. They must first ensure that their science is of a high standard, and that the technical solutions it suggests will be proof against a very variable and varied environment. But they also need to see themselves and their proposals in a cultural context. Different situations require different techniques, not only in regard to the means available to tackle problems, but also in relation to how the problems are perceived and evaluated and their role in local cultural and economic processes. As the scale of the enquiry expands so will the scale of these complexities.

Further reading

A number of books cover some of the topics discussed in this Chapter. Cooke and Doornkamp (1990) include a Chapter on aeolian hazards in a more general text on applied geomorphology. The books by Cooke *et al.* (1982) and Goudie (1990) are concerned specifically with the drylands, and include contributions on aeolian geomorphology. Papers by Watson (1985) and Kerr and Nigra (1952) give detailed lists of measures for desert areas. Ranwell and Boar (1986) provide a handbook for the management of coastal dune systems.

GLOSSARY

Aarhus trap: a simple, non-rotating sand trap (*qv*) composed of a series of chambers arranged above the surface.

Abrasion: erosion by the wind of cohesive materials (as opposed to deflation *qv*).

Absolute dating: dating in years, generally achieved for aeolian sediments by the use of the ^{14}C or thermoluminescence (*qv*) techniques (*qv* relative dating).

Aeolian geomorphology: the study of the work of the wind in creating landforms.

Aeolianite: a cemented sandstone of aeolian origin. Most aeolianites are rich in carbonate (from whence their cement), and most are associated with coastal sources of sediment, and are thus close to modern or ancient shorelines.

Alios: a term applied in France to the hardened B horizons in podzols developed in stabilized coastal dunes.

Anemometer: a device for measuring the speed of the wind. Aeolian geomorphologists most commonly use cup anemometers (*qv*) or hot-wire anemometers (*qv*).

Angle of initial yield: the angle of the surface of a deposit of granular material (such as dune sand) at which avalanching (*qv*) begins.

Angle of repose (or the residual angle after shearing): the angle at which the surface of a slope made of granular materials comes to rest. On most sand dunes in dry conditions this is between 30° and 33°.

Anti-wetting (or water-repellency): a property acquired by many dune soils in semi-arid and humid areas, whereby they repel water. Anti-wetting is provided by a chemical coating of sand grains, probably derived from plants.

Armoured hot-wire (film) anemometer: an hot-wire anemometer (*qv*) whose active surface has been protected to survive bombardment by saltating sand.

Attrition: the wear of clastic particles as they collide with each other in transport.

Avalanching: used in aeolian geomorphology to refer to the movement of loose sand on slip faces (*qv*).

Bagnold's kink: the inflexion of the wind velocity profile (*qv*) caused by saltation (*qv*).

Barchan dune: a free dune with a crescentic plan-shape in which the crescent opens downwind.

Barchanoid ridge: a barchan-shaped section of a continuous transverse dune.

Barrier island: an elongated island, separated from the coast by a lagoon, generally covered by sand dunes.

Beach-nourishment: the artificial 'feeding' of sand onto a beach for coastal protection.

Bedform climbing: a situation in which each subsequent bedform, such as a migrating transverse dune, is deposited onto remnants of its predecessors, and thus 'climbs' over it.

Bedload: sediment that moves along or in close proximity to or in frequent contact with the bed of flow. Usually taken as the sediment transported by saltation, reptation and creep (*qv*).

Bimodal grain-size pattern: a sediment composed of two main size modes (occurring in some natural dusts and dune sands).

Bimodal wind regime: a wind regime in which winds blow generally from two directions in a yearly (or daily) cycle.

Blow: colloquial English for a sand- and dust-moving event.

Blowout: an elliptical hollow of bare sand eroded into an otherwise vegetated dune landscape.

Bombardment: the effect of grains in saltation (*qv*) as they hit the surface on return.

Boundary layer: the depth of fluid affected by the contact of the fluid with a stationary boundary. The contact creates a velocity gradient (*qv* velocity profile).

Bounding surface: the surface marking the break between the older and newer deposits, often representing an intervening episode of erosion.

Brink: the sharp upper edge of a slip face (*qv*).

Bulk transport: the transport of sand by the rolling over of dunes (as opposed to transport by saltation across intervening surfaces).

^{137}Cs: an isotope of caesium produced exclusively by nuclear bomb testing, which, because maximum production was in the 1960s, can be used as a tracer for measuring rates of soil erosion.

Carolina Bays: rounded lakes found in the Carolinas and nearby states of the USA, thought to have formed by wind action in late Pleistocene times.

Cascading: the process whereby sediment transport increases progressively from the edge of a loose surface, such as an agricultural field.

Chiflones: wind-parallel streams of coarse sand at the ripple scale (of the order of a metre wide and a few metres long).

Clay dunes: dunes (such as lunettes *qv*) formed of clay, which has travelled to the dunes as pellets (*qv*).

Cliffing: the erosion by the sea of the seaward margins of coastal dune ridges.

Cliff-top dune: an anchored dune formed in the calm zones created by persistent eddies at the sharp crest of a hill or cliff.

Climbing dune: an anchored dune climbing up the windward face of a hill.

Complex dune: a dune on which one fundamental dune type is superimposed on another different type.

Compound dune: a dune on which dune elements of one type are superimposed on larger forms of the same type.

Cooper-Thom model: a hypothesis which suggests that coastal erosion at the time of rising sea levels releases sand which creates dunefields.

Coulisses: closely spaced rows of crop stalks, designed as a defence against wind erosion.

Coversands: layers of aeolian sand, with few well-developed dune forms. The term is used principally for now-stabilized accumulations of sand in north-western Europe.

Creep: the slow movement of grains of sand in contact with the surface under the influence of bombardment (*qv*).

Crest: the summit of a dune.

Crust: a hardened surface layer (generally less than a centimetre thick) on the surface of a soil.

Cup anemometer: an anemometer (*qv*) in which the speed of the wind is measured by the rotation of arms on the end of which are cups to catch the wind.

Deflation: the removal by the wind of loose clastic particles.

Desert pavement: a pebbly covering of a finer soil, which may be formed partly in some circumstances by the deflation of finer material.

Desert varnish: a coating rich in iron and manganese covering most stable rock surfaces in deserts (and some other locations). The iron and manganese may often be deposited as dust.

Drag: forces exerted by the wind on the bed, usually resolved into: 'form drag' created by the difference in pressure between the upwind and downwind sides of a particle; and 'surface drag' created by shear across the surface.

Drift potential (DP): the potential amount of sand that can be transported at a site (derived from wind data using a sand transport equation).

Dune: a depositional mound of sand-sized particles, sometimes pellets (*qv*) or snow, created by grain-by-grain movement.

Dunefield: a collection of dunes covering between about 0.5 and 30 000 km^2.

Dune network: a system of overlapping transverse dune ridges, each adjusted to winds from a different direction.

Dust Bowl: parts of western Kansas and nearby states of the USA which suffered very heavy wind erosion in the 1930s.

Dust devil: a narrow circular pattern of rising concentrations of dust, created by thermal and/or wind-shear effects.

Dust plume: a narrow concentration of dust flowing parallel to the surface.

Dynamic (or impact) threshold: the threshold (*qv*) needed to maintain movement in already mobile sediment.

Echo dune: a dune upwind of an escarpment and 'echoing' its shape, created by the pattern of flow against the escarpment.

Entrainment: the lifting of sand or dust into the wind.

Eolation: a term used by Keyes to refer to the wind-erosion of large areas, producing level plains.

Episodically active dune: a dune on which sand movement may be at a low or negligible level for long periods, but on which the indicators of lengthy inactivity are also absent.

Erg: a sand sea (*qv*).

Erodibility: the degree to which a surface can be eroded (depending on its properties, such as its grain size, cohesion, roughness, etc.).

Erosivity: the power of an agent such as the wind to erode a surface (depending on its velocity, turbulence, viscosity, etc.).

Falling dune: an anchored dune on the downwind side of a hill formed as sand is driven over the hill into calmer areas in the lee.

Flow separation: occurs when flow streamlines leave the boundary surface (separate) creating a closed cell.

Foredune: the main and outer ridge of a coastal dune system, parallel to the beach.

Foresets: high-angle deposits dipping downwind, typical of dune slip faces (*qv*).

Froude number: $Fr = u_*/(gh)^{1/2}$, where h can be defined as the height of an obstacle and u_* is the friction velocity.

Grainfall deposition: deposition in the lee of the dune crest as a result of a rapid decrease in the transporting capacity of the wind (also termed *sedimentation*).

Grain shape: grain shape can be resolved into sphericity, roundness, and surface texture (*qv*).

Haboob: a cloud of dust raised by downdraft systems associated with thunderstorm squalls.

Harmattan: a winter wind in West Africa, which transports large quantities of dust south-westward from the Sahara.

Hot-wire (film) anemometer: an anemometer (*qv*) in which the speed of the wind is measured by the decrease in temperature of a heated probe.

Katabatic wind: a wind induced by topographic differences in temperature and pressure between mountain massifs and plains. Katabatic winds, which blow from the mountains to the plains, are generally stronger than anabatic winds, which blow in the opposite direction.

Kosa: outbreaks of dust originating in the Chinese deserts which move out over Japan and the northern Pacific.

Lag deposit: a covering of coarse particles on the surface left by the removal of fines by wind of surface wash (*qv* desert pavement).

Laminar flow: (*qv* turbulent flow) flow in which there is no significant mixing between sub-layers of the fluid.

Landsat: satellite data resolved at about 20 m (MSS), in several wavebands, used for interpreting earth-resource information.

Law of the Wall: the proposition that the velocity distribution with height of flow over a smooth, level surface can be described as a straight line on a velocity : log height graph (*qv* velocity profile).

Leatherman trap: a sand trap (*qv*) which rotates with the wind, but in which sand is not separated according to height of transport.

Lee dune: an anchored dune trailing downwind from a fixed obstacle. Lee dunes can occur singly or in pairs (and occasionally in threes).

Lift: the aerodynamic upward force exerted on loose particles by the effect of low pressure created by acceleration of flow near the bed or over protrusions from the bed.

Linear dune: a dune on which length greatly exceeds width, and where net sand transport is parallel to the dune crest.

Linguoid: a term derived from the study of ripples, referring to projections of transverse ridges in the windward direction.

Lithification: the rendering of a loose sediment into hard, cemented rock.

Loess: a deposit originating as wind-blown dust, which has since been lightly lithified (or changed in other ways, as by redistribution by surface wash or streams).

Loessite: ancient, highly lithified loess (*qv*).

Logarithmic profile: the velocity : log height profile of the wind above a smooth level surface, this being a straight line (*qv* velocity profile).

Log-hyperbolic transformation: a transformation of grain-size data requiring both grain-size and grain-frequency scales to be transformed logarithmically, which produces a hyperbolic curve when plotted.

Lunette: a crescentic dune on the downwind side of a pan (*qv*), formed in places of sand, but more frequently of silt and clay. The crescent shape, which is formed largely by wave action in an ephemeral lake, opens into the wind.

Machair: shelly dune sands deposits with irregular topography found on the north-western coasts of Scotland and Ireland.

Magnetic susceptibility: the ability of a sediment to accept magnetic signals, a property which can used to relate it to its source and pedogenic processes.

Managed retreat: a policy that accepts that coastal erosion (into coastal dunes and other coastal features) will occur, for whatever reason, and plans for the consequences.

Marram (*Ammophila arenaria*): a grass native to north-western European dunes, now widely introduced to stabilize coastal sand dunes. It can withstand a certain degree of burial by aeolian sand and salinity.

Mega-ripple: a ripple with a wavelength of a metre or more, and reaching up to 0.3 m in height, generally formed of coarse sand.

Mega-yardang: a yardang in a system in which the 'wavelength' or mean distance between yardangs is of the order of 1 km.

Meteosat: a geostationary satellite collecting coarsely resolved data mostly for meteorological forecasting.

Milankovitch rhythms: changes in the solar input to the Earth's surface produced by variations in its solar orbital eccentricity, and precession and its axial obliquity.

Modernism: (very briefly) the belief in the inevitability of progress and in science as its principal instrument.

Monsoons: wind and weather systems forced by differences in temperature and pressure between major continental land masses and the oceans.

Morphodynamic: the interaction of form and process.

Multimodal wind regime: a wind regime in which winds blow from many directions in a yearly cycle.

Nabkha: *qv* vegetated sand mound.

Oblique dune: a dune ridge that is apparently oblique to the prevailing or resultant direction of sand movement.

Palaeosol: a soil formed at some period in the past, and now not actively being formed, often being buried by later deposits.

Pan: a shallow, ephemeral lake, generally with a smooth elliptical outline, especially on the downwind side, and generally with the long axis at right angles to the prevailing wind direction (*qv* lunette).

Parabolic dune: a dune with a 'hairpin' shape (a crescent with elongated parallel arms), the crescent opening into the prevailing wind. Parabolic dunes are associated with the presence of vegetation.

Parna: a term used in Australia to describe clayey loess (*qv*).

Pellet: a sand-sized aggregate of silt or clay particles. Pellets can travel in the wind as sand, most in saltation (*qv*).

Peri-desert loess: loess (*qv*) formed in deserts and deposited on their margins.

Phi scale: a logarithmic grain-size scale such that grain size in phi (ϕ) $= -\log_2 d$, where d is the grain size in millimetres. Larger grain sizes are represented by smaller phi values.

Photometer: device for measuring light extinction (as by dust).

Pisoliths: rounded concretions in soils, cemented by various substances such as iron or calcium carbonate.

Pivot point: the point on a slip face where there is the maximum deposition rate of sand, and at which, therefore, failure begins.

Playa: an ephemeral, shallow lake in semi-arid or arid areas, often, but not always saline.

Plume: a narrow concentration (often of dust) elongated in the wind direction.

Primary dunes: dunes close to the sea in coastal dune systems.

Proximal source: the immediate source of a sediment (as opposed to its origin in the distant past).

Relative dating: dating of a deposit by its stratigraphic position (relative to overlying or underlying material) (*qv* absolute dating).

Reptation: the movement of grains of sand thrown up by a descending saltating grain.

Residual angle after shearing: *qv* angle of repose.

Resultant direction of movement: the direction of movement of sand or dust during a full annual cycle.

Resultant drift direction (RDD): The compass direction of the RDP (*qv*).

Resultant drift potential (RDP): the quantified (though generally relative) potential for sand movement in the resultant direction.

Reynolds number: an empirical expression for distinguishing between laminar (*qv*) and turbulent flow (*qv*).

Ripples: low ridges of sand, generally less than 0.01 m high, mostly aligned at right angles to the wind, and repeated over large areas of bare, loose sand.

Roughness length (z_0): the depth of the zone of near zero-velocity flow close to a surface across which the wind is blowing. The roughness length is a function of the roughness of the surface.

Roundness: the angularity of the edges of sedimentary particles (such as sand).

Saltation: the movement of grains under the influence of wind shear, in which grains rise a few centimetres into the wind, are propelled forward by it and descend again to the surface after a path of many centimetres.

Sand fences: fences designed to accumulate sand and prevent it burying valuable installations.

Sandflow cross-stratum: a single lamina deposited due to the avalanching (*qv*) of a slip face (*qv*) (also termed *encroachment deposition*).

Sand (or dust) rose: a diagram showing the movement of sand (or dust) from different directions in an annual cycle, generally derived from long-term wind data converted into sand movement using a standard formula.

Sand sea: a collection of dunes covering an area of over 30 000 km^2.

Sand sheet: an expanse of sand with few or very low dune forms.

Sand trap: a device used to trap sand and thereby measure the amount of sand being transported by the wind.

Sea breezes: wind systems on coasts. The onshore wind, blowing during the latter part of the day, is generally stronger than the offshore wind at night.

Secondary dunes: generally stabilized dunes, far from the sea in coastal dune systems.

Seif (sayf): a sinuous, sharp-crested, active linear dune (*qv*).

Separation bubble: the zone in the lee of a transverse dune in which there is return flow towards the dune on the surface, but onward flow above (*qv* flow separation).

Serir: a North African term for a surface with a desert- or stone-pavement.

Shamal: the strong summer wind that blows from the north-west down the Arabian Gulf, carrying large quantities of dust.

Shear stress: shear occurs when one body (such as the air) slides over another (such as the ground). The shear force per unit area (in $N\,m^{-2}$) is termed τ_0 ('tau-zero').

Shear velocity (u_*) ('u star'): $u_* = (\tau_0/\rho_a)$ where ρ_a ('rho-a') is the density of the air (in $kg\,m^{-3}$). Said to be related to the velocity profile (*qv*) such that $u_z/u_* = 1/\kappa \ln z/z_0$ where u_z is wind velocity at height z; and κ ('kappa') is Kármán's constant, usually taken as 0.4.

Shelterbelt: *qv* windbreak.

Slip face: the slope to the lee of the dune crest on which sand is avalanching (*qv*) to maintain the angle of repose (*qv*).

Sphericity: the nearness of the shape of a clastic particle to that of a sphere.

Splash effect: by analogy with splashing water, the throwing up of loose grains by the impact of a returning saltating grain.

Squall line: a moving line of intense thunderstorms typical of places like Arizona and the Sahel in summer. The squalls raise large quantities of dust.

Stabilised dune: a dune inactivated because of the growth of vegetation or the development of a cohesive soil.

Star dune: a large pyramidal or dome-like dune.

Static (or fluid) threshold: the threshold (*qv*) of initiation of movement in a surface of particles.

Stone pavement: *qv* desert pavement.

Stoss slope: the windward slope of a dune (or other feature).

Surface texture: the microscopic form of clastic particles.

Suspension: the movement of particles by suspension in air (or water). Suspension in air generally occurs only with silts and clays, rather than sands. Where flow is turbulent (as it generally is), the suspended particles follow the turbulent paths.

Thermoluminescence (TL) dating: the absolute dating (*qv*) of sediments in which the release of electrons under the influence of heat or light is measured. Electrons are stored up in sedimentary particles at a steady time-dependent rate once they are covered and protected from the light (or heat).

Threshold (or critical) shear stress τ, threshold velocity (u_t) and threshold shear velocity (u_{*t}): stress parameters measuring the initiation or maintenance of movement in sedimentary particles (being mostly a function of their size).

Toe: the base of the windward slope of a dune.

Tractional deposition: deposition of grains on rippled surfaces in the lee of ripples or other grains (also termed *accretion deposits*).

Transgressive dunes: coastal dunes which migrate inland unrestrained by vegetation, occurring generally either where there is little vegetation (as on arid coasts) or where there is a superabundance of sand.

Transverse dune: dune ridges with crests transverse to the dominant wind, which migrate, for the most part, in the direction of the dominant wind.

Turbulent flow: (*qv* laminar flow) flow in which there is significant mixing between sub-layers of the fluid as a result of turbulent eddies (indicated in air by Reynolds number (*qv*) values >6000). The flow comprises eddies, which, although making overall downwind progress, also have a complex internal motion.

Unimodal sand: sand whose size distribution has a strongly developed, single mode.

Unimodal wind regime: a wind regime in which winds blow largely from one direction throughout a yearly cycle.

United States Geological Survey (USGS) sand trap: a large sand trap (*qv*) which rotates with the wind and which differentiates sand according to height of travel.

Vegetated sand mound: a dune deposited around a core of vegetation, sometimes known as a nabkha (*qv*) or nebkha.

Velocity profile: the change in velocity with height above the surface (*qv* Law of the Wall; logarithmic profile).

Ventifact: a loose stone or boulder whose surface has been abraded by the wind.

Viscosity: the capacity of a fluid to resist changes of shape. Perfectly viscous materials (fluids) move at any level of applied stress. Water is 100 times more viscous than air.

Windbreak: a line (or multiple lines) of trees and bushes designed to break the force of the wind in an attempt to control wind erosion (or other ill-effects of the wind).

Wind Erosion Equation (WEQ): a system for predicting wind erosion on agricultural fields developed by the United States Department of Agriculture.

Wind Erosion Prediction System (WEPS): a new system for predicting wind erosion on agricultural fields, due to replace the WEQ and promised for preliminary release in 1995.

Wind rose: a circular diagram showing the directions and strengths of winds in the mean annual cycle at a station, the basis of a sand rose (*qv*).

Yardang: a mound composed of generally cohesive sediment, whose surface has been abraded (and perhaps deflated) by the wind into an aerodynamic shape.

Zeugen: a yardang with a hard cap-rock.

Zibar: a dune, without a slip face, formed of coarse sand. Most zibar are transverse to the prevailing wind, and most are less than a few metres high.

REFERENCES

References to papers or chapters in edited collections make cross-reference to the editor(s).

Abrahams, A. D. and Parsons, A. J. (eds) 1944. *Geomorphology of desert environments*, Chapman and Hall, London.

Achtnich, W. and Homeyer, B. 1980. Protective measures against desertification in oasis farming, as demonstrated by the example of the oasis Al Hasa, Saudi Arabia, in Meckelein (1980) 93–105.

Adetunji, J. and Ong, C. K. 1980. Quantitative analysis of the Harmattan haze by X-ray diffraction, *Atmospheric Environment*, **14**, 857–858.

Adriaanse, L. A. and Coosen, J. 1991. Beach and dune nourishment and environmental aspects, *Coastal Engineering*, **16**, 129–146.

Agnew, C. T. 1988. Soil hydrology in the Wahiba Sands, in Dutton (1988) 191–200.

Ahlbrandt, T. S. 1979. Textural parameters of eolian deposits, in McKee (1979a) 21–51.

Ahlbrandt, T. S. and Fryberger, S. G. 1980. Eolian deposits in the Nebraska Sand Hills, *USGS Professional Paper*, **1120A**, 1–24.

Ahlbrandt, T. S., Swinehart, J. B. and Maroney, D. G. 1983. The dynamic Holocene dune fields of the Great Plains and Rocky Mountain Basins, in Brookfield and Ahlbrandt (1983) 379–406.

Ahlcrona, E. 1988. The impact of climate and man on land transformation in central Sudan. Applications of remote Sensing, *Meddelanden fran Lunds Universitetets Geografiska Institutioner, Avhandlinger*, **103**.

Akiner, S., Cooke, R. U. and French, R. A. 1992. Salt damage to islamic monuments in Uzbekistan, *Geographical Journal*, **158**, 257–272.

Albritton, C. C. Brooks, J. E., Issawi, B. and Swedan, A. 1990. Origin of the Qattara Depression, Egypt, *Bulletin of the Geological Society of America*, **102**, 952–960.

Al-Hinai, K. G. 1988. *Quaternary aeolian sand mapping in Saudi Arabia, using remotely-sensed imagery*, PhD thesis, Imperial College, University of London.

Al-Hinai, K. G., Moore, J. M. and Bush, P. R. 1987. Landsat image enhancement study of possible submerged sand dunes in the Arabian Gulf, *International Journal of Remote Sensing*, **2**, 251–258.

Alimen, M.-H., Faure, H., Chevaillon, M., Taleb, M. and Battistini, R. 1969. Les études françaises sur le Quaternaire de l'Afrique, *Supplement au Bulletin de l'Association français pour l'étude du Quaternaire*, 201–214.

Al-Janabi, K. Z., Jawad Ali, A., Al-Taie, F. H. and Jack, F. J. 1988. Origin and nature of sand dunes in the alluvial plain of southern Iraq, *Journal of Arid Environments*, **14**, 27–34.

Allen, J. R. L. 1968. The nature and origin of bedform hierarchies, *Sedimentology*, **10**, 161–182.

Allen, J. R. L. 1969. The maximum slope angle attainable by surfaces underlain by bulked equal spheroids with variable dimensional ordering, *Bulletin of the Geological Society of America*, **80**, 1923–1930.

Allen, J. R. L. 1970. The avalanching of granular solids on dunes and similar slopes, *Journal of Geology*, **78**, 326–351.

Allen, J. R. L. 1971. Intensity of deposition from avalanches and loose packing of avalanche deposits, *Sedimentology*, **18**, 105–111.

Allen, J. R. L. 1974. Reaction, relaxation and lag in natural sedimentary systems: general principles, examples and lessons, *Earth Science Reviews*, **10**, 263–342.

Allen, J. R. L. 1982. Simple models for the shape and symmetry of tidal sand waves. 1. Statistically stable equilibrium forms, *Marine Geology*, **48**, 31–49.

Allen. J. R. L. 1985. *Principles of physical sedimentology*, Allen and Unwin, London.

Amiel, A. J. 1975. Progressive pedogenesis of eolianite sandstone, *Journal of Sedimentary Petrology*, **45**, 513–519.

Amit, R. and Gerson, R. 1986. The evolution of Holocene reg (gravelly) soils in deserts – an example from the Dead Sea Region, *Catena*, **13**, 59–80.

Anderson, R. S. 1986. Erosion profiles due to particles entrained by wind: application of an eolian sediment transport model, *Bulletin of the Geological Society of America*, **97**, 1270–1278.

Anderson, R. S. 1987a. Eolian sediment transport as a stochastic process: the effects of a fluctuating wind on particle trajectories, *Journal of Geology*, **95**, 497–512.

Anderson, R. S. 1987b. A theoretical model for aeolian impact ripples, *Sedimentology*, **34**, 943–956.

Anderson, R. S. 1988. The pattern of grainfall deposition in the lee of aeolian dunes, *Sedimentology*, **34**, 175–188.

Anderson, R. S. 1989. Saltation of sand: a qualitative review with biological analogy, in Gimmingham *et al.* (1989) 149–165.

Anderson, R. S. and Bunas, K. L. 1993. Grain size segregation and stratigraphy in aeolian ripples modelled with a cellular automaton, *Nature*, **365**, 740–743.

Anderson, R. S. and Haff, P. K. 1988. Simulation of eolian saltation, *Science*, **241**, 820–823.

Anderson, R. S. and Haff, P. K. 1991. Wind modification and bed response during saltation of sand in air, *Acta Mechanica Supplementum*, **1**, 21–52.

Anderson, R. S. and Hallet, B. 1986. Sediment transport by wind: toward a general model, *Bulletin of the Geological Society of America*, **97**, 523–535.

Anderson, R. S., Sørensen, M. and Willetts, B. B. 1991. A review of recent progress in our understanding of aeolian sediment transport, *Acta Mechanica Supplementum*, **1**, 1–20.

Andrews, S. G. 1981. Sedimentology of Great Sand Dunes, Colorado, in Ethridge, F. G. and Flores, R. M. (eds) *Ancient and modern non-marine depositional environments: models for exploration, Special Publication*, **31**, Society for Economic Mineralogists and Palaeontologists, 279–291.

Angus, S. and Elliott, M. M. 1992. Erosion in Scottish machair with particular reference to the Outer Hebrides, in Carter *et al.* (1992) 93–111.

Anonymous 1955. *Views of the Soviet Union*, Foreign Languages Publishing House, Moscow, 271 pp.

Anton, D. and Ince, F. 1986. A study of sand color and maturity in Saudi Arabia, *Zeitschrift für Geomorphologie NF*, **30**, 339–356.

Anton, D. and Vincent, P. 1986. Parabolic dunes of the Jafurah Desert, Eastern province, Saudi Arabia, *Journal of Arid Environments*, **11**, 187–198.

Arens, S. M. 1994. *Aeolian processes in the Dutch foredunes*, Landscape and Environmental Research Group, University of Amsterdam/Technical Advisory Committee for Water Defences, Amsterdam.

Arens, S. M. 1995. Actual and potential aeolian transport rates on a beach in a temperate humid climate, *Geomorphology*, *in press*.

Arens, S. M. and van der Lee, G. E. M. 1995. Saltation traps for the measurement of aeolian transport into the foredunes, *Soil Technology*, *in press*.

Arens, S. M. and Wiersma, J. 1994. The Dutch foredunes: inventory and classification, *Journal of Coastal Research*, **10**, 189–202.

Argabright, S. 1991. Evolution in use and development of the wind erosion equation, *Journal of Soil and Water Conservation*, **46**, 104–105.

Armbrust, D. V., Chepil, W. S. and Siddoway, F. H. 1964. Effects of ridges on erosion of soil by wind, *Proceedings of the Soil Science Society of America*, **28**, 557–560.

Armon, J. W. and McCann, S. B. 1979. Longshore variations in shoreline erosion, Malpeque barrier system, Prince Edward Island, *Canadian Geographer*, **23**, 18–31.

Ash, J. E. and Wasson, R. H. 1983. Vegetation and sand mobility in the Australian desert dunefield, *Zeitschrift für Geomorphologie Supplementband*, **45**, 7–25.

Aufrère, M. L. 1928. L'orientation des dunes et la direction des vents, *Comptes Rendus de l'Academie Scientifique à Paris*, **187**, 833–835.

Azizov, A., Ismailov, I. M., Kadyrov, K. G., Mirzazhanov, K. M. and Toshov, B. R. 1979. Relation between the amount of loamy sand soil removed by wind and soil moisture (in Russian), *Pochvovedeniye*, **4**, 105–107.

Bagnold, R. A. 1937. The size-grading of sand by wind, *Proceedings of the Royal Society of London*, **163A**, 250–264.

Bagnold, R. A. 1941. *The physics of blown sand and desert dunes*, Methuen, London.

Bagnold, R. A. 1953. The surface movement of blown sand in relation to meteorology, *Research Council of Israel Special Publication*, **2**, 89–96.

Bagnold, R. A. 1966. The shearing and dilation of dry sand and the 'singing mechanism', *Proceedings of the Royal Society of London*, **A295**, 219–232.

Bagnold, R. A. and Barndorff-Nielsen, O. E. 1980. The pattern of natural size distributions, *Sedimentology*, **27**, 199–207.

Bailey, S. I. 1907. Sand dunes in the Peruvian desert, *Geographical Journal*, **49**, 53–56.

Baker, V. R. 1978. The Spokane flood controversy and the Martian outflow channels, *Science*, **202**, 1249–1259.

Bakker, Th. W. M., Jungerius, P. D. and Klijn, J. A. (eds) 1990. *Dunes of the European coasts*, *Catena Supplement*, **18**, Catena Verlag, Brockenblick.

Ball, J. 1927. Problems of the Libyan Desert, *Geographical Journal*, **70**, 21–38, 105–128, 209–224.

Ballais, J.-L. 1994. Aeolian activity, desertification and the 'Green Dam' in the Ziban Range, Algeria, in Millington and Pye (1994) 177–198.

Ballantyne, C. K. and Whittington, G. 1987. Niveo-aeolian sand deposits on An Teallach, Wester Ross, Scotland, *Philosophical Transactions of the Royal Society of Edinburgh: Earth Sciences*, **78**, 51–63.

Barbey, C. and Couté, A. 1976. Croûtes à cyanophycères sur les dunes du Sahél mauritanéen, *Bulletin de l'Institut fondemental de l'Afrique Noire*, **A38**, 732–736.

Barenblatt, G. I. and Golitsyn, G. S. 1974. Local structure of mature dust storms, *Journal of Atmospheric Science*, **31**, 1917–1933.

Barndorff-Nielsen, O. E. and Willetts, B. B. (eds) 1991. Aeolian grain transport, *Acta Mechanica Supplementum*, **1/2**.

Barndorff-Nielsen, O. E., Dalsgaard, K., Halgren, C., Kuhlman, G., Møller, J.-T. and Schon, G. 1982. Variations in particle size over a small dune, *Sedimentology*, **29**, 53–65.

Barndorff-Nielsen, O. E., Møller, J.-T., Rasmussen, K. R. and Willetts, B. B. (eds) 1985. *Proceedings of the International Workshop on the Physics of Blown Sand*, Aarhus, 28–31 May 1985, *Memoir*, **8**, Department of Theoretical Statistics, Institute of Mathematics, University of Aarhus.

Barrère, P. 1992. Dynamics and management of the coastal dunes of the Landes, Gascony, France, in Carter *et al.* (1992) 25–31.

Barrett, P. J. 1980. The shape of rock particles, a critical review, *Sedimentology*, **27**, 291–303.

Barry, R. G. and Chorley, R. J. 1992. *Atmosphere, weather and climate*, 6th edn, Methuen, London.

Bauer, B. O., Sherman, D. J. and Wolcott, J. F. 1992. Sources of uncertainty in shear stress and roughness length estimates derived from velocity profiles, *Professional Geographer*, **44**, 453–464.

Beadnell, H. J. L. 1909. *An Egyptian oasis: an account of the oasis of Kharga in the Libyan desert*, John Murray, London.

Beadnell, H. J. L. 1910. The sand dunes of the Libyan Desert, *Geographical Journal*, **35**, 379–395.

Beget, J. E. and Hawkins, D. B. 1989. Influence of orbital parameters on Pleistocene loess deposition in central Alaska, *Nature*, **337**, 151–153.

Belknap, R. L. 1928. Some Greenland sand dunes, *Papers of the Michigan Academy of Science, Arts and Letters*, **10**.

Benbow, M. C. 1990. Tertiary coastal dunes of the Eucla basin, Australia, *Geomorphology*, **3**, 9–29.

Ben Mohamed, A. and Frangi, J.-P. 1986. Results from ground-based monitoring of spectral aerosol optical turbidity and horizontal extinction: some specific characteristics of dusty Sahelian atmospheres, *Journal of Climatology and Applied Meteorology*, **25**, 1807–1815.

Ben Mohamed, A., Frangi, J.-P., Fontan, J. and Druilhet, A. 1992. Spatial and temporal variations in atmospheric turbidity and related parameters in Niger, *Journal of Applied Meteorology*, **13**, 1286–1294.

Bergametti, G., Gomes, L., Coudé-Gaussen, G., Rognon, P. and Coustumer, M. N. 1989a. African dust over Canary Islands: source-regions identification and transport pattern for some summer situations, *Journal of Geophysical Research*, **94** (D212), 14 855–14 864.

Bergametti, G., Gomes, L., Remoudaki, E., Desbois, M., Matin, D. and Buat-Ménard, P. 1989b. Present transport and deposition patterns of African dusts to the northwestern Mediterranean, in Leinen and Sarnthein (1989) 227–252.

Bergametti, G., Chatenet, B., Marticorena, B. and Gomes, L. 1994. Assessing the actual grain-size distributions of desert erodible soils for atmospheric transport modelling, in Desert Research Institute (1994) 11–12.

Berger, A. and Loutre, M. F. 1991. Insolation values for the climate of the last million years, *Quaternary Science Reviews*, **10**, 297–317.

Besler, H. 1980. Die Dünen-Namib: Entstehung und Dynamik eines Ergs, *Stuttgarter Geographische Studien*, **96**.

Besler, H. 1982. The north-eastern Rub' al Khāli within the borders of the United Arab Emirates, *Zeitschrift für Geomorphologie NF*, **26**, 495–505.

Besler, H. and Pfeiffer, L. 1992. Sand encroachment at Oursi (Burkina Faso): reactivation of fixed draa or deflation of Mare sands? *Zeitschrift für Geomorphologie Supplementband*, **91**, 185–195.

Betzer, P. R., Cardes, K. L., Duce, R. A., Merrill, J. T., Tindale, N. W., Uematsu, K., Costello, D. K., Young, R. W., Feley, R. A., Breland, J. A., Bernstein, R. E. and Greco, A. M. 1988. Long-range transport of giant mineral aerosol particles, *Nature*, **336**, 568–571.

Bigarella, J. J. 1965. Sand-ridge structures from Parana Coastal Plain, *Marine Geology*, **3**, 269–278.

Bigarella, J. J. 1979. Ancient sandstones considered to be eolian, in McKee (1979a) 187–240.

Bigarella, J. J. and Salamuni, R. 1961. Early Mesozoic wind patterns as suggested by dune bedding in the Botucatu sandstone of Brazil and Uruguay, *Bulletin of the Geological Society of America*, **72**, 1089–1106.

Bigarella, J. J., Becker, R. D. and Duarte, G. M. 1969. Coastal dune structures from Paraná (Brazil), *Marine Geology*, **7**, 5–55.

Binda, P. L. and Hildred, P. R. 1973. Bimodal grain-size distributions of some Kalahari-type sands from Zambia, *Sedimentary Geology*, **10**, 233–237.

Bird, E. C. F. 1974. Dune stability on Fraser Island, *Queensland Naturalist*, **21**, 15–21.

Bird, E. C. F. (ed.) 1985. *Coastline changes: a global view*, John Wiley and Son, Chichester.

Bird, E. C. F. 1990. Classification of European dune coasts, in Bakker *et al.* (1990) 15–23.

Black, R. F. and Barksdale, W. L. 1949. Oriented lakes of northern Alaska, *Journal of Geology*, **33**, 134–140.

Blackwelder, E. 1928. The origin of desert basins of southwestern USA (abstract), *Bulletin of the Geological Society of America*, **39**, 262–263.

Blackwelder, E. 1934. Yardangs, *Bulletin of the Geological Society of America*, **24**, 159–166.

Blake, W. P. 1855. On the grooving and polishing of hard rocks and minerals by dry sand, *American Journal of Science*, **20**, 178–179.

Blanchard, W. O. 1926. The Landes: reclaimed waste lands of France, *Economic Geography*, **2**, 249–272.

Blanford, W. T. 1876. On the physical geography of the Great Indian Desert with especial reference to the former existence of the sea in the Indus Valley; and on the origin and mode of formation of the sand hills, *Journal of the Asiatic Society of Bengal (Calcutta)*, **45**, 86–103.

Blatt, H. 1987. Oxygen isotopes and the origin of quartz, *Journal of Sedimentary Petrology*, **57**, 373–377.

Blount, G. and Lancaster, N. 1990. Development of the Gran Desierto sand sea, northwestern Mexico, *Geology*, **18**, 724–728.

Borówka, R. K. 1990. The Holocene development and present morphology of the Łeba dunes, Baltic coast of Poland, in Nordstrom *et al.* (1990) 289–314.

Borsy, Z. 1993. Blown sand territories in Hungary, *Zeitschrift für Geomorphologie Supplementband*, **90**, 1–14.

Böse, M. 1993. A palaeoclimatalogical interpretation of frost wedge casts and aeolian sand deposits in the lowlands between Rhine and Vistula in the Upper Peneglacial and Late Glacial, *Zeitschrift für Geomorphologie Supplementband*, **90**, 15–28.

Bosworth, T. 1922. *Geology of the Tertiary and Quaternary periods in the northwest parts of Peru*, Macmillan, London.

Bourcart, J. 1928. L'Action du vent á la surface de la terre, *Revue de Géographie physique et Géologie dynamique*, **1**, 26–54.

Bourman, R. P. 1986. Aeolian sand transport along beaches, *Australian Geographer*, **17**, 30–35.

Bowden, A. R. 1983. Relict terrestrial dunes: legacies of a former climate in coastal northeastern Tasmania, *Zeitschrift für Geomorphologie Supplementband*, **45**, 153–174.

Bowden, L. W., Huning, J. R., Hutchinson, C. F. and Johnson, C. W. 1974. Satellite photograph presents first comprehensive view of local wind: the Santa Ana, *Science*, **184**, 1077–1078.

Bowler, J. M. 1968. Australian landform example: lunette, *Australian Geographer*, **10**, 402–404.

Bowler, J. M. 1973. Clay dunes: their occurrence, formation and environmental significance, *Earth Science Reviews*, **9**, 315–338.

Bowler, J. M. 1980. Quaternary chronology and palaeohydrology in the evolution of the Mallee landscapes, in Storrier and Stannard (1980) 17–36.

Bowler, J. M. 1983. Lunettes as indices of hydrologic change: a review of Australian evidence, *Proceedings of the Royal Society of Victoria*, **95**, 147–168.

Bowler, J. M. 1986. Spatial variability and hydrologic evolution of Australian lake basins: analogue for Pleistocene hydrologic change and evaporite formation, *Palaeogeography, Palaeoclimatology, Palaeoecology*, **54**, 21–41.

Bowler, J. M. and Magee, J. W. 1978. Geomorphology of the Mallee region in semi-arid northern Victoria and western New South Wales, *Proceedings of the Royal Society of Victoria*, **90**, 5–21.

Bowler, J. M. and Wasson, R. J. 1984. Glacial age environments of inland Australia, in Late Cainozoic palaeoclimates of the southern hemisphere, in Vogel (1984) 183–208.

Bowman, D. 1982. Iron coating in a recent terrace sequence under extremely arid conditions, *Catena*, **9**, 353–360.

Brazel, A. J. 1989. Dust and climate in the American Southwest, in Leinen and Sarnthein (1989) 65–95.

Brazel, A. J. and Nickling, W. G. 1986. The relationship of weather types to dust storm generation in Arizona (1955–1980), *Journal of Climatology*, **6**, 255–275.

Breed, C. S. and Grow, T. 1979. Morphology and distribution of dunes in sand seas observed by remote sensing, in McKee (1979a) 253–303.

Breed, C. S. and McCauley, J. F. 1986. Use of dust storm observations on satellite images to identify areas vulnerable to severe wind erosion, *Climatic Change*, **9**, 243–251.

Breed, C. S., Fryberger, S. C., Andrews, S., McCauley, C., Lennartz, F., Gebel, D. and Horstman, K. 1979a. Regional studies of sand seas using LANDSAT (ERTS) imagery, in McKee (1979a) 305–398.

Breed, C. S., Grolier, M. J., and McCauley, J. F. 1979b. Eolian features in the Western Desert of Egypt and some applications to Mars, *Journal of Geophysical Research*, **84**, 8205–8221.

Breed, C. S., Embabi, N. S., El-Etr, H. and Grolier, M. J. 1980. Wind deposits in the Western Desert, *Geographical Journal*, **146**, 88–90.

Breed, C. S., McCauley, J. F. and Grolier, M. J. 1982. Relic drainages, conical hills, and the aeolian veneer in southwest Egypt – applications to Mars, *Journal of Geophysical Research* **87**, 9929–9950.

Breed, C. S., McCauley, J. F. and Davis, P. A. 1987. Sand sheets of the eastern Sahara and ripple blankets on Mars, in Frostick and Reid (1987) 337–360.

Breed, C. S., McCauley, J. F. and Whitney, M. I. 1989. Wind erosion forms, in Thomas (1989a) 284–307.

Breuninger, R. H., Gillette, D. A. and Kihl, R. 1989. Formation of wind-erodible aggregates for salty soils and soils with less than 50% sand composition in natural terrestrial environments, in Leinen and Sarnthein (1989) 31–64.

Bridge, B. J. and Ross, P. J. 1983. Water erosion in vegetated sand dune at Cooloola, southeast Queensland, *Zeitschrift für Geomorphologie Supplementband*, **45**, 227–244.

Briggs, D. J. and France, J. 1982. Mapping soil erosion by wind for regional environmental planning, *Journal of Environmental Management*, **15**, 158–168.

Brimhall, G. H., Lewis, C. J., Ague, J. J., Dietrich, W. E., Hampel, J., Teague, T. and Rix, P. 1988. Metal enrichment in bauxites by deposition of chemically mature aeolian dust, *Nature*, **333**, 819–824.

Brookfield, M. E. 1970. Dune trends and wind regime in central Australia, *Zeitschrift für Geomorphologie Supplementband*, **10**, 121–153.

Brookfield, M. E. 1977. The origin of bounding surfaces in ancient eolian sandstones, *Sedimentology*, **24**, 303–332.

Brookfield, M. E. 1980. Permian intermontane basin sedimentation in southern Scotland, *Sedimentary Geology*, **27**, 167–194.

Brookfield, M. E. and Ahlbrandt, T. S. (eds) 1983. *Eolian sediments and processes*, Developments in Sedimentology, **38**, Elsevier, Amsterdam.

Brugmans, F. 1983. Wind ripples in an active drift sand area in the Netherlands: a preliminary report, *Earth Surface Processes and Landforms*, **8**, 527–534.

Bruins, H. J. 1976. *The origin, nature and stratigraphy of palaeosols in the loessal deposits of the north-west Negev (Netivot, Israel)*, MSc dissertation, Hebrew University of Jerusalem.

Bruins, H. J. and Yaalon, D. H. 1979. Stratigraphy of the Netivot section in the desert loess of the Negev (Israel), *Acta Geologica Academiae Scientiarum Hungaricae*, **22**, 161–169.

Brunt, D. 1937, Natural and artificial clouds, *Quarterly Journal of the Royal Meteorological Society*, **63**, 277–288.

Buat-Ménard, P. and Duce, R. A. 1986. Precipitation scavenging of aerosol particles over remote marine regions, *Nature*, **321**, 508–510.

Buckley, R. C. 1987. The effect of sparse vegetation on the transport of dune sand by wind, *Nature*, **325**, 426–428.

Buckley, R. 1989. Grain-size characteristics of linear dunes in central Australia, *Journal of Arid Environments*, **16**, 23–28.

Bullard, J. E. 1994. *An analysis of the morphological variation of linear sand dunes and of their relationship with environmental parameters in the southwest Kalahari*, Unpublished PhD thesis, University of Sheffield.

Bullard, J. E., Thomas, D. S. G., Livingstone, I. and Wiggs, G. F. S. 1995. Analysis of linear sand dune morphological variability, southwestern Kalahari Desert, *Geomorphology* **11**, 189–203.

Burkinshaw, J. R. and Rust, I. C. 1993. Aeolian dynamics on the windward slope of a reversing transverse dune, Alexandria coastal dunefield, South Africa, in Pye and Lancaster (1993) 13–21.

Burkinshaw, J. R., Illenberger, W. K. and Rust, I. C. 1993. Wind speed profiles over a reversing transverse dune, in Pye (1993a) 25–36.

Buritt, B. and Hyers, A. D. 1981. Evaluation of Arizona's highway dust warning system, *Geological Society of America, Special Paper* 186, 281–292.

Busche, D., Draga, M. and Hagedorn, H. 1984. *Les sables éoliens – modelés et dynamique – la menace éolienne et son contrôle, bibliographie annotée*, Deutsche Gesellschaft für technische Zusammensarbeit (GTZ), T2 Verlagsgesellschaft, mbh, Rossdorf.

Butler, B. E. and Hutton, J. T. 1956. Parna in the Riverine Plain of southeastern Australia and the soils thereon, *Australian Journal of Agricultural Research*, **7**, 536–553.

Butt, C. R. M. 1985. Granite weathering and silcrete formation on the Yilgarn block, Western Australia, *Australian Journal of Earth Sciences*, **32**, 415–432.

Butterfield, G. R. 1991. Grain transport rates in steady and unsteady turbulent airflows, *Acta Mechanica Supplementum*, **1**, 97–122.

Butterfield, G. R. 1993. Sand transport response to fluctuating wind velocity, in Clifford, N. J., French, J. R. and Hardisty, J. (eds) *Turbulence: perspectives on sediment transport*, John Wiley and Sons, Chichester, 305–334.

Butzer, K. W., Stuckenrath, R., Bruzewick, A. J. and Helgren, D. M. 1978. Late Cenozoic palaeoclimates of the Gaop Escarpment, Kalahari margin, South Africa, *Quaternary Research*, **10**, 310–339.

Cahill, T. A., Gill, T. E., Reid, J. S. and Gearhart, E. A. 1995. Saltating particles, playa crusts, and dust aerosols from Owens (dry) Lake, California, *Earth Surface Processes and Landforms*, in press.

Cailleux, A. 1967. Periglacial of McMurdo Strait (Antarctica), *Biuletyn Periglacjalnly*, **17**, 57–90.

Callen, R. A. and Nanson, G. C. 1992. Discussion – Formation and age of dunes in the Lake Eyre depocentres, *Geologische Rundschau*, **81**, 589–593.

Capot-Rey, R. 1945. Dry and humid morphology in the Western Erg, *Geographical Review*, **35**, 391–407.

Capot-Rey, R. 1957. Le vent et le modelé éolien au Borkou, *Traveaux de l'Institut de Recherches sahariennes*, **15**, 155–157.

Capot-Rey, R. 1963. Contribution à l'étude et la représentation des barkanes, *Traveaux de l'Institut de Recherches sahariennes*, **22**, 37–60.

Carlson, T. N. and Prospero, J. M. 1972. Large-scale movement of Saharan air outbreaks over the northern equatorial Atlantic, *Journal of Applied Meteorology*, **11**, 283–297.

Carrigy, M. A. 1970. Experiments on the angles of repose of granular materials, *Sedimentology*, **14**, 147–158.

Carroll, J. J. and Ryan, J. A. 1970. Atmospheric vorticity and dust devil rotation, *Journal of Geophysical Research*, **75**, 5179–5184.

Carruthers, R. A. 1987. Aeolian sedimentation from the Galtymore Formation (Devonian), Ireland, in Frostick and Reid (1987) 251–269.

Carson, C. E. and Hussey, K. M. 1960. Hydrodynamics of three Arctic lakes, *Journal of Geology*, **68**, 585–600.

Carson, C. E. and Hussey, K. M. 1962. The oriented lakes of Alaska, *Journal of Geology*, **70**, 417–439.

Carter, R. W. G. 1988. *Coastal environments: an introduction to the physical, ecological and cultural systems of coastlines*, Academic Press, London.

Carter, R. W. G. 1990. The geomorphology of Irish coastal dunes, in Bakker *et al.* (1990) 31–39.

Carter, R. W. G. 1991. The impact of near future sea level rise on coastal dunes, *Landscape Ecology*, **6**, 29–39.

Carter, R. W. G. and Wilson, P. 1993. Aeolian processes and deposits in northwest Ireland, in Pye (1993a) 173–190.

Carter, R. W. G., Hesp, P. A. and Nordstrom, K. F. 1990. Erosional landforms in coastal dunes, in Nordstrom *et al.* (1990) 217–249.

Carter, R. W. G., Bauer, B. O., Sherman, D. J., Davidson-Arnott, R. G. D., Gares, P. A., Nordstrom, K. F. and Orford, J. D. 1992a. Dune development in the aftermath of stream outlet closure: examples from Ireland and California, in Carter *et al.* (1992b) 57–69.

Carter, R. W. G., Curtis, T. F. G. and Sheehy-Skeffington, M. (eds). 1992b. *Coastal dunes: geomorphology, ecology and management for conservation*, Balkema, Rotterdam.

Castro, I. P. and Wiggs, G. F. S. 1994. Pulsed-wire anemometry on rough surfaces, with application to desert sand dunes, *Journal of Wind Engineering and Industrial Aerodynamics*, **52**, 53–71.

Catt, J. A. 1978. Loess and coversands, in Shotton, F. W. (ed.) *British Quaternary studies: recent advances*, Oxford University Press, Oxford, 221–229.

Catt, J. A. 1979. Distribution of loess in Britain, *Proceedings of the Geologists' Association*, **90**, 93–95.

Chadwick, H. W. and Dalke, P. D. 1965. Plant succession on dune sands in Fremont County, Idaho, *Ecology*, **46**, 765–780.

Chadwick, O. A. and Davis, J. O. 1990. Soil-forming intervals caused by eolian sediment pulses in the Lohontan Basin, northwestern Nevada, *Geology*, **18**, 243–246.

Chakraborty, C. 1992. Morphology, internal structure and mechanisms of small longitudinal (seif) dunes in an aeolian horizon of the Proterozoic Dhandraul Quartzite, India, *Sedimentology*, **40**, 79–86.

Chappell, A. 1995. Modelling the spatial variation of processes in the redistribution of soil: digital terrain models and ^{137}Cs in SW Niger, *Geomorphology, in press.*

Chepil, W. S. 1945. Dynamics of wind erosion: II. Initiation of soil movement, *Soil Science*, **60**, 397–411.

Chepil, W. S. 1957. Width of field strips to control wind erosion, *Technical Bulletin*, **92**, Kansas State College of Agriculture and Applied Science.

Chepil, W. S., Siddoway, F. H. and Armbrust, D. V. 1962. Climatic factor for estimating wind erodibility of farm fields, *Journal of Soil and Water Conservation*, **17**, 162–165.

Chester, R. 1990. The atmospheric transport of clay minerals to the world ocean, in Farmer, V. C. and Tardy, Y. (eds) *Proceedings of the 9th International Clay Conference*, Strasbourg, 1989, *Science Géologiques, Mémoires*, **88**, 23–32.

Chorley, R. J., Schumm, S. A. and Sugden, D. E. 1984. *Geomorphology*, Methuen, London.

Chrintz, T. and Clemmensen, L. B. 1993. Draa reconstruction, the Permian Yellow Sands, northeast England, in Pye and Lancaster (1993) 151–161.

Christiansen, C. and Hartmann, D. 1988. On using the log-hyperbolic distribution to describe the textural characteristic of aeolian sediments – discussion, *Journal of Sedimentary Petrology*, **58**, 159–160.

Clark, S. S. 1975. The effect of sand mining on the coastal heath vegetation in New South Wales, *Proceedings of the Ecological Society of Australia*, **9**, 1–16.

Clarke, M. L., Wintle, A. G., and Lancaster, N. 1995. Infra-red stimulated luminescence dating of sands from the Cronese Basins, Mojave Desert, *Geomorphology, in press.*

Clements, T., Stone, R. O., Mann, J. F. Jr. and Eymann, J. L. 1963. *A study of windborne sand and dust in desert areas, Technical Report*, ES-8, United States Army, Natick Laboratories.

Clemmensen, L. B. 1986. Storm-generated eolian sand shadows and their sedimentary structures, Vejers Strand, Denmark, *Journal of Sedimentary Petrology*, **56**, 520–527.

Clemmensen, L. B. 1987. Complex star dunes and associated aeolian bedforms, Hopeman Sandstone (Permo-Triassic), Moray Firth Basin, Scotland, in Frostick and Reid (1987) 213–231.

Clifton, H. E. 1977. Rain impact ripples, *Journal of Sedimentary Petrology*, **47**, 678–679.

Coffey, G. N. 1909. Clay dunes, *Journal of Geology*, **17**, 754–755.
Conacher, A. J. 1971. The significance of vegetation fire and man in the stabilization of sand dunes near the Warburton Ranges, central Australia, *Earth Science Journal*, **5**, 92–94.
Connally, G. G., Krinsley, D. H. and Sirkin, L. A. 1972. Late Pleistocene erg in the upper Hudson Valley, New York, *Bulletin of the Geological Society of America*, **83**, 1537–1542.
Consortium of Indian Scientists for Sustainable Development c. 1992. *Regeneration of Pushkar (Ajmer) Lake Valley ecosystem. Final technical report*, Sponsored by Ministry of Environment and Forests, Government of India, New Delhi.
Cook, P. J., Colwell, J. B., Firman, J. B., Lindsay, J. M., Schwebel, D. A. and Von Der Borch, C. C. 1977. Late Cainozoic sequence of south east of South Australia and Pleistocene sea level changes, *BMR Journal of Australian Geology and Geophysics*, **2**, 81–88.
Cooke, R. U. 1970. Stone pavements in deserts, *Annals of the Association of American Geographers*, **60**, 560–577.
Cooke, R. U. and Doornkamp, J. C. 1990. *Geomorphology in environmental management: a new introduction*, 2nd edn, Clarendon, Oxford.
Cooke, R. U., Brunsden, D., Doornkamp, J. C. and Jones, D. K. C. 1982. *Urban geomorphology in drylands*, United Nations University, Oxford University Press, London.
Cooke, R. U., Warren, A. and Goudie, A. S. 1993. *Desert Geomorphology*, UCL Press, London.
Cooper, W. S. 1958. Coastal sand dunes of Oregon and Washington, *Geological Society of America Memoir*, **72**.
Coque, R. and Jauzein, A. 1967. The geomorphology and Quaternary Geology of Tunisia, *9th Annual Field Conference of the Petroleum Exploration Society of Libya*, 227–257.
Corbett, I. 1993. The modern and ancient pattern of sandflow through the southern Namib deflation basin, in Pye and Lancaster (1993) 45–60.
Cornish, V. 1897. On the formation of sand dunes, *Geographical Journal*, **9**, 278–309.
Cornish, V. 1908. On the observation of desert sand-dunes, *Geographical Journal*, **31**, 400–402.
Cotton, C. A. 1947. *Climatic accidents in landscape-making*, Whitcombe and Tombs, Christchurch, New Zealand.
Coudé-Gaussen, G. 1985. Présence de grains éolisés de palygorskite dans les poussières actuelles et les sédiments récents d'origine désertique, *Bulletin de la Société géologique de France*, **1**, 571–579.
Coudé-Gaussen, G. 1987. The presaharan loess sedimentological characterisation and palaeoclimatological significance, *GeoJournal*, **15**, 177–183.
Coudé-Gaussen, G. 1990. The loess and loess-like deposits on either side of the western Mediterranean Sea: genetic and paleoclimatic significance, *Quaternary International*, **5**, 1–8.
Coudé-Gaussen, G. 1991. *Les poussières sahariennes: cycle sédimentaire et place dans les environments et paléoenvironments désertiques*, John Libby Eurotext, Montrouge.
Coudé-Gaussen, G., Riser, J. and Rognon, P. 1983. Tri éolien et évolution du matériel dunaire par vanage et fragmentation: l'Erg In Koussamène (Nord Mali), *Comptes rendus de l'Académie des Sciences, Paris*, **II, 296**, 291–296.
Coudé-Gaussen, G., Rognon, P., Rapp, A. and Nilhén, T. 1987. Dating of peridesert loess in Matmata, south Tunisia, by radio-carbon and thermoluminescence methods, *Zeitschrift für Geomorphologie NF*, **31**, 129–144.
Coursin, A. 1964. Observations et expériences fait en avril et mai 1956 sur les barkhanes du Souhel et Abiodh (région est de Port Etiènne), *Institut Française de l'Afrique Noire, Bulletin*, Serie **A26**, 989–1022.
Crocker, R. L. 1946. The Simpson Desert expedition, 1939, scientific reports: 8 – The soils and vegetation of the Simpson Desert and its borders, *Transactions of the Royal Society of South Australia*, **70**, 235–258.
D'Almeida, G. A. 1989. Desert aerosol: characteristics and effects on climate, in Leinen and Sarnthein (1989) 311–337.
D'Almeida, G. A., Jaenicke, R., Roggendorf, P. and Richter, D. 1983. New sunphotometer for network operation, *Applied Optics*, **22**, 3796–3801.
Dan, J. 1990. The effect of dust deposition on the soils of the Land of Israel, *Quaternary International*, **5**, 107–113.
Danin, A. and Ganor, E. 1991. Trapping of airborne dust by mosses in the Negev Desert, Israel, *Earth Surface Processes and Landforms*, **16**, 153–162.
Danin, A. and Gerson, R. 1983. Weathering patterns on hard limestone and dolomite by endolithic lichens and bacteria. Supporting evidence for eolian contribution to terra rossa soil, *Soil Science*, **136**, 213–217.
Dare-Edwards, A. J. 1983. Loessic clays of south Eastern Australia, *Loess Letter Supplement*, **2**.
Dare-Edwards, A. J. 1984. Aeolian clay deposits of south-eastern Australia: parna or loessic clay? *Transactions of the Institute of British Geographers*, **9**, 227–344.
Darwin, C. 1846. An account of the fine dust which often falls on vessels in the Atlantic Ocean, *Quarterly Journal of the Geological Society of London*, **2**, 26–30.
Daveau, S. 1965. Dunes ravinnées et dépôts du Quaternaire récent dans le Sahel mauritanéen, *Revue de Géographie de l'Afrique orientale*, **1**, 7–47.
Davis, W. M. 1905. The geographical cycle in an arid climate, *Journal of Geology*, **13**, 381–407 (reprinted as Davis 1954).
Davis, W. M. 1930. Rock floors in arid and in humid climates, *Journal of Geology*, **38**, 1–27, 136–138.
Davis, W. M. 1954. The geographical cycle in an arid climate, in Johnson, D. W. (ed.) *Geographical essays*, Constable, London, 296–322.
Dawson, A. G. 1992. *Ice Age Earth: Late Quaternary geology and climate*, Routledge, London.
Dayan, U., Hefter, J., Miller, J. and Gutman, G. 1991. Dust intrusion events into the Mediterranean basin, *Journal of Applied Meteorology*, **30**, 1185–1199.
Deacon, S. G., Lancaster, N. and Scott, L. 1984. Evidence for Late Quaternary climatic change in southern Africa, in Vogel (1984) 391–404.
De Félice, P. 1955. Etude de la formation des rides de sable, *Comptes rendus de l'Académie des Sciences, Paris*, **240**, 1253–1255.
DeKimpe, N. M., Dolan, R. and Hayden, B. P. 1991. Predicted dune recession on the outer banks of North Carolina, *Journal of Coastal Research*, **7**, 53–84.
Dekker, L. W. and Jungerius, P. D. 1990. Water repellency in the dunes with special reference to the Netherlands, in Bakker *et al.* (1990) 185–193.

Dekker, L. W. and Ritsema, C. J. 1994. Fingered flow: the creator of sand columns in dune and beach sands, *Earth Surface Processes and Landforms*, **19**, 153–164.

De Lancey Forth, N. B. 1930. More journeys in search of Zerura, *Geographical Journal*, **75**, 49–64.

de Lima, J. L. M. P., van Dijk, P. M. and Spaan, W. P. 1993. Splash-saltation transport under wind-driver rain, *Soil Technology*, **5**, 151–166.

De Ploey, J. 1977. Some experimental data on slopewash and wind action with reference to Quaternary morphogenesis in Belgium, *Earth Surface Processes*, **2**, 101–115.

De Raeve, F. 1989. Sand dune vegetation and management dynamics, in van der Meulen *et al.* (1989) 99–110.

Derbyshire, E. 1983. Origin and characteristics of some Chinese loess at two locations in China, in Brookfield and Ahlbrandt (1983) 69–90.

Desert Research Institute 1994. *Response of eolian processes to global change: abstracts of a workshop*, Desert Studies Center, Zzyzx, CA, USA, 24–29 March 1994, *Occasional Paper*, **2**, Desert Research Institute, Quaternary Sciences Center, University of Nevada.

Ding Zhougli, Rutter, N. and Liu Tungsheng 1992. Pedostratigraphy of Chinese loess deposits and climatic cycles in the last 2.5 Myr, *Catena*, **20**, 73–91.

Diver, C. 1933. The physiography of the South Haven Peninsula, Dorset, *Geographical Journal*, **81**, 404–427.

Dobson, M. 1781. An account of the Harmattan, a singular African wind, *Philosophical Transactions of the Royal Society of London*, **71**, 46–57.

Dodonov, A. E. 1991. Loess of central Asia, *GeoJournal*, **24**, 185–194.

Donkin, R. A. 1981. The 'manna lichen': *Lecanora esculenta*, *Anthropos*, **76**, 562–576.

Doody, J. P. 1989. Conservation and development of the coastal dunes of Great Britain, in van der Meulen *et al.* (1989) 53–67.

Dorn, R. I. and Oberlander, T. M. 1982. Rock varnish, *Progress in Physical Geography*, **6**, 317–367.

Dovland, H. and Eliassen, A. 1976. Dry deposition on a snow surface, *Atmospheric Environment*, **10**, 783–785.

Drees, L. R., Manu, A. and Wilding, L. P. 1993. Characteristics of aeolian dusts in Niger, West Africa, *Geoderma*, **59**, 213–233.

Dulac, F., Tanré, D., Bergametti, G., Buat-Ménard, P., Desbois, M. and Sutton, D. 1992. Assessment of the African airborne dust over the western Mediterranean Sea using Meteosat data, *Journal of Geophysical Research*, **97** (D2), 2489–2506.

Dutton, R. (ed.) 1988. *Scientific results of the Royal Geographical Society's Oman Wahiba Sands Project, Journal of Oman Studies, Special Report*, **3**.

East Anglian Coastal Research Programme 1977. *Beach profiles, form and change*, University of East Anglia, Norwich.

Eden, D. N. 1980. The loess of north-east Essex, *Boreas*, **9**, 1.

Eden, D. N., Wen Qizhong, Hunt, J. L. and Whiton, J. S. 1994. Mineralogical and geochemical trends across the Loess Plateau, North China, *Catena*, **21**, 73–90.

Edgett, K. S. and Lancaster, N. 1993. Volcaniclastic aeolian dunes: terrestrial examples and application to martian sands, *Journal of Arid Environments*, **25**, 271–297.

Edlin, H. L. 1976. The Culbin Sands, in Lenihan, J. and Fletcher, W. W. (eds) *Reclamation*, Blackie, Glasgow, 1–31.

Edwards, M. B. 1979. Late Precambrian loessites from North Norway and Svalbard, *Journal of Sedimentary Petrology*, **49**, 85–91.

Edwards, S. R. 1993. Luminescence dating of sand from the Kelso Dunes, California, in Pye (1993a) 59–68.

El-Baz, F. 1984a. The desert in the space age, in El-Baz (1984b) 1–29.

El-Baz, F. (ed.) 1984b. *Deserts and arid lands*, Martinus Nijhof, The Hague.

El-Baz, F. and Hassan, M. H. A. (eds) 1986. *Physics of desertification*, Martinus Nijhof, The Hague.

Eldridge, F. R. 1980. *Wind machines*, 2nd edn, Van Nostrand Reinhold, New York.

Ellis, J. E. and Swift, D. M. 1988. Stability of African pastoral ecosystems: alternate paradigms and implications for development, *Journal of Range Management*, **41**, 450–459.

Ellwood, B. B. and Howard, J. H. 1981. Magnetic fabric development in an experimentally produced barchan dune, *Journal of Sedimentary Petrology*, **51**, 97–100.

Ellwood, J. M., Evans, P. D. and Wilson, I. G. 1975. Small scale aeolian bedforms, *Journal of Sedimentary Petrology*, **45**, 554–561.

El-Shobokshy, M. S. and Al-Saedi, Y. G. 1993. The impact of the Gulf War on the Arabian environment. I. Particulate pollution and reduction of solar irradiance, *Atmospheric Environment*, **27A**, 95–108.

Embabi, N. S. 1970/71. Structures of barchan dunes at the Kharga Oasis Depression, The Western Desert, Egypt (and a comparison with structures of two aeolian microforms from Saudi Arabia), *Bulletin de la Société de Géographie d'Egypte*, **43/44**, 53–71.

Erchinger, H. F. 1992. Conservation of barrier dunes as a smooth, natural method of coastal protection on the East Friesian Islands, Germany, in Carter *et al.* (1992b) 389–396.

Eriksson, P. G., Nixon, N., Snyman, C. P. and Bothma, J. DuP. 1989. Ellipsoidal parabolic dune patches in the southern Kalahari desert, *Journal of Arid Environments*, **16**, 111–124.

Evans, J. W. 1911. Dreikanter, *Geological Magazine*, **8**, 334–335.

Exner, F. M. 1927. Über Dünen und Sandwellen, *Geografiska Annaler*, **9**, 81–89.

Fagotto, F. 1987. Sand-dune fixation in Somalia, *Environmental Conservation*, **14**, 157–163.

Felix-Henningson, P. 1984. Zur Relief und Bodentwicklung der Goz-Zone Nordkordofans im Sudan, *Zeitschrift für Geomorphologie NF*, **28**, 285–304.

Fenley, J. M. 1948. Sand dune control in Les Landes, France, *Journal of Forestry*, **46**, 514–520.

Fieller, N. R. J., Gilbertson, D. D. and Olbricht, W. 1984. A new method for environmental analysis of particle size distribution data from shoreline sediments, *Nature*, **311**, 648–651.

Filion, L. and Morisset, P. 1983. Eolian landforms along the eastern coast of Hudson Bay, Northern Quebec, *Nordicana*, **47**, 73–94.

Filion, L., Saint-Laurent, D., Desponts, M. and Payette, S. 1991. The Late Holocene record of eolian and fire activity in northwestern Quebec, *The Holocene*, **1**, 201–208.

Finkel, H. J. 1959. The barchans of southern Peru, *Journal of Geology*, **67**, 614–647.

Finnigan, J. J., Raupach, M. R., Bradley, E. F. and Aldis, G. K. 1990. A wind tunnel study of turbulent flow over a two-dimensional ridge, *Boundary-Layer Meteorology*, **50**, 277–317.

Fisher, R. V. and Schmincke, H.-U. 1994. Volcaniclastic sediment transport and deposition, in Pye (1994) 351–388.

Flenley, E. C., Fieller, N. R. J. and Gilbertson, D. D. 1987. The statistical analysis of 'mixed' grain size distributions from aeolian sands in the Libyan Pre-Desert using log skew Laplace models, in Frostick and Reid (1987) 271–280.

Flohn, H. 1969. Local wind systems, in Flohn, H. (ed.) *General climatology*, vol 2. World survey of climatology, Elsevier, Amsterdam, 139–172.

Folk, R. L. 1971a. Genesis of longitudinal and oghurd dunes elucidated by rolling upon grease, *Bulletin of the Geological Society of America*, **82**, 3461–3468.

Folk, R. L. 1971b. Longitudinal dunes of the North-western edge of the Simpson Desert, Northern Territory, Australia. 1. Geomorphology and grain size relationships, *Sedimentology*, **16**, 5–54.

Folk, R. L. 1976. Rollers and ripples in sand, streams and sky: rhythmic alteration of transverse and longitudinal vortices in three orders, *Sedimentology*, **23**, 649–669.

Folk, R. L. 1977. Longitudinal ridges with tuning-fork junctions in the laminated interval of flysch beds: evidence of low-order helical flow in turbidites, *Sedimentary Geology*, **19**, 1–6.

Folk, R. L. and Ward, W. C. 1957. Brazos River bar: a study in the significance of grain size parameters, *Journal of Sedimentary Petrology*, **27**, 3–26.

Food and Agriculture Organization of the United Nations (FAO) 1985. *Sand dune stabilization, shelterbelts and afforestation, FAO Conservation Guide* 10, FAO, Rome.

Fookes, P. G. and Best, R. 1969. Consolidation characteristics of some Late Pleistocene periglacial metastable soils of east Kent, *Quarterly Journal of Engineering Geology*, **2**, 103–128.

Forman, S. L., Goetz, A. F. H. and Yuhas, R. H. 1992. Large-scale stabilized dunes on the High Plains of Colorado: understanding the landscape response to Holocene climates with the aid of images from space, *Geology*, **20**, 145–148.

Frank, A. and Kocurek, G. 1994. Models for airflow velocity profiles in natural settings: accounting for atmospheric convection, and secondary flow over dunes (on both the windward and lee slopes), in Desert Research Institute (1994) 37–38.

Frazee, C. J., Fehrenbacher, J. B. and Krumbein, W. C. 1970. Loess distribution from source, *Proceedings of the Soil Science Society of America*, **34**, 296–301.

French, R. A. 1967. Contemporary landscape change in the USSR, in Lawton, R. W. and Steel, D. W. (eds) *Essays in geography*, University of Liverpool Press, Liverpool, 547–564.

Frère, H. B. E. 1870. On the Ran of Cutch and neighbouring regions, *Journal of the Royal Geographical Society*, **40**, 181–207.

Friedman, G. M. 1961. Distinction between dune beach and river sands from textural characteristics, *Journal of Sedimentary Petrology*, **31**, 514–529.

Froidefond, J. M. and Prud'homme, R. 1991. Coastal erosion and aeolian sand transport on the Aquitiane coast, *Acta Mechanica Supplementum*, **2**, 147–160.

Frostick, L. E. and Reid, I. (eds) 1987. *Desert sediments, ancient and modern, Special Publication*, **35**, Geological Society of London, Blackwell, Oxford.

Fryberger, S. G. 1979. Dune forms and wind regime, in McKee (1979a) 305–397.

Fryberger, S. G. and Ahlbrandt, T. S. 1979. Mechanisms for the formation of eolian sand seas, *Zeitschrift für Geomorphologie NF*, **23**, 440–460.

Fryberger, S. and Goudie, A. S. 1981. Arid geomorphology, *Progress in Physical Geography*, **5**, 420–428.

Fryberger, S. G. and Schenk, C. 1981. Wind sedimentation tunnel experiments on the origins of aeolian strata, *Sedimentology*, **28**, 805–821.

Fryberger, S. G. and Schenk, C. 1988. Pin stripe lamination: a distinctive feature of modern and ancient eolian sediments, *Sedimentary Geology*, **55**, 1–16.

Fryberger, S. G., Ahlbrandt, T. S. and Andrews, S. 1979. Origin, sedimentary features and significance of low-angle eolian 'sand sheet' deposits, Great Sand Dunes National Monument and vicinity, Colorado, *Journal of Sedimentary Petrology*, **49**, 733–746.

Fryberger, S. G., Al-Sari, A. M., Clisham, T. J., Rizvi, S. A. R. and Al-Hinai, K. G. 1984. Wind sedimentation in the Jafurah sand sea, Saudi Arabia, *Sedimentology*, **31**, 413–431.

Fryrear, D. W. 1976. Windbarriers for erosion control in Texas, in Tinus (1976) 37–40.

Fryrear, D. W. 1984. Soil ridges, clods and wind erosion, *Transactions of the American Society of Agricultural Engineers*, **27**, 445–448.

Fryrear, D. W. 1986. A field dust sampler, *Journal of Soil and Water Conservation*, **41**, 117–120.

Fryrear, D. W., Krammes, C. A., Williamson, D. L. and Zobeck, T. M. 1994. Computing the wind erodible fraction of soils, *Journal of Soil and Water Conservation*, **49**, 183–180.

Fullen, M. A. 1985. Wind erosion of arable soils in East Shropshire (England) during spring 1983, *Catena*, **12**, 111–120.

Gabriel, A. 1938. The southern Lut and Iranian Baluchistan, *Geographical Journal*, **92**, 195–210.

Ganor, E. and Mamane, Y. 1982. Transport of Saharan dust across the eastern Mediterranean, *Atmospheric Environment*, **16**, 581–587.

Gardner, G. J., Mortlock, A. J., Price, D. M., Readhead, M. L. and Wasson, R. J. 1987. Thermoluminescence and radiocarbon dating of Australian coastal dunes, *Australian Journal of Earth Science*, **34**, 343–357.

Gardner, R. A. M. 1981. Reddening of dune sands – evidence from southeast India, *Earth Surface Processes and Landforms*, **6**, 459–468.

Gardner, R. A. M. 1983. Aeolianite, in Goudie, A. S. and Pye, K. (eds) *Chemical sediments in geomorphology*, Academic Press, London, 265–300.

Gardner, R. A. M. 1988. Aeolianites and marine deposits of the Wahiba Sands: character and palaeoenvironments, in Dutton (1988) 75–94.

Gardner, R. A. M. and McLaren, S. J. 1993. Progressive vadose diagenesis in late Quaternary aeolianite deposits? in Pye (1993a) 219–234.

Gardner, R. A. M. and Pye, K. 1981. Nature, origin and palaeoenviromental significance of red coastal and desert sand dunes, *Progress in Physical Geography*, **5**, 514–534.

Gares, P. A. 1990. Eolian processes and dune changes at developed and undeveloped sites, Island Beach, New Jersey, in Nordstrom *et al.* (1990) 361–380.

Gares, P. A. and Nordstrom, K. F. 1991. Coastal dune blowouts – dynamics and management implications, *Proceedings of Coastal Zone '91*, American Society of Civil Engineers, 2851–2862.

Gaylord, D. R. and Dawson, P. J. 1987. Airflow-terrain interactions through a mountain gap, with an example of eolian activity beneath and atmospheric hydraulic jump, *Geology*, **15**, 789–792.

Gerlach, A. 1992. Dune cliffs: a buffered system, in Carter *et al.* (1992b) 51–55.

Gibbard, P. L., Wintle, A. G. and Catt, J. A. 1987. Age and origin of clayey silt 'brickearth' in west London, England, *Journal of Quaternary Science*, **2**, 3–9.

Gibbens, R. P., Tromble, J. N., Hennessey, J. T. and Cardenas, M. 1983. Soil movement in mesquite dunelands and former grasslands of southern New Mexico from 1933 to 1980, *Journal of Range Management*, **36**, 145–148.

Gilbertson, D. D. 1981. The impact of the past and present land use on a major coastal barrier system, *Applied Geography*, **1**, 97–119.

Gile, L. H. 1975. Holocene soils and soil-geomorphic relations in an arid region of southern New Mexico, *Quaternary Research*, **5**, 321–360.

Gill, T. E. 1995. Geomorphic impacts of human-induced desiccation of playas, *Geomorphology, in press.*

Gillette, D. A. 1980. Major contributions of natural primary continental aerosols: source mechanisms, in Kneip, J. J. and Lioy, P. J. (eds) *Aerosols: anthropogenic and natural, sources and transport, Annals of the New York Academy of Sciences*, **338**, 348–358.

Gillette, D. A. 1981. Production of dust that may be carried great distances, *Geological Society of America, Special Paper*, **186**, 11–26.

Gillette, D. A. 1986. Production of dust, in El-Baz and Hassan (1986) 251–260.

Gillette, D. A. 1987. Recommendations for measurements and expanded instrumentation at the USGS Geomet sites, in McCauley, J. F. and Rinker, J. N. (eds) *A workshop on desert processes, United States Geological Survey, Circular* C-0989, 10–12.

Gillette, D. 1988. Threshold friction velocities for dust production for agricultural soils, *Journal of Geophysical Research*, **93**, 12 622–12 645.

Gillette, D. A. and Hanson, K. J. 1989. Spatial and temporal variability of dust production caused by wind erosion in the United States, *Journal of Geophysical Research*, **94**, 2197–2206.

Gillette, D. A. and Passi, R. 1988. Modelling dust emission caused by wind erosion, *Journal of Geophysical Research*, **93**, 14 233–14 242.

Gillette, D. A. and Stockton, P. H. 1989. The effect of nonerodible particles on wind erosion on erodible surfaces, *Journal of Geophysical Research*, **94** (D10), 12 885–12 893.

Gillette, D. A. and Walker, T. R. 1977. Characteristics of airborne particles produced by wind erosion of sandy soil, High Plains of West Texas, *Soil Science*, **123**, 97–110.

Gillette, D. A., Blifford, I. H. and Fenster, C. R. 1972. Measurements of aerosol size-distribution and vertical fluxes on land subject to wind erosion, *Journal of Applied Meteorology*, **11**, 977–987.

Gillette, D. A., Adams, J., Endo, A., Smith, D. and Kihl, R. 1980. Threshold velocities for input of soil particles into the air by desert soils, *Journal of Geophysical Research*, **85** (C10), 5621–5630.

Gillette, D. A., Adams, J., Muhs, D. and Kihl, R. 1982. Threshold friction velocities and rupture moduli for crusted desert soils for the input of soil particles to the air, *Journal of Geophysical Research*, **87C**, 9003–9015.

Gillette, D. A., Stensland, G. J., Williams, A. L., Barnard, W., Gatz, D., Sinclair, P. C. and Johnson, T. C. 1992. Emissions of alkaline elements Calcium, Magnesium, Potassium and Sodium from open sources in the contiguous United States, *Global Biogeochemical Cycles*, **6**, 437–457.

Gillette, D. A., Herbert, G., Stockton, P. H. and Owen, P. R. 1995. Causes of the fetch effect in wind erosion, *Earth Surface Processes and Landforms, in press.*

Gillies, J. A. and Nickling, W. G. 1994. The relationship between surface roughness and the origin of the aerodynamic zero reference plane, in Desert Research Institute (1994) 43–44.

Gimmingham, C. H., Ritchie, W., Willetts, B. B. and Willis, A. J. (eds) 1989. Coastal sand dunes, *Proceedings of the Royal Society of Edinburgh*, **B96**.

Glaccum, R. A. and Prospero, J. M. 1980. Saharan aerosols over the tropical North Atlantic, mineralogy, *Marine Geology*, **37**, 295–321.

Glazovski, N. F. 1990. The Aral crisis: the source, the current situation and the ways of solving it, in *International Symposium on the Aral crisis: origins and solutions*, 2–5, October, Nukus.

Glennie, K. W. 1970. *Desert sedimentary environments*, Developments in Sedimentology **14**, Elsevier, Amsterdam.

Glennie, K. W. 1972. Permian Rotliegendes of northwest Europe interpreted in the light of modern desert sedimentation studies, *Bulletin of the American Association of Petroleum Geologists*, **56**, 1048–1071.

Glennie, K. W. 1983. Lower Permian Rotliegende desert sedimentation in the North Sea area, in Brookfield and Ahlbrandt (1983) 521–541.

Glennie, K. W. 1985. Early Permian (Rotliegendes) palaeowinds of the North Sea – reply, *Sedimentary Geology*, **54**, 297–313.

Goldsmith, V. 1985. Coastal dunes, in Davis, R. A. (ed.) *Coastal sedimentary environments*, Springer-Verlag, New York, 171–236.

Gomes, L., Bergametti, G., Coudé-Gaussen, G. and Rognon, P. 1990. Submicron desert dusts: a sand-blasting process? *Journal of Geophysical Research*, **95D**, 13 927–13 935.

Goodman, G. T., Inskip, M. J., Smith, S., Parry, G. D. R. and Burton, M. A. S. 1979. The use of moss-bags in aerosol monitoring, in Morales (1979) 71–91.

Goosens, D. 1985. The granulometric characteristics of a slowly moving dust cloud, *Earth Surface Processes and Landforms*, **10**, 353–362.

Goosens, D. 1988. Scale model simulations of the deposition of loess in hilly terrain, *Earth Surface Processes and Landforms*, **13**, 533–544.

Goosens, D. 1991. Superposition of aeolian dust ripple patterns as a result of changing wind directions, *Earth Surface Processes and Landforms*, **16**, 689–700.

Goosens, D. and Offer, Z. I. 1990. A wind-tunnel simulation and field verification of desert dust deposition, *Sedimentology*, **37**, 7–22.

Goosens, D. and Offer, Z. I. 1994. An evaluation of the efficiency of some aeolian dust collectors, *Soil Technology*, **7**, 25–36.

Goring-Morris, A. N. and Goldberg, P. 1990. Late Quaternary dune incursions in the southern Levant: archaeology, chronology, and palaeoenvironments, *Quaternary International*, **5**, 115–137.

Goudie, A. S. 1970. Notes on some major dune types in southern Africa, *South African Geographical Journal*, **52**, 93–101.

Goudie, A. S. 1972. Climate, weathering, crust formation, dunes and fluvial features of the central Namib Desert, near Gobabeb, South West Africa, *Madoqua*, series II, **1**, 15–31.

Goudie, A. S. 1978. Dust storms and their geomorphological implications, *Journal of Arid Environments*, **1**, 291–310.

Goudie, A. S. 1983. Dust storms in space and time, *Progress in Physical Geography*, **7**, 502–530.

Goudie, A. S. 1989. Wind erosion in deserts, *Proceedings of the Geologists' Association*, **100**, 83–92.

Goudie, A. S. (ed.) 1990. *Techniques for desert reclamation*, John Wiley and Sons, Chichester.

Goudie, A. S. 1991. Pans, *Progress in Physical Geography*, **15**, 221–237.

Goudie, A. S. 1992. *Environmental change*, 3rd Edn, Clarendon, Oxford.

Goudie, A. S. and Sperling, C. H. B. 1977. Long distance transport of formaniferal tests by wind in the Thar Desert, north-west India, *Journal of Sedimentary Petrology*, **47**, 630–633.

Goudie, A. S. and Thomas, D. S. G. 1985. Pans in southern Africa with particular reference to South Africa and Zimbabwe, *Zeitschrift für Geomorphologie NF*, **29**, 1–19.

Goudie, A. S. and Thomas, D. S. G. 1986. Lunette dunes in southern Africa, *Journal of Arid Environments*, **10**, 1–12.

Goudie, A. S. and Watson, A. 1981. The shape of desert sand dune grains, *Journal of Arid Environments*, **4**, 185–190.

Goudie, A. S., Rendell, H. M. and Bull, P. A. 1984. The loess of Tajik SSR, in Miller, K. J. (ed.) *International Karakoram Project*, Cambridge University Press, Cambridge, 399–412.

Goudie, A. S., Warren, A., Jones, D. K. C. and Cooke, R. U. 1987. The character and possible origins of the aeolian sediments of the Wahiba Sand Sea, Oman, *Geographical Journal*, **153**, 231–256.

Goudie, A. S., Stokes, S., Livingstone, I., Baliff, I. K. and Allison, R. J. 1993. The nature, age and sedimentology of the relict dunes of West Kimberley, N. W. Australia, *Geographical Journal*, **159**, 306–317.

Goumandakoye, M. 1990. Un cordon de défense pour triompher des cordons dunaires, in *Interdune*, Bulletin de réseau sahélien de lutte contre l'érosion éolienne et de fixation des dunes, *Reflets sahéliens, Supplement*, **1**, (April) 3–4.

Goździk, J. 1993. Sedimentological record of aeolian processes from the Upper Plenivistulian and the turn of Pleni- and Late-Vistulian Loesses in central Poland, *Zeitschrift für Geomorphologie Supplementband*, **90**, 51–61.

Graf, W. H. 1971. *Hydraulics of sediment transport*, McGraw-Hill, New York.

Grant, K. E. 1975. Erosion in 1973–4: the record and the challenge, *Journal of Soil and Water Conservation*, **30**, 29–32.

Grass, A. J. 1971. Structural features of turbulent flow over smooth and rough boundaries, *Journal of Fluid Mechanics*, **50**, 233–255.

Greeley, R. 1986. Aeolian landforms: laboratory simulations and field studies, in Nickling (1986) 195–211.

Greeley, R. and Iversen, J. D. 1985. *Wind as a geological process on Earth, Mars, Venus and Titan*, Cambridge University Press, Cambridge.

Greeley, R., Williams, S. H., White, B. R. and Pollack, J. B. 1984. Wind abrasion on Earth and Mars, in Woldenberg, M. J. (ed.) *Models in geomorphology*, Allen and Unwin, Boston, 373–422.

Greeley, R., Blumberg, D. G. and Williams, S. H. 1994. Field measurements of active windblown sand, in Desert Research Institute (1994) 47–48.

Griffith, P. 1976. Introduction of the problem, in Tinus (1976) 3–7.

Grolier, M. J., McCauley, J. F., Breed, C. S. and Embabi, N. S. 1980. Yardangs of the Western Desert, *Geographical Journal*, **135**, 191–212.

Grousset, F. E., Rognon, P., Coudé-Gaussen, G. and Pedemay, P. 1992. Origins of peri-Saharan dust deposits traced by their Nd and Sr isotopic composition, *Palaeogeography, Palaeoclimatology, Palaeoecology*, **93**, 203–212.

Grove, A. T. 1960. Geomorphology of the Tibesti region with special reference to the western Tibesti, *Geographical Journal*, **126**, 18–31.

Grove, A. T. 1969. Landforms and climatic change in the Kalahari and Ngamiland, *Geographical Journal*, **135**, 191–212.

Grove, A. T. 1985. The physical evolution of the river basins, in Grove, A. T. (ed.) *The Niger and its neighbours: environmental history and hydrobiology, human use and health hazards of the major West African rivers*, Balkema, Rotterdam, 21–60.

Hack, J. T. 1941. The dunes of western Navajo county, *Geographical Review*, **31**, 240–362.

Haff, P. K. and Werner, B. T. 1986. Computer simulation of the mechanical sorting of grains, *Powder Technology*, **48**, 239–245.

Hagedorn, H. 1968. Über äolische Abtragung und Formung in der Südest-Sahara, *Erdekunde*, **22**, 257–269.

Hagen, L. J. 1991. A wind erosion prediction system to meet user needs, *Journal of Soil and Water Conservation*, **46**, 106–111.

Hagen, L. J. and Armbrust, D. V. 1985. Effects of field ridges on soil transport by wind, in Barndorff-Nielsen *et al.* (1985) 563–586.

Halevy, G. and Steinberger, E. H. 1974. Inland penetration of the summer inversion from the Mediterranean coast of Israel, *Israel Journal of Earth Science*, **23**, 47–54.

Halimov, M. and Fezer, F., 1989, Eight yardang types in central Asia, *Zeitschrift für Geomorphologie NF*, **33**, 205–217.

Hall, K. 1989. Wind blown particles as weathering agents? An Antarctic example, *Geomorphology*, **2**, 405–410.

Hallet, J. and Hoffer, T. 1971. Dust devil systems, *Weather*, **24**, 247–250.

Hammond, F. D. C. and Heathershaw, A. D. 1981. A wave theory for sandwaves in shelf seas, *Nature*, **293**, 208–210.

Handey, R. H. 1973. Collapsible loess in Iowa, *Proceedings of the Soil Science Society of America*, **37**, 281–284.

Hansen, J. E. and Lacis, A. A. 1990. Sun and dust versus greenhouse gases: an assessment of their relative roles in global climatic change, *Nature*, **346**, 713–719.
Hardisty, J. and Whitehouse, R. J. S. 1988. Evidence for a new sand transport process from experiments on Saharan dunes, *Nature*, **332**, 532–534.
Harmse, J. T. 1982. Geomorphologically effective winds in the northern part of the Namib sand desert, *South African Geographer*, **10**, 43–52.
Hartmann, D. and Christiansen, C. 1988. Settling-velocity distributions and sorting processes on a longitudinal dune: a case study, *Earth Surface Processes and Landforms*, **13**, 649–656.
Hastenrath, S. L. 1967. The barchans of the Arequipa region, southern Peru, *Zeitschrift für Geomorphologie NF*, **11**, 300–331.
Hastenrath, S. L. 1978. Mapping and surveying – dune shape and multiannual displacement, in Lettau, H. H. and Lettau, K. (eds) *Exploring the world's driest climates*, Report 101, Institute for Environmental Studies, University of Wisconsin-Madison, Institute for Environmental Studies, 74–88.
Haynes, C. V. Jr. 1989. Bagnold's barchan: a 57-year record of dune movement in the eastern Sahara and implications for dune origin and palaeoclimate since Neolithic times, *Quaternary Research*, **32**, 153–167.
Hedin, S. 1903. *Central Asia and Tibet*, Charles Scribners and Sons, New York.
Heidinga, H. A. 1984. Indications of severe drought during the 10th century A. D. from an inland dune area in the central Netherlands, *Geologie en Mijnbouw*, **63**, 241–248.
Heisler, G. M. and Dewalle, D. R. 1988. Effects of windbreak structure on wind flow, *Journal of Agriculture, Ecosystems and Environment*, **22/23**, 41–69.
Held, R. B. and Clawson, M. 1965. *Soil conservation in perspective*, Johns Hopkins University Press for Resources for the Future, Baltimore.
Helm, P. J. and Breed, C. S. 1994. Sediment transport by wind under monitored climatic conditions in Arizona and New Mexico, in Desert Research Institute (1994) 51–52.
Hennessey, J. T., Kies, B., Gibbens, R. P. and Tromble, J. M. 1986. Soil sorting by forty-five years of wind erosion on a southern New Mexico range, *Soil Science Society of America, Journal*, **50**, 391–394.
Hesp, P. A. 1979. Sand trapping ability of culms of marram grass (*Ammophila arenaria*), *Journal of the Soil Conservation Service of New South Wales*, **35**, 156–160.
Hesp, P. A. 1987. Morphology, dynamics and internal stratification of some established foredunes in southeast Australia, *Sedimentary Geology*, **55**, 17–42.
Hesp, P. A. 1988. Surfzone, beach and foredune interactions on the Australian south east coast, *Journal of Coastal Research*, (Special Issue) **3**, 15–25.
Hesp, P. A., Illenberger, W., Rust, I., McLachlan, A. and Hyde, R. 1988. Some aspects of transgressive dunefields and transverse ridge geomorphology and dynamics, south coast, South Africa, *Zeitschrift für Geomorphologie Supplementband*, **73**, 111–123.
Hesp, P., Hyde, R., Hesp, V. and Zhengyu, Q. 1989. Longitudinal dunes can move sideways, *Earth Surface Processes and Landforms*, **14**, 447–451.
Hickox, C. F. 1959. Formation of ventifacts in a moist temperate climate, *Bulletin of the Geological Society of America*, **70**, 1489–1490.
Hidore, J. J. and Albokhair, Y. 1982. Sand encroachment in Al-Hasa Oasis, Saudi Arabia, *Geographical Review*, **72**, 350–356.
Higgins, C. G. 1956. Formation of small ventifacts, *Journal of Geology*, **64**, 506–516.
Högbom, I. 1923. Ancient inland dunes of north and middle Europe, *Geografiska Annaler*, **5**, 113–143.
Holben, B. N., Eck, T. and Fraser, R. S. 1991. Temporal and spatial variability of aerosol optical depth in the Sahel region, *International Journal of Remote Sensing*, **12**, 1147–1163.
Holliday, V. T. 1989. The Blackwater Draw Formation (Quaternary): a 1.4±my record of eolian sedimentation and soil formation on the southern High Plains, *Bulletin of the Geological Society of America*, **101**, 1598–1607.
Holm, D. A. 1960. Desert geomorphology in the Arabian peninsula, *Science*, **132**, 1369–1379.
Hoogheimstra, H. 1989. Variations of the NE African trade wind regime during the last 140,000 years: changes in pollen flux evidenced by marine sediment records, in Leinen and Sarnthein (1989) 733–769.
Horikawa, K., Hotta, S. and Kraus, N. 1986. Literature review of sand transport by wind on a dry sand surface, *Coastal Engineering*, **9**, 503–526.
Hörner, N. G. 1932. Lop Nor: topographical amd geological summary, *Geografiska Annaler*, **14**, 297–321.
Hotta, S., Kubota, S., Katori, S. and Horikawa, K. 1984. Blown sand on a wet sand surface, *Proceedings of the 19th Coastal Engineering Conference*, American Society of Civil Engineers, New York, 1265–1281.
Hovan, S. A., Rea, D. K., Pisias, N. G. and Shackleton, N. J. 1989. A direct link between the China loess and marine ^{18}O records: aeolian flux into the north Pacific, *Nature*, **340**, 296–298.
Howard, A. D. 1977. The effect of slope on the threshold of motion and its application to the orientation of wind ripples, *Bulletin of the Geological Society of America*, **88**, 853–856.
Howard, A. D. 1985. Interaction of sand transport with topography and local winds in the northern Peruvian coastal desert, in Barndorff-Nielsen *et al.* (1985) 511–544.
Howard, A. D., Morton, J. B., Gad-el-Hak, M. and Pierce, D. B. 1977. Simulation model of erosion and deposition on a barchan dune, *Contractor Report*, NASA CR-2838, Contract NGR-47-005-172, National Aeronautics and Space Administration (NASA), Washington, DC; also School of Engineering and Applied Science, University of Virginia, Charlottesville.
Howard, A. D., Morton, J. B., Gad-el-Hak, M. and Pierce, D. B. 1978. Sand transport model of barchan dune equilibrium, *Sedimentology*, **25**, 307–338.
Howard, A. D. and Walmsley, J. L. 1985. Simulation model of isolated dune sculpture by wind, in Barndorff-Nielsen *et al.* (1985) 377–392.
Hoyt, J. H. 1966. Air and sand movement to the lee of dunes, *Sedimentology*, **7**, 137–143.
Hsu, S.-A. 1971. Measurement of shear stress and roughness length on a beach, *Journal of Geophysical Research*, **76**, 2880–2885.

Hunt, J. C. R. and Barrett, C. F. 1989. Wind loss from stockpiles and transport of dust, *Warren Spring Laboratory, Laboratory Report*, LR 688 (PA), Her Majesty's Stationery Office, London, Warren Spring Laboratory, Gunnels Wood Road, Stevenage, Hertfordshire.

Hunt, J. C. R. and Nalpanis, P. 1985. Saltating and suspended particles over flat and sloping surfaces. I. Modelling concepts, in Barndorff-Nielsen *et al.* (1985) 9–36.

Hunt, J. C. R., Richards, K. J. and Brighton, P. M. W. 1988. Stably stratified flow low hills, *Quarterly Journal of the Royal Meteorological Society*, **114**, 859–886.

Hunter, R. E. 1977a. Basic types of stratification in small eolian dunes, *Sedimentology*, **24**, 361–387.

Hunter, R. E. 1977b. Terminology of cross-stratified sedimentary layers and climbing-ripple structures, *Journal of Sedimentary Petrology*, **47**, 697–706.

Hunter, R. E. 1980. Quasi-planar adhesion stratification – an eolian structure formed in wet sand, *Journal of Sedimentary Petrology*, **50**, 263–266.

Hunter, R. E. 1985. A kinematic model for the structure of lee-side deposits, *Sedimentology*, **32**, 409–422.

Hunter, R. E. and Richmond, B. M. 1988. Daily cycles in coastal dunes, *Sedimentary Geology*, **55**, 43–67.

Hunter, R. E. and Rubin, D. M. 1983. Interpreting cyclic cross-bedding, with an example from the Navajo Sandstone, in Brookfield and Ahlbrandt (1983) 429–454.

Hunter, R. E., Richmond, B. M. and Alpha, T. R. 1983. Storm-controlled oblique dunes of the Oregon coast, *Bulletin of the Geological Society of America*, **94**, 1450–1465.

Huszar, P. C. 1988. Nature and causes of the blowout problem in Colorado, in Whitehead *et al.* (1988) 659–664.

Idso, S. B. 1974. Tornado or dust devil: enigma of desert whirlwinds, *American Scientist*, **62**, 530–541.

Idso, S. B., Ingram, R. S. and Pritchard, J. M. 1972. An American haboob, *Bulletin of the American Meterological Society*, **53**, 930–955.

Illenberger, W. K. and Rust, I. C. 1988. A sand budget for the Alexandria coastal dunefield, South Africa, *Sedimentology*, **35**, 513–521.

Imbrie, J. 1985. A theoretical framework for the Pleistocene ice ages, *Journal of the Geological Society of London*, **142**, 417–432.

Inman, D. L., Ewing, G. C. and Corliss, J. B. 1966. Coastal sand dunes of Guerrero Negro, Baja California, Mexico, *Bulletin of the Geological Society of America*, **77**, 787–802.

Inoue, K. and Naruse, T. 1991. Accumulation of Asian long-range eolian dust in Japan and Korea from the late Pleistocene to the Holocene, *Catena Supplement*, **20**, 25–42.

Issar, A., Tsoar, H. and Levin, D. 1989. Climatic changes in Israel during historical times and their impact on hydrological, pedological and socio-economic systems, in Leinen and Sarnthein (1989) 525–541.

Iversen, J. D. 1983. Saltation threshold and deposition rate modelling, in Brookfield and Ahlbrandt (1983) 103–114.

Iversen, J. D. 1986a. Aeolian processes in the environmental wind tunnel and in the atmosphere, in El-Baz and Hassan (1986) 318–321.

Iversen, J. D. 1986b. Small scale wind tunnel modelling of particle transport – Froude number effect, in Nickling (1986) 19–33.

Iversen, J. D. and White, B. R. 1982. Saltation threshold on Earth, Mars and Venus, *Sedimentology*, **29**, 111–119.

Iversen, J. D., Greeley, R., Marshall, J. R. and Pollack, J. 1987. Aeolian saltation threshold: effect of density ratio, *Sedimentology*, **34**, 699–706.

Iversen, J. D., Wang, W. P., Rasmussen, K. R., Mikkelsen, H. E. and Leach, R. N. 1990. Roughness element effect on local and universal saltation transport, *Acta Mechanica Supplementum*, **2**, 65–75.

Jackson, D. W. T. and Nevin, G. H. 1992. Sand transport in a cliff-top dune system at Fonte de Telha, Portugal, in Carter *et al.* (1992b) 81–91.

Jackson, P. S. 1981. On the displacement height in the logarithmic velocity profile, *Journal of Fluid Mechanics*, **111**, 15–25.

Jaenicke, R. 1979. Monitoring and critical review of the estimated source strength of mineral dust from the Sahara, in Morales (1979) 233–242.

Jennings, J,N. 1967. Cliff-top dunes, *Australian Geographical Studies*, **5**, 40–49.

Jennings, J. N. 1975. Desert dunes and estuarine fill in the Fitzroy estuary (north-western Australia), *Catena*, **2**, 215–262.

Jensen, F. 1994. Dune management in Denmark: application of the Nature Protection Act of 1992, *Journal of Coastal Research*, **10**, 263–269.

Jensen, J. L., Rasmussen, K. R., Sørensen, M. and Willetts, B. B. 1984. *The Hanstholm Experiment, 1982. Sand grain saltation on a beach, Research Reports* **125**, Department of Theoretical Statistics, Institute of Mathematics, University of Aarhus.

Jensen, J. L. and Sørensen, M. 1986. Estimation of some aeolian saltation transport parameters: a reanalysis of William's data, *Sedimentology*, **33**, 547–558.

Jensen, M. 1954. *Shelter effect: investigations into the aerodynamics of shelter and its effects on climate and crops*, Danish Technical Press, Copenhagen.

Jensen, N.-O. and Zeman, O. 1985. Perturbations to mean wind and turbulence in flow over topographic forms, in Barndorff-Nielsen *et al.* (1985) 351–368.

Jones, D. K. C., Cooke, R. U. and Warren, A. 1986. Geomorphological investigation, for engineering purposes, of blowing sand and dust hazard, *Quarterly Journal of Engineering Geology*, **19**, 251–270.

Jones, R. L. and Beavers, A. H. 1964. Aspects of catenary and depth distribution of opal phytoliths in Illinois soils, *Proceedings of the Soil Science Society of America*, **28**, 413–416.

Joussaume, S. 1990. Three dimensional simulations of the atmospheric cycle of desert dust particles using a general circulation model, *Journal of Geophysical Research*, **95**, 1909–1941.

Jungerius, P. D. 1984. A simulation model of blowout development, *Earth Surface Processes and Landforms*, **9**, 509–512.

Jungerius, P. D. 1989. Geomorphology, soils and dune management, in van der Meulen (1989) 91–98.

Jungerius, P. D. and van der Meulen, F. 1988. Erosion processes in a dune landscape along the Dutch coast, *Catena*, **15**, 217–228.

Jungerius, P. D. and van der Meulen, F. 1989. The development of dune blowouts, as measured with erosion pins and sequential air photos, *Catena*, **16**, 369–376.

Jutson, J. T. 1934. The physiography (geomorphology) of Western Australia, *Bulletin of the Geological Survey of Western Australia*, 2nd edn, **95**.

Kaiser, E. 1926. *Die Diamentenwüste Sudwest-Afrikas*, Dietrich Reimer, Berlin.

Kaldi, J., Krinsley, D. H. and Lawson, D. 1978. Experimentally produced aeolian surface textures on quartz sand grains from various experiments, in Whalley (1978) 261–274.

Kar, A. 1990. The megabarchanoids of the Thar: their environment, morphology and relationship with longitudinal dunes, *Geographical Journal*, **156**, 51–61.

Kaul, R. N. 1983. Some silvicultural aspects of sand dune afforestation, *International Tree Crops Journal*, **2**, 133–146.

Kebin, Z. and Kaigo, Z. 1989. Afforestation and sand dune fixation in China, *Journal of Arid Environments*, **16**, 3–10.

Kennedy, J. F. 1969. The formation of sediment ripples, dunes and antidunes, *Annual Review of Fluid Mechanics*, **1**, 147–169.

Kerr, R. C. and Nigra, J. O. 1952. Eolian sand control, *Bulletin of the American Association of Petroleum Geologists*, **36**, 1541–1573.

Keyes, C. R. 1909. Base level of eolian erosion, *Journal of Geology*, **17**, 659–663.

Keyes, C. R. 1912. Deflative scheme of the geographic cycle in an arid climate, *Bulletin of the Geological Society of America*, **23**, 537–562.

Khalaf, F. I. 1989. Textural characteristics and genesis of aeolian sediments in the Kuwaiti desert, *Sedimentology*, **36**, 253–271.

Khalaf, F. I., Gharib, I. M. and Al-Hashash, M. Z. 1984. Types and characteristics of the recent surface deposits of Kuwait, Arabian Gulf, *Journal of Arid Environments*, **7**, 9–33.

Killigrew, L. P. and Gilkes, R. J. 1974. Development of playa lakes in south western Australia, *Nature*, **247**, 454–455.

Kimberlin, L. W., Hidelbaugh, A. L. and Grunewald, A. R. 1977. The potential wind erosion problem in the United States, *Transactions of the American Society of Agricultural Engineers*, **20**, 873–879.

King, D. 1956. The Quaternary stratigraphic record at Lake Eyre North and the evolution of existing topographic forms, *Transactions of the Royal Society of South Australia*, **83**, 93–103.

King, D. 1960. The sand ridge deserts of South Australia and related aeolian landforms of the Quaternary arid cycles, *Transactions of the Royal Society of South Australia*, **83**, 99–108.

Klijn, J. A. 1990. Dune forming factors in a geographical context, *Catena Supplement*, **18**, 1–14.

Knott, P. 1979. *The structure and pattern of dune-forming winds*, Unpublished PhD dissertation, University of London.

Knott, P. and Warren, A. 1990. Aeolian processes, in Goudie, A. S. (ed.) *Geomorphological techniques*, 2nd edn, Allen and Unwin, London, 226–246.

Knottnerus, D. J. C. 1980. Relative humidity of the air and critical wind velocity in relation to erosion, in De Boodt, M. and Gabriels, D. (eds) *Assessment of erosion*, John Wiley and Sons, Chichester, 531–539.

Knutson, E. O., Sood, S. K. and Stockham, J. D. 1977. Aerosol collection by snow and ice crystals, *Atmospheric Environment*, **11**, 395–402.

Kobayshi, S. and Ishihara, T. 1979. Interaction between wind and snow surface, *Boundary-Layer Meteorology*, **16**, 35–47.

Kocurek, G. 1981. Significance of interdune deposits and bounding surfaces in aeolian dune sands, *Sedimentology*, **28**, 753–780.

Kocurek, G. 1988. First-order and super bounding surfaces in eolian sequences – bounding surfaces revisited, *Sedimentary Geology*, **56**, 193–206.

Kocurek, G. 1991 Interpretation of ancient eolian sand dunes. *Annual Review of Earth and Planetary Sciences*, **19**, 43–75.

Kocurek, G. and Nielson, J. 1986. Conditions favourable for the formation of warm-climate aeolian sand sheets, *Sedimentology*, **33**, 795–816.

Kocurek, G., Knight, J. and Havholm, K. 1991. Outcrop and semi-regional three-dimensional architecture and reconstruction of a portion of the eolian Page Sandstone (Jurassic), in Miall, A. and Tyler, N. (eds) *Three-dimensional facies architecture*, Society of Economic Palaeontologists and Mineralogists, 25–43.

Kocurek, G., Townsley, M., Yeh, E., Sweet, M. and Havholm, K. 1992. Dune and dune-field development stages on Padre Island, Texas: effects of lee airflow and sand saturation levels and implications for interdune deposition, *Journal of Sedimentary Petrology*, **62**, 622–635.

Kocurek, G., Murphy, K., Frank, A. and Lake, L. 1994. High-resolution wind velocity profiles in nature using hot-wire probes, in Desert Research Institute (1994) 61.

Kolm, K. E. 1985. Predicting surface wind characteristics of southern Wyoming from remote sensing and eolian geomorphology, in Barndorff-Nielsen *et al.* (1985) 421–481.

Koster, E. A., Castel, I. I. Y. and Nap, R. L. 1993. Genesis and sedimentary structures of the Late Holocene eolian drift sands in north-west Europe, in Pye (1993a) 247–267.

Krinsley, D. H. and Smalley, I. J. 1972. Sand, *American Scientist*, **60**, 286–291.

Krumbein, W. C. 1938. Size frequency distributions and the normal phi curve, *Journal of Sedimentary Petrology*, **8**, 84–90.

Kuenen, Ph. 1960. Experimental abrasion, 4: eolian action, *Journal of Geology*, **68**, 427–449.

Kutzbach, J. E. 1989. Possible effects of orbital variations on past sources and transports of eolian material: estimates from general circulation model experiments, in Leinen and Sarnthein (1989) 513–521.

Laity, J. E. 1987. Topographic effects on ventifact development, Mojave Desert, California, *Physical Geography*, **8**, 113–132.

Laity, J. E. 1994. Landforms of aeolian erosion, in Abrahams and Parsons (1994) 506–535.

Lancaster, I. N. 1978. Composition and formation of southern Kalahari pan margin dunes, *Zeitschrift für Geomorphologie NF*, **22**, 148–149.

Lancaster, J., Lancaster, N. and Seely, M. K. 1984. The climate of the central Namib Desert, *Madoqua*, **14**, 5–61.

Lancaster, N. 1980. The formation of seif dunes from barchans – supporting evidence for Bagnold's model from the Namib Desert, *Zeitschrift für Geomorphologie NF*, **24**, 160–167.

Lancaster, N. 1981. Palaeoenvironmental implications of fixed dune systems in southern Africa, *Palaeogeography, Palaeoclimatology, Palaeoecology*, **33**, 327–346.

Lancaster, N. 1982a. Dunes on the Skeleton Coast, Namibia (South West Africa): Geomorphology and grain size relationships, *Earth Surface Processes and Landforms*, **7**, 575–587.
Lancaster, N. 1982b. Spatial variations in linear dune morphology and sediments in the Namib sand sea, *Palaeoecology of Africa*, **15**, 173–182.
Lancaster, N. 1983a. Controls of dune morphology in the Namib sand sea, in Brookfield and Ahlbrandt (1983) 261–289.
Lancaster, N. 1983b. Linear dunes of the Namib sand sea, *Zeitschrift für Geomorphologie Supplementband*, **45**, 27–49.
Lancaster, N. 1984. Characteristics and occurrence of wind erosion features in the Namib Desert, *Earth Surface Processes and Landforms*, **9**, 469–478.
Lancaster, N. 1985a. Wind and sand movements in the Namib sand sea, *Earth Surface Processes and Landforms*, **10**, 607–619.
Lancaster, N. 1985b. Variations in wind velocity and sand transport on the windward flanks of desert sand dunes, *Sedimentology*, **32**, 581–593.
Lancaster, N. 1986. Grain-size characteristics of linear dunes in the southwestern Kalahari, *Journal of Sedimentary Petrology*, **56**, 395–400.
Lancaster, N. 1987a. Variations in wind velocity and sand transport on the windward flanks of desert sand dunes – reply, *Sedimentology*, **43**, 516–520.
Lancaster, N. 1987b. Dunes of the Gran Desierto sand sea, Sonora, Mexico, *Earth Surface Processes and Landforms*, **12**, 277–288.
Lancaster, N. 1988a. A bibliography of dunes: Earth, mars and Venus, *NASA Contractor Report* 4149.
Lancaster, N. 1988b. Development of linear dunes in the southwestern Kalahari, southern Africa, *Journal of Arid Environments*, **14**, 233–244.
Lancaster, N. 1988c. Controls of eolian dune size and spacing, *Geology*, **16**, 972–975.
Lancaster, N. 1989a. *The Namib sand sea: dune forms, processes and sediments*, Balkema, Rotterdam.
Lancaster, N. 1989b. Star dunes, *Progress in Physical Geography*, **13**, 67–91.
Lancaster, N. 1989c. The dynamics of star dunes: an example from the Gran Desierto, Mexico, *Sedimentology*, **36**, 273–289.
Lancaster, N. 1994. Dune morphology and dynamics, in Abrahams and Parsons (1994) 475–505.
Lancaster, N. 1995. *Geomorphology of desert dunes*, Routledge, London.
Lancaster, N. and Greeley, R. 1990. Sediment volume in the North Polar sand seas of Mars, *Journal of Geophysical Research*, **95**, 10 921–10 927.
Lancaster, N. and Nickling, W. G. 1994. Aeolian transport systems, in Abrahams and Parsons (1994) 447–473.
Lancaster, N. and Ollier, C. D. 1983. Sources of sand for the Namib Sand Sea, *Zeitschrift für Geomorphologie Supplementband*, **45**, 71–83.
Lancaster, N., Nickling, W. G., McKenna Neuman, C., and Wyatt, V. 1995. Sediment flux and airflow on the stoss side of a barchan dune, *Geomorphology*, *in press*.
Langbein, W. B. 1961. Salinity and hydrology of closed lakes, *United States Geological Survey Professional Paper* **412**.
Langford, R. P. 1989. Fluvial-aeolian interaction, Part I. Modern Systems, *Sedimentology*, **36**, 1023–1036.
Leatherman, S. P. 1978. A new aeolian sand trap design, *Sedimentology*, **25**, 303–306.
Leatherman, S. P. and Zaremba, R. E. 1987. Overwash and eolian processes on a U.S. northeast coast barrier, *Sedimentary Geology*, **52**, 183–206.
Leathers, C. R. 1981. Plant components of desert dust in Arizona and their significance for man, *Geological Society of America, Special Paper* **186**, 191–206.
Lebret, P. and Lautridou, J.-P. 1991. The loess of western Europe, *GeoJournal*, **24**, 151–156.
Lee, J. A., Wigner, K. A. and Gregory, J. M. 1993. Drought, wind and blowing dust on the southern High Plains of the United States, *Physical Geography*, **14**, 56–67.
Lees, D. J. 1982. The sand dunes of Gower as potential indicators of climatic change in historical time, *Cambria*, **9**, 25–35.
Leger, M. 1990. Loess landforms, *Quaternary International*, **7/8**, 53–61.
Legrand, M., Bertrand, J. J., Desbois, M., Menenger, L. and Fouquart, Y. 1989. The potential of infrared satellite data for the retrieval of Saharan-dust optical depth over Africa, *Journal of Applied Meteorology*, **28**, 309–318.
Leinen, M. 1989. The Late Quaternary record of atmospheric transport to the northwest Pacific from Asia, in Leinen and Sarnthein (1989) 693–731.
Leinen, M. and Sarnthein, M. (eds). 1989. *Palaeoclimatology and palaeometeorology: modern and past patterns of global atmospheric transport*, Kluwer Academic, Dordrecht.
Le Ribault, L. 1978. The exoscopy of quartz sand grains, in Whalley (1978) 319–327.
Lettau, K. and Lettau, H. 1969. Bulk transport of sand by the barchans of La Pampa La Hoja in southern Peru, *Zeitschrift für Geomorphologie NF*, **13**, 182–195.
Lettau, K. and Lettau, H. H. 1978. Experimental and micrometeorological studies of dune migration, in Lettau, H. H. and Lettau, K. (eds) *Exploring the world's driest climates, Institute of Environmental Science Report* 101, Center for Climatic Research, University of Wisconsin, Madison, 110–147.
Li, P.-Y. and Zhou, L.-P. 1993. Occurrence and palaeoenvironmental implications of the Late Pleistocene loess along the eastern coasts of the Bohai Sea, China, in Pye (1993a) 293–309.
Li, Z. and Komar, P. D. 1986. Laboratory measurements of pivoting angles for application to selective entrainment of gravel in cement, *Sedimentology*, **33**, 413–423.
Lindquist, S. J. 1988. Practical characterisation of eolian reservoirs for development: Nugget Sandstone, Utah–Wyoming Thrust Belt, *Sedimentary Geology*, **56**, 315–339.
Lindsay, J. F. 1973. Ventifact evolution in Wright Valley, Antactica, *Bulletin of the Geological Society of America*, **84**, 1791–1798.
Littmann, T. 1991a. Dust storm frequency in Asia: climatic control and variability, *International Journal of Climatology*, **11**, 393–412.
Littmann, T. 1991b. Rainfall, temperature and dust storm anomalies in the African Sahel, *Geographical Journal*, **157**, 136–160.
Littmann, T. 1991c. Recent African dust deposition in West Germany: sediment characteristics and climatological aspects, *Catena Supplement*, **20**, 57–73.

Livingstone, I. 1986. Geomorphological significance of wind flow patterns over a Namib linear dune, in Nickling (1986) 97–112.
Livingstone, I. 1987. Grain-size variation on a 'complex' linear dune in the Namib Desert. in Frostick and Reid (1987) 281–291.
Livingstone, I. 1988. New models for the formation of linear sand dunes, *Geography*, **73**, 105–115.
Livingstone, I. 1989a. Monitoring surface change on a Namib linear dune, *Earth Surface Processes and Landforms*, **14**, 317–332.
Livingstone, I. 1989b. Temporal trends in grain-size measures on a linear sand dune, *Sedimentology*, **36**, 1017–1022.
Livingstone, I. 1993. A decade of surface change on a Namib linear dune, *Earth Surface Processes and Landforms*, **18**, 661–664.
Livingstone, I. and Thomas, D. S. G. 1993. Modes of linear dune activity and their palaeoenvironmental significance: an evaluation with reference to southern African examples, in Pye (1993a) 91–101.
Logie, M. 1982. Influence of roughness elements and soil moisture on the resistance of sand to wind erosion, *Catena Supplement*, **1**, 161–173.
Long, J. T. and Sharp, R. P. 1964. Barchan-dune movement in the Imperial Valley, California, *Bulletin of the Geological Society of America*, **75**, 149–156.
Loope, D. B. 1985. Episodic deposition and preservation of eolian sands: a late Paleozoic example from southeastern Utah, *Geology*, **13**, 73–76.
Loopey, J. W. 1991. Land degradation in Australia: the search for a legal remedy, *Journal of Soil and Water Conservation*, **46**, 256–259.
Lowdermilk, W. C. 1944. Les Landes, where over three quarters of a century France has transformed vast mobile sand dunes and waste marshland into a rich pine-producing area, *American Forests*, **50**, 380–382, 412–415.
Lowe, D. R. 1976. Grain flow and grain-flow deposits, *Journal of Sedimentary Petrology*, **46**, 188–199.
Loÿe-Pilot, M. D., Martin, J. M. and Morelli, J. 1986. Influence of Saharan dust on rain acidity and atmospheric input to the Mediterranean, *Nature*, **321**, 427–428.
Lucchitta, B. K. 1982. Ice sculpture in Martian outflow channels, *Journal of Geophysical Research*, **87**, 9951–9973.
Lundqvist, J. and Bentsson, K. 1970. The red snow – a meteorological and pollen analytical study of long-transported material from snow falls in Sweden, *Geologiska Föreningens i Stockholm Förhandlingar*, **92**, 288–301.
Lyles, L. 1975. Possible effects of wind erosion on soil productivity, *Journal of Soil and Water Conservation*, **30**, 279–283.
Lyles, L. 1985. Predicting and controlling wind erosion, *Agricultural History*, **59**, 205–214.
Lyles, L. and Schrandt, R. L. 1972. Wind erodibility as influenced by rainfall and soil salinity, *Soil Science*, **114**, 367–372.
Maat, P. B. and Johnson, W. C. 1995. Thermoluminescence and new ^{14}C age estimates for late Quaternary loesses in southwestern Nebraska, *Geomorphology, in press*.
Mabbutt, J. A. 1968. Aeolian landforms in central Australia, *Australian Geographical Studies*, **6**, 139–150.
Mabbutt, J. A. 1977. *Desert landforms*, Australian National University Press, Canberra.
Mabbutt, J. A. 1980. Some general characteristics of the aeolian landscapes, in Storrier and Stannard (1980) 1–16.
Mabbutt, J. A. 1984. Factors determining desert dune type, *Nature*, **309**, 92.
Mabbutt, J. A. and Sullivan, M. E. 1968. The formation of longitudinal dunes: evidence from the Simpson Desert, *Australian Geographer*, **10**, 483–487.
Machette, M. N. 1985. Calcic soils of the southwestern United States, in Weide, D. L. (ed.) *Soils and Quaternary geology of the southwestern United States, Special Paper*, **203**, Geological Society of America, 1–21.
MacKay, J. R. 1956. Notes on oriented lakes of the Liverpool Bay area, Northwestern Territories, *Revue Canadienne de Géographie*, **10**, 169–173.
Madigan, C. T. 1936. The Australian sand-ridge deserts, *Geographical Review*, **26**, 205–227.
Madigan, C. T. 1946. The Simpson Desert expedition 1939; scientific reports, 6: geology – the sand formations, *Transaction of the Royal Society of South Australia*, **70**, 45–63.
Magaritz, M. and Enzel, Y. 1990. Standing-water deposits as indicators of Late Quaternary dune migration in the northwestern Negev, Israel, *Climatic Change*, **16**, 307–318.
Mainguet, M. 1968. Le Borkou, aspect d'une modelé éolien, *Annales de Géographie*, **77**, 296–322.
Mainguet, M. 1978. The influence of trade winds, local air masses and topographic obstacles on the aeolian movement of sand particles and the origin and distribution of ergs in the Sahara and Australia, *Geoforum*, **9**, 17–28.
Mainguet, M. 1983. Dunes vives, dunes fixées, dunes vêtues: une classification selon le bilan d'alimentation, le régime éolien et la dynamique des édifices sableux, *Zeitschrift für Geomorphologie Supplementband*, **45**, 265–285.
Mainguet, M. 1984. A classification of dunes based on aeolian dynamics and the sand budget, in El-Baz (1984a) 31–58.
Mainguet, M. and Chemin, M.-C. 1983. Sand seas of the Sahara and Sahel: an explanation of their thickness and sand dune type by the sand budget principle, in Brookfield and Ahlbrandt (1983) 353–363.
Mainguet, M., Canon-Cossus, L. and Chemin, M.-C. 1980. Le Sahara: géomorphologie et paléogéomorphologie éoliennes, in Williams, M. A. J. and Faure, H. (eds) *The Sahara and the Nile*, Balkema, Rotterdam, 17–35.
Maizels, J. K. 1990. Raised channel systems as indicators of paleohydrologic change: a case study from Oman, *Palaeogeography, Palaeoclimatology, Palaeoecology*, **76**, 241–277.
Maley, J. 1982. Dust clouds, rain types and climatic variations in tropical North Africa, *Quaternary Research*, **18**, 1–16.
Manohar, M. and Bruun, P. 1970. Mechanics of dune growth by sand fences, *Dock and Harbour Authority*, **51**, 243–252.
Marsh, W. M. and Marsh, B. D. 1987. Wind erosion and sand dune formation on high Lake Superior bluffs, *Geografiska Annaler*, **A69**, 379–391.
Marshall, J. K. 1973. Drought, land use and soil erosion, in Lovett, J. V. (ed.) *Drought*, Angus and Robertson, Sydney, 55–80.
Marshall, T. R. 1987. Morphotectonic analysis of the Wesselsborn panveld, *South African Journal of Geology*, **90**, 209–218.

Marston, R. A. 1986. Manoeuvre-caused wind-erosion impacts, South Central New Mexico, in Nickling (1986) 273–290.

Mason, G.-M. 1961. L'emploi des murs anti-sable ou 'draa' sur les pistes du Sahara septentrional, *Annales de Géographie*, **60**, 127–128.

Matchinski, M. 1962. Sur la distribution des petits mares de l'Île de France, *Comptes Rendus, Academie des Sciences à Paris*, **254**, 331–334.

Mattsson, J. O. 1976. Wind-tilted pebbles in sand – some field observations and simple experiments, *Nordic Hydrology*, **7**, 181–208.

Maxwell, T. A. and Haynes, C. V. 1989. Large-scale, low amplitude bedforms (chevrons) in the Selima Sand Sheet, *Science*, **243**, 1179–1182.

Mayer, M., McFadden, L. D. and Harden, J. W. 1988. Distribution of calcium carbonate in desert soils: a model, *Geology*, **16**, 303–306.

McCauley, J. F., Grolier, M. J. and Breed, C. S. 1977a. Yardangs, in Doehring, D. O. (ed.) *Geomorphology in arid regions*, Allen and Unwin, London, 233–269.

McCauley, J. F., Grolier, M. J. and Breed, C. S. 1977b. Yardangs of Peru and other desert regions, *United States Geological Survey Interagency Report, Astrogeology*, **81**.

McClure, H. A. 1976. Radiocarbon chronology of late-Quaternary lakes in the Arabian desert, *Nature*, **263**, 755–756.

McCoy, F. W. Jr., Nokleberg, W. J. and Norris, R. M. 1967. Speculations on the origin of the Algodones Dunes, California, *Bulletin of the Geological Society of America*, **78**, 1039–1044.

McDonald, R. R. and Anderson, R. S. 1994. Controls on aeolian dune grain flow dynamics and geometry through laboratory experiments on sand slopes, in Desert Research Institute (1994) 81–82.

McEwan, I. K. 1993. Bagnold's kink: a physical feature of a wind velocity profile modified by blown sand, *Earth Surface Processes and Landforms*, **18**, 145–156.

McEwan, I. K. and Willetts, B. B. 1991. Numerical model of the saltation cloud, *Acta Mechanica Supplementum*, **1**, 53–66.

McEwan, I. K. and Willetts, B. B. 1993. Sand transport by wind: a review of the current physical model, in Pye (1993a) 7–16.

McFadden, L. D., Wells, S. G. and Jercinovich, M. J. 1987. Influences of eolian and pedogenic processes on the origin and evolution of desert pavements, *Geology*, **15**, 504–508.

McGee, A. W., Bull, P. A. and Goudie, A. S. 1988. Chemical textures on quartz grains: an experimental approach, *Earth Surface Processes and Landforms*, **13**, 665–670.

McKee, E. D. 1933. The Coconino Sandstone – its history and origin, *Publication*, Carnegie Institute of Washington, 78–125.

McKee, E. D. 1944. Tracks that go uphill, *Plateau*, **16**, 61–72.

McKee, E. D. 1966. Structures of dunes at White Sands National Monument, New Mexico (and a comparison with structures of dunes from other selected areas), *Sedimentology*, **7**, 1–69.

McKee, E. D. (ed.) 1979a. *A study of global sand seas, United States Geological Survey, Professional Paper* **1052**.

McKee, E. D. 1979b. Introduction to a study of global sand seas, in McKee (1979a) 1–19.

McKee, E. D. 1979c. Sedimentary structures in dunes, in McKee (1979a) 83–113.

McKee, E. D. and Tibbitts, G. C. Jr. 1964. Primary structures of a seif dune and associated deposits in Libya, *Journal of Sedimentary Petrology*, **34**, 5–17.

McKee, E. D. and Ward, W. C. 1983. Eolian environment, in Scholle, P. A., Bebont, D. G. and Moore, C. H. (eds) *Carbonate depositional environments, Memoir* 33, American Association of Petroleum Geologists, 131–170.

McKee, E. D., Douglass, J. R. and Rittenhouse, S. 1971. Deformation of lee-side laminae in eolian dunes, *Bulletin of the Geological Society of America*, **82**, 359–378.

McKenna Neuman, C. and Gilbert, R. 1986. Aeolian processes and landforms in glaciofluvial environments of southeastern Baffin Island, N.W.T., Canada, in Nickling (1986) 213–235.

McKenna Neuman, C. and Nickling, W. G. 1989. A theoretical and wind-tunnel investigation of the effect of capillary water on the entrainment of soil by wind, *Canadian Journal of Soil Science*, **69**, 79–96.

McLachlan, A. and Burns, M. 1992. Headland bypass dunes on the South African coast: 100 years of (mis)management, in Carter *et al.* (1992b) 71–79.

McLaren, P. 1981. An interpretation of trends in grain size measures, *Journal of Sedimentary Petrology*, **51**, 611–624.

McLean, S. R. 1990. The stability of ripples and dunes, *Earth Science Reviews*, **29**, 131–144.

McTainsh, G. H. 1980. Harmattan dust deposition in northern Nigeria, *Nature*, **286**, 587–588.

McTainsh, G. H. 1985. Dust processes in Australia and West Africa: a comparison, *Search*, **16**, 104–106.

McTainsh, G. H. 1986. A dust monitoring programme for desertification control in West Africa, *Environmental Conservation*, **13**, 17–25.

McTainsh, G. H. 1987. Desert loess in northern Nigeria, *Zeitschrift für Geomorphologie NF*, **31**, 145–165.

McTainsh, G. H. and Pitblado, J. R. 1987. Dust storms and related phenomena measured from meteorological records in Australia, *Earth Surface Processes and Landforms*, **12**, 415–424.

McTainsh, G. H., Burgess, R. and Pitblado, J. R. 1989. Aridity, drought and duststorms in Australia, *Journal of Arid Environments*, **16**, 11–22.

McTainsh, G. H., Lynch, A. W. and Burgess, R. C. 1990. Wind erosion in eastern Australia, *Australian Journal of Soil Research*, **28**, 323–339.

Melton, F. A. 1940. A tentative classification of sand dunes: its application to dune history in the southern High Plains, *Journal of Geology*, **48**, 113–174.

Michalsky, J. J., Pearson, E. W. and LeBarron, B. A. 1990. An assessment of the impact of volcanic eruptions on the northern hemisphere aerosol burden during the last decade, *Journal of Geophysical Research*, **95**, 5677–5688.

Michel, P. 1973. Les bassins des fleuves Sénégal et Gambie: études géomorphologiques, Mémoire, **63**, *Office de la Recherche Scientifique et Technique d'Outre Mer (ORSTOM), Paris*, 3 volumes, 752 pp.

Middleton, N. J. 1986. Dust storms in the Middle East, *Journal of Arid Environments*, **10**, 83–96.

Middleton, N. J. 1989. Desert dust, in Thomas (1989a) 262–283.

Middleton, N. J. 1991. Dust storms in the Mongolian People's Republic, *Journal of Arid Environments*, **20**, 287–298.

Middleton, N. J., Goudie, A. S. and Wells, G. L. 1986. The frequency and source areas of dust storms, in Nickling (1986) 237–260.

Miller, G. H., Paskoff, R. and Stearns, Ch. E. 1986. Amino-geochronology of Pleistocene littoral deposits in Tunisia, *Zeitschrift für Geomorphologie Supplementband*, **62**, 197–207.

Miller, R. P. 1937. Drainage in bas-relief, *Journal of Geology*, **45**, 432–438.

Millington, A. C. and Pye K. (eds) 1994. *Environmental change in drylands: biogeographical and geomorphological perspectives*, Wiley, Chichester.

Miotke, F. 1982. Formation and rate of formation of ventifacts in Victoria land, Antarctica, *Polar Geography and Geology*, **6**, 98–113.

Mitha, S., Tran, M. Q., Werner, B. T. and Haff, P. K. 1986. The grain-bed impact process in aeolian saltation, *Acta Mechanica* **63**, 267–278.

Moiola, R. J. and Weiser, D. 1968. Textural parameters: an evaluation, *Journal of Sedimentary Petrology*, **38**, 45–53.

Møller, J.-T. 1986. Soil degradation in a north European region, in Fantechi, R. and Margaris, N. S. (eds) *Desertification in Europe*, Reidel, Dordrecht, 214–230.

Monod, Th. 1958. Le Majabât al-Koubrâ, *Institut Français de l'Afrique Noire, Memoirs* 52.

Morales, C. (ed.) 1979. *Saharan dust*, John Wiley and Sons, Chichester.

Mortimore, M. J. 1989. *Adapting to drought: farmers, famines and desertification in West Africa*, Cambridge University Press, Cambridge.

Mulligan, K. R. 1988. Velocity profiles on the windward slope of a transverse dune, *Earth Surface Processes and Landforms*, **13**, 573–582.

Mullins, C. E., Mcleod, D. A., Northcote, K. H., Tisdall, J. M. and Young, I. M. 1990. Hard-setting soils: behaviour occurrence and management, *Advances in Soil Science*, **7**, 37–108.

Musick, H. B. and Gillette, D. A. 1990. Field evaluation of relationships between a vegetation structural parameter and sheltering against wind erosion, *Land Degradation and Rehabilitation*, **2**, 87–94.

Nakata, J. K., Wilshire, H. G. and Barnes, C. G. 1976. Origin of Mojave Desert dust plumes photographed from space, *Geology*, **4**, 644–648.

Nägeli, W. 1946. Weitere Untersuchungen über die Windverhältnisse im Bereich von Windschutzanlagen, *Mitteilungen der schweizerein Ant. Forstl. Versuchswesen*, **24**, 660–737.

Nalpanis, P. 1985. Saltating and suspended particles over flat and sloping surfaces. II. Experiments and numerical simulations, in Barndorff-Nielsen *et al.* (1985) 37–66.

Nanson, G. C., Chen, X. Y. and Price, D. M. 1992. Lateral migration, thermoluminescence chronology and colour variation of longitudinal dunes near Birdsville in the Simpson desert, central Australia, *Earth Surface Processes and Landforms*, **17**, 807–820.

Newell, R. E., Gould-Stewart, S. and Chung, J. C. 1981. A possible interpretation of palaeoclimatic reconstructions for 18 000 BP for the region 60° N to 60° S, 60° W to 100° E, *Palaeoecology of Africa*, **13**, Balkema, Rotterdam, 1–19.

Nickling, W. G. 1978. Eolian sediment transport during dust storms: Slims River Valley, Yukon Territory, *Canadian Journal of Earth Science*, **15**, 1069–1084.

Nickling, W. G. (ed.) 1986. *Aeolian geomorphology*, Allen and Unwin, Boston.

Nickling, W. G. 1988. The initiation of particle movement by wind, *Sedimentology*, **35**, 499–511.

Nickling, W. G. and Gillies, J. A., 1993. Dust emissions and transport in Mali, West Africa, *Sedimentology*, **40**, 859–863.

Nickling, W. G. and McKenna Neuman, C. K. 1995. Development of deflation lag surfaces, *Sedimentology*, **42**, 403–414.

Nickling, W. G. and Wolfe, S. A. 1994. The morphology and origin of nabkhas, region of Mopti, Mali, West Africa, *Journal of Arid Environments*, **28**, 13–30.

Nielson, J. and Kocurek, G. 1986. Climbing zibars of the Algodones, *Sedimentary Geology*, **48**, 1–15.

Nielson, J. and Kocurek, G. 1987. Surface processes, deposits, and development of star dunes, Dumont dune field, California, *Bulletin of the Geological Society of America*, **99**, 177–186.

Nordstrom, K. F. and Gares, P. A. 1990. Changes in the volume of coastal dunes in New Jersey, USA, *Ocean and Shoreline Management*, **14**, 1–10.

Nordstrom, K. F. and McCluskey, J. M. 1985. The effects of houses and sand fences on the eolian sediment budget at Fire island, New York, *Journal of Coastal Research*, **1**, 39–46.

Nordstrom, K. F., McCluskey, J. M. and Rosen, P. S. 1986. Aeolian processes and dune characteristics of a developed shoreline: Westhampton Beach, New York, in Nickling (1986) 131–147.

Nordstrom, K., Psuty, N. and Carter, R. W. G. (eds). 1990. *Coastal dunes: processes and morphology*, John Wiley and Sons, Chichester.

Norris, R. M. 1966. Barchan dunes of Imperial Valley, California, *Journal of Geology*, **74**, 292–306.

Norris, R. M. 1969. Dune reddening and time, *Journal of Sedimentary Petrology*, **39**, 7–11.

Norris, R. M. and Norris, K. S. 1961. Algodones dunes of southeastern California, *Bulletin of the Geological Society of America*, **72**, 605–620.

Obruchev, V. A. 1945. Loess types and their origin, *American Journal of Science*, **243**, 256–262.

Offer, Z. I. and Goosens, D. 1990. Airborne dust in the northern Negev Desert (January–December 1987) general occurrence and dust content measurement, *Journal of Arid Environments*, **18**, 1–20.

Ohmori, H., Iwasaki, K. and Takeuchi, K. 1983. Relationships between the recent dune activities and the rainfall fluctuations in the southern part of Australia, *Geographical Review of Japan*, **56**, 131–150.

Oke, T. R. 1990. *Boundary layer climates*, 2nd edn, Routledge, London.

Oldfield, F., Thompson, R. and Barber, K. E. 1978. The changing atmospheric fall-out of magnetic particles recorded in ombrotrophic peat sections, *Science*, **199**, 679–680.

Oldfield, F., Brown, A. F. and Thompson, R. 1979. The effect of microtopography and vegetation on the catchment of airborne particles measured by remnant magnetism, *Quaternary Research*, **12**, 326–332.

Oliver, F. W. 1945. Dust storms in Egypt and their relation to the war period, as noted in Maryut, 1939–45, *Geographical Journal*, **106**, 26–49.

Orange, D. and Gac, J.-Y. 1990. Bilan géochimique des apports atmosphériques en domaines sahéliens et soudano-guinéen d'Afrique de l'Ouest, *Géodynamique*, **5**, 51–65.

Orange, D., Gac, J.-Y., Probst, J.-L. and Tanré, D. 1990. Mesure du dépôt au sol des aerosols désertiques – une méthode simple de prelévement: le capteur pyramidal, *Comptes rendus de Académie des Sciences, Paris*, série 2, **311**, 167–172.

Orme, A. R. and Tchakerian, V. P. 1986. Quaternary dunes of the Pacific coast of the Californias, in Nickling (1986) 149–175.

Osterkamp, W. R. and Wood, W. W. 1987. Playa-lake basins on the southern High plains of Texas and New Mexico. Part I: Hydrologic, geomorphic and geologic evidence for their development, *Bulletin of the Geological Society of America*, **99**, 215–233.

Owen, P. R. 1964. Saltation of uniform grains in air, *Journal of Fluid Mechanics*, **20**, 225–242.

Pachur, H.-J., Röper, H. P., Kröpelin, S. and Groschin, M. 1987. Late Quaternary hydrography of the eastern Sahara, *Berliner geowissenschaftliche Abhandlungen*, **A75**, 331–384.

Parrish, J. T. and Peterson, F. 1988. Wind directions predicted from global circulation models and wind directions determined from eolian sandstones of the western United States – a comparison, *Sedimentary Geology*, **56**, 261–282.

Passarge, S. 1904. *Die Kalahari*, Reimer, Berlin.

Passarge, S. 1930. Ergebnisse einer Studienreise nach Sudtunisien im Jahre 1928, *Mitteilungen der Geographische Gesellschaft, Hamburg*, **61**, 96–122.

Pécsi, M. (ed.). 1987. *Loess and environment, Catena Supplement*, **9**, Catena Verlag, Brockenblick.

Pécsi, M. 1990. Loess is not just the accumulation of dust, *Quaternary International*, **7/8**, 1–21.

Pécsi, M. and Lóczy, D. (eds). 1990. Loess and the palaeoenvironment, *Quaternary International*, **7/8**.

Peel, R. F. 1968. Landscape sculpture by wind, *Papers of the 21st International Geographical Congress*, India, vol. I, 99–104.

Penck, A. 1905. Climatic features of the land surface, *American Journal of Science*, **19**, 165–174.

Peoples' Republic of China, Sinkiang Uighur Autonomous Region, Office of Environmental Protection. 1977. *China: tame the wind, harness the sand and transform the Gobi. Summary report on the experiences of the Sinkiang Turfan people in combatting Desertification*, United Nations Conference on Desertification (UNCOD), August 29–September 9, Nairobi, A/CONF. 74/16, also in: Biswas, M. R. and Biswas, A. K. (eds) *Desertification*, Pergamon, Oxford, 163–177.

Pesce, A. 1968. *Gemini space photographs of Libya and Tripoli: a geological and geographical analysis*, Petroleum Exploration Society of Libya, Tripoli.

Petrov, M. P. 1976. *Deserts of the world*, Halsted, New York.

Péwé, T. L., Péwé, E. A., Péwé, R. H., Journaux, A. and Slatt, R. 1981. Desert dust: characteristics and rates of deposition in central Arizona, *Geological Society of America, Special Paper* **186**, 169–190.

Phillips, C. J. and Willetts, B. B. 1979. Predicting sand deposition at porous fences, *Proceedings of the American Society of Civil Engineers, Journal of the Waterways, Port, Coastal, Oceanic Division*, **105**, 15–31.

Piper, S. 1989. Measuring the particulate pollution damage from wind erosion in the western United States, *Journal of Soil and Water Conservation*, **44**, 70–75.

Pollard, E. and Miller, A. 1968. Wind erosion in the East Anglian fens, *Weather*, **23**, 415–417.

Porter, M. L. 1986. Sedimentary record of erg migration, *Geology*, **14**, 497–500.

Porter, M. L. 1987. Sedimentology of an ancient erg margin: the Lower Jurassic Aztec Sandstone, southern Nevada and southern California, *Sedimentology*, **34**, 661–680.

Price, W. A. 1963. Physico-chemical and environmental factors in clay-dune genesis, *Journal of Sedimentary Petrology*, **33**, 766–778.

Price, W. A. 1968. Carolina Bays, in Fairbridge, R. W. (ed.) *The encyclopaedia of geomorphology*, Reinhold, New York, 102–109.

Prospero, J. M. and Nees, R. T. 1977. Dust concentration in the atmosphere of the equatorial North Atlantic: possible relationship to the Sahelian Drought, *Science*, **196**, 1196–1198.

Prouty, W. F. 1933. The Carolina Bays and elliptical lake basins, *Journal of Geology*, **43**, 200–207.

Prouty, W. F. 1952. Carolina Bays and their origin, *Bulletin of the Geological Society of America*, **63**, 167–224.

Psuty, N. P. (ed). 1988. Beach–dune interaction, *Journal of Coastal Research, Special Issue*, **3**.

Psuty, N. P. 1989. An application of science to management problems in dunes along the Atlantic coast of the USA, in Gimmingham *et al.* (1989) 289–307.

Psuty, N. P. 1992. Spatial variation in coastal foredune development, in Carter *et al.* (1992b) 3–13.

Psuty, N. P. 1993. Foredune morphology and sediment budget: Perdido Key, Florida, USA, in Pye (1993a) 145–157.

Pye, K. 1982. Morphological development of coastal dunes in a humid tropical environment, Cape Bedford and Cape Flattery, North Queensland, *Geografiska Annaler*, **A64**, 212–227.

Pye, K. 1983. Early post-depositional modification of aeolian dune sands, in Brookfield and Ahlbrandt (1983) 197–221.

Pye, K. 1984. Models of transgressive dune building episodes and their relationship to Quaternary sea-level changes: a discussion with reference to evidence from eastern Australia, in Clark, M. (ed.) *Coastal research: UK perspectives*, Geobooks, Norwich, 81–104.

Pye, K. 1987. *Aeolian dust and dust deposits*, Academic Press, London.

Pye, K. 1989. The process of fine particle formation, dust source regions, and climatic changes, in Leinen and Sarnthein (1989) 3–30.

Pye, K. 1992. Aeolian dust transport and deposition over Crete and adjacent parts of the Mediterranean Sea, *Earth Surface Processes and Landforms*, **17**, 271–288.

Pye, K. (ed.) 1993a. *The dynamics and environmental context of aeolian sedimentary systems*, Special Publication **72**, Geological Society of London.

Pye, K. 1993b. Late Quaternary development of coastal parabolic megadune complexes in northeastern Australia, in Pye and Lancaster (1993) 123–144.

Pye, K. (ed.) 1994. *Sediment transport and depositional processes*, Blackwell, Oxford.

Pye, K. and Johnson, R. 1988. Stratigraphy, geochemistry, and thermoluminescence ages of Lower Mississippi Valley loess, *Earth Surface Processes and Landforms*, **13**, 103–124.

Pye, K. and Lancaster, N. (eds) 1993. *Aeolian sediments: ancient and modern*, Special Publication **16**, International Association of Sedimentologists, Blackwell, Oxford.

Pye, K. and Neal, A. 1993. Late Holocene dune formation on the Sefton coast, northwest England, in Pye (1993a) 201–217.

Pye, K. and Sperling, C. H. 1983. Experimental investigation of silt formation by static breakage processes: the effect of temperature, moisture and salt on quartz dune sand and granitic regolith, *Sedimentology*, **30**, 49–62.

Pye, K. and Tsoar, H. 1987. The mechanics and geological implications of dust transport and deposition in deserts, with particular reference to loess formation and dune sand diagenesis in the northern Negev, Israel, in Frostick and Reid (1987) 139–156.

Pye, K. and Tsoar, H. 1990. *Aeolian sand and sand deposits*, Unwin Hyman, London.

Queiroz, J., Southard, A. R. and Woolridge, G. L. 1982. Characteristics of soils in a stone-free surficial deposit in central Utah, *Journal of the Soil Science Society of America*, **46**, 777–781.

Radley, J. and Simms, C. 1967. Wind erosion in East Yorkshire, *Nature*, **216**, 20–23.

Ranwell, D. S. 1972. *The ecology of salt marshes and sand dunes*, Chapman and Hall, London.

Ranwell, D. S. and Boar, R. 1986. *Coast dune management guide*, Natural Environment Research Council (NERC), Institute of Terrestrial Ecology (ITE).

Rasmussen, K. R. 1989. Some aspects of flow over coastal dunes, in Gimmingham *et al.* (1989) 129–147.

Rasmussen, K. R., Sørensen, M. and Willetts, B. B. 1985. Measurement of saltation and wind strength on beaches, in Barndorff-Nielsen *et al.* (1985) 301–326.

Raudkivi, A. J. 1976. *Loose boundary hydraulics*, Pergamon, Oxford.

Raupach, M. R. 1991. Saltation layers, vegetation canopies and roughness lengths, *Acta Mechanica Supplementum*, **1**, 83–96.

Raupach, M. R., Gillette, D. A. and Leys, J. F. 1993. The effect of roughness on wind erosion threshold, *Journal of Geophysical Research*, **98-D2**, 3023–3029.

Rea, D. K. 1989. Geologic record of atmospheric circulation on tectonic time scales, in Leinen and Sarnthein (1989) 841–855.

Rea, D. K. 1990. Aspects of atmospheric circulation: the late Pleistocene (0–950,000 yr) record of eolian deposition in the Pacific Ocean, *Palaeogeography, Palaeoclimatology, Palaeoecology*, **78**, 217–227.

Reheis, M. C. 1987. Soils in granitic alluvium in humid and semi-arid climates along Rock Creek, Carbon County, Montana, *Bulletin of the United States Geological Survey*, 1590-D.

Reid, D. G. 1985. Wind statistics and the shape of sand dunes, in Barndorff-Nielsen *et al.* (1985) 393–419.

Rendell, H. M. 1995. Luminescence dating of sand ramps in the eastern Mojave Desert, *Geomorphology, in press.*

Rendell, H. M. and Townsend, P. D. 1988. Thermoluminescence dating of a 10 m loess profile in Pakistan, *Quaternary Science Review*, **7**, 251–255.

Rendell, H. M., Yair, A. and Tsoar, H. 1993. Thermoluminescence dating of periods of sand movement and linear dune formation in the northern Negev, Israel, in Pye (1993a) 69–74.

Richards, P. 1985. *Indigenous agricultural revolution*, Hutchinson, London.

Richter, G. 1980. On the soil erosion problem in the temperate humid area of central Europe, *GeoJournal*, **4**, 279–287.

Riebsame, W. E. 1987. Human transformation of the United States Great Plains: patterns and causes, in Turner, B. L. II, Clark, W., Kates, R. W., Richards, J. F., Mathews, J. T. and Meyer, W. B. (eds) *The Earth as transformed by human action*, Cambridge University Press, Cambridge, 561–575.

Ritchie, W. 1989. Restoration of coastal dunes breached by pipeline landfalls in north-east Scotland, in Gimmingham *et al.* (1989) 247–265.

Ritchie, W. and Mather, A. 1971. Conservation and use: case study of the beaches of Sutherland, Scotland, *Biological Conservation*, **3**, 199–207.

Ritchie, W. and Penland, S. 1988. Cyclical changes in the coastal dunes of southern Louisiana, *Journal of Coastal Research, Special Issue*, **3**, 111–114.

Robertson-Rintoul, M. J. 1990. A quantitative analysis of the near-surface wind flow pattern over coastal parabolic dunes, in Nordstrom *et al.* (1990) 57–78.

Robertson-Rintoul, M. and Ritchie, W. 1990. The geomorphology of coastal dunes in Scotland: a review, *Catena Supplement*, **18**, 41–49.

Robinson, D. N. 1969. Soil erosion by wind in Lincolnshire, March 1968, *East Midlands Geographer*, **4**, 351–362.

Rögner, K. and Smykatz-Kloss, W. 1991. The deposition of eolian sediments in lacustrine and fluvial environments of central Sinai (Egypt), *Catena Supplement*, **20**, 75–92.

Rognon, P. 1987. Late Quaternary climatic reconstruction for the Maghreb (North Africa), *Palaeogeography, Palaeoclimatology, Palaeoecology*, **58**, 11–34.

Rognon, P. 1991. Un projet japonais de lutte contre la sécheresse au Sahel, *Sécheresse*, **2**, 135–138.

Rohrlich, V. and Goldsmith, V. 1984. Sediment transport along the southeastern Mediterranean: a geologic perspective, *Geo-Marine Letters*, **4**, 99–103.

Romeril, M. G. 1989. Dune management on Les Quennevais, Jersey, Channel Islands, GB, in van der Meulen *et al.* (1989) 255–260.

Ross, G. M. 1983. Bigbear Erg: a Proterozoic intermontane eolian sand sea in the Hornby Bay Group, Northwest Territories, Canada, in Brookfield and Ahlbrandt (1983) 483–519.

Rossby, C. G. 1941. The scientific basis of modern meteorology, in *Climate and man yearbook of agriculture*, United States Department of Agriculture, 599–655.

Rubin, D. M. 1984. Factors determining desert dune type – discussion, *Nature*, **309**, 91–92.

Rubin, D. M. 1990. Lateral migration of linear dunes in the Strzelecki Desert, Australia, *Earth Surface Processes and Landforms*, **15**, 1–14.

Rubin, D. M. and Hunter, R. E. 1982. Bedform climbing in theory and nature, *Sedimentology*, **29**, 121–138.

Rubin, D. M. and Hunter, R. E. 1985. Why deposits of longitudinal dunes are rarely recognized in the geological record, *Sedimentology*, **32**, 147–157.

Rubin, D. M. and Hunter, R. E. 1987. Bedform alignment in directionally varying flows, *Science*, **237**, 276–278.

Rubin, D. M. and Ikeda, H. 1990. Flume experiments on the alignment of transverse, oblique, and longitudinal dunes in directionally varying flows, *Sedimentology*, **37**, 673–684.

Rubin, D. M. and McCulloch, D. S. 1980. Single and superimposed bedforms: a synthesis of San Francisco Bay and flume observations, *Sedimentary Geology*, **26**, 207–231.

Rude, A. 1959. Les galets éolisées de l'Isle Herd (Australie) et les galets éolisées fossiles de Pont-Le-Chateau (Limogne d'Auvergne), *Revue de Géomorphologie dynamique*, **10**, 33–34.

Rutin, J. 1983. *Erosional processes on a coastal sand dune, De Blink, Noordwijkerhiut, The Netherlands*, Dissertatie Universiteit van Amsterdam, Publicaties **35**, Fysische Geografisch en Bodenkundig Laboratorium van de Universiteit van Amsterdam, Kaal BV, Amsterdam.

Rutin, J. 1992. Geomorphic activity of rabbits on a coastal sand dune, De Blink Dunes, the Netherlands, *Earth Surface Processes and Landforms*, **17**, 85–94.

Said, R. 1960. New light on the origin of the Qattara Depression, *Société Géographique d'Egypt, Bulletin*, **33**, 37–44.

Said, R. 1962. *The geology of Egypt*, Elsevier, Amsterdam.

Sakamoto-Arnold, C. M. 1981. Eolian features produced by the December, 1977 windstorm, southern San Joaquin Valley, California, *Journal of Geology*, **89**, 129–137.

Salisbury, E. J. 1922. The soils of Blakeney Point: a study of soil reaction and succession in relation to the plant covering, *Annals of Botany*, **36**, 391–431.

Sandford, K. S. 1933. Geology and geomorphology of the southern Libyan Desert, *Geographical Journal*, **82**, 213–219.

Sarnthein, M. 1978. Sand deserts during the last glacial maximum and climatic optimum, *Nature*, **272**, 43–46.

Sarnthein, M. and Walger, K. 1974. Der äolische Sandstrom aus der W-Sahara zur Atlantikküste, *Geologische Rundschau*, **63**, 1065–1087.

Sarnthein, M., Tetzlaff, G., Koopmann, B., Wolter, K. and Pflaumann, U. 1981. Glacial and interglacial wind regimes over the eastern tropical Atlantic and North-West Africa, *Nature*, **293**, 193–196.

Sarre, R. D. 1987. Aeolian sand transport, *Progress in Physical Geography*, **11**, 157–182.

Sarre, R. D. 1988. Evaluation of aeolian sand transport equations using intertidal zone measurements, Saunton Sands, England, *Sedimentology*, **35**, 671–679.

Sarre, R. D. 1990. Evaluation of aeolian sand transport equations using intertidal-zone measurements, Saunton Sands, England – reply, *Sedimentology*, **37**, 389–392.

Sarre, R. D. and Chancey, C. C. 1990. Size segregation during aeolian saltation on sand dunes, *Sedimentology*, **37**, 357–365.

Schenk, C. J. 1983. Textural and structural characteristics of some experimentally formed eolian strata, in Brookfield and Ahlbrandt (1983) 41–49.

Schenk, C. J. and Fryberger, S. G. 1988. Early diagenesis of eolian dune and interdune sands at White Sands, New Mexico, *Sedimentary Geology*, **55**, 109–120.

Schenk, C. J., Gautier, D. L., Olhoeft, G. R. and Lucius, J. E. 1993. Internal structure of an aeolian dune using ground-penetrating radar, in Pye and Lancaster (1993) 61–69.

Schoewe, W. H. 1932. Experiments on the formation of wind-faceted pebbles, *American Journal of Science*, **24**, 111–134.

Schwein, J. D., Willis, W. V. and Grable, A. R. 1983. Specter of another Dust Bowl seems laid to rest, in *Using our natural resources: Yearbook*, United States Department of Agriculture, Washington, DC, 422–429.

Scott, W. D. 1994. Wind erosion of residue waste. I. Using the wind profile to characterise wind erosion, *Catena*, **21**, 291–305.

Seablom, E. W. and Wiedemann, A. M. 1994. Distribution and effects of *Ammophila breveligulata* Fern, (American beachgrass) on the foredunes of the Washington coast, *Journal of Coastal Research*, **10**, 178–188.

Seevers, P. M., Lewis, D. T. and Drew, J. V. 1975. Use of ERTS-1 imagery to interpret the wind erosion in Nebraska's Sand Hills, *Journal of Soil and Water Conservation*, **30**, 181–184.

Seginer, I. 1975. Flow round a windbreak in an oblique wind, *Boundary-Layer Meteorology*, **9**, 133–141.

Selby, M. J., Palmer, R. W. P., Smith, C. J. R. and Rains, R. B. 1973. Ventifact distribution and wind directions in the Victoria Valley, Antarctica, *New Zealand Journal of Geology and Geophysics*, **16**, 303–306.

Seliskar, D. M. 1994. The effect of accelerated sand accretion on growth, carbohydrate reserves and ethylene production in *Ammophyla breveligulata* (Poaceae), *American Journal of Botany*, **81**, 536–541.

Semeniuk, V. and Glassford, D. K. 1988. Significance of aeolian limestone lenses in quartz sand formations: an interdigitation of coastal and continental facies, Perth Basin, southwestern Australia, *Sedimentary Geology*, **57**, 199–210.

Seppälä, M. and Lindé, K. 1978. Wind tunnel studies of ripple formation, *Geografiska Annaler*, **60A**, 29–40.

Sexton, W. J. and Hays, M. O. 1992. The geologic impact of Hurricane Hugo and post-storm shoreline recovery along the undeveloped coastline of South Carolina, Dewees Island to the Santee Delta, *Journal of Coastal Research*, **8**, 275–290.

Sharp, R. P. 1949. Pleistocene ventifacts east of the Bighorn Mountains, Wyoming, *Journal of Geology*, **57**, 175–195.

Sharp, R. P. 1963. Wind ripples, *Journal of Geology*, **71**, 617–636.

Sharp, R. P. 1964. Wind-driven sand in the Coachella valley, California, *Bulletin of the Geological Society of America*, **75**, 785–804.

Sharp, R. P. 1966. Kelso Dunes, Mojave Desert, California, *Bulletin of the Geological Society of America*, **77**, 1045–1074.

Sharp, R. P. 1979. Intradune flats of the Algodones chain, Imperial Valley, California, *Bulletin of the Geological Society of America*, **90**, 908–916.

Shaw, P. A. and Cooke, H. J. 1986. Geomorphic evidence for the Late Quaternary palaeoclimates of the middle Kalahari and northern Botswana, *Catena*, **13**, 349–359.

Shepard, F. P. and Young, R. 1961. Distinguishing between beach and dune sands, *Journal of Sedimentary Petrology*, **31**, 196–214.

Sherman, D. J. 1992. An equilibrium relationship for shear velocity and apparent roughness length in aeolian saltation, *Geomorphology*, **5**, 419–431.

Sherman, D. J. and Bauer, B. O. 1993. Dynamics of beach–dune systems, *Progress in Physical Geography*, **17**, 413–447.

Sherman, D. J. and Hotta, S. 1989. Eolian sediment transport: theory and measurement, in Nordstrom *et al.* (1989) 17–33.

Short, A. D. 1988. Holocene coastal dune formation in South Australia: a case study, *Sedimentary Geology*, **55**, 121–143.

Short, A. D. and Hesp, P. A. 1982. Wave, beach and dune interactions in southeastern Australia, *Marine Geology*, **48**, 259–284.

Simonett, D. S. 1949. Sand dunes near Castlereagh, New South Wales, *Australian Geographer*, **5**, 3–10.

Simons, D. B., Richardson, E. V. and Nordin, C. F. Jr. 1965. Bedload equation for ripples and dunes, *United States Geological Survey, Professional Paper*, **462-H**, H1–H9.

Simons, F. S. 1956. A note on the Pur-Pur dune, Viru Valley, Peru, *Journal of Geology*, **64**, 517–521.

Simons, F. S. and Eriksen, G. E. 1953. Some desert features of northwest central Peru, *Sociedad Geologica del Peru, Boletin*, **26**, 229–245.

Simpson, J. E. 1994. *Sea breeze and local winds*, Cambridge University Press, Cambridge.

Sinclair, P. C. 1969. General characteristics of dust devils, *Journal of Applied Meteorology*, **8**, 32–45.

Singhvi, A. K., Sharma, P. and Agarwal, D. P. 1982. Thermoluminescence dating of sand dunes in Rajasthan, India, *Nature*, **295**, 313–315.

Sirocko, F. and Sarnthein, M. 1989. Wind-borne deposits in the northwestern Indian Ocean: record of Holocene sediments versus modern satellite data, in Leinen and Sarnthein (1989) 401–433.

Sirocko, F., Lange, H. and Erlenkeuser, H. 1991. Atmospheric summer circulation and coastal upwelling in the Arabian sea during the Holocene and the last glaciation, *Quaternary Research*, **36**, 72–93.

Skarregaard, P. 1989. Stabilisation of coastal dunes in Denmark, in van der Meulen *et al.* (1989) 151–162.

Skidmore, E. L. 1986a. Wind erosion control, *Climatic Change*, **9**, 209–218.

Skidmore, E. L. 1986b. Soil erosion by wind: an overview, in El-Baz and Hassan (1986) 261–273.

Skidmore, E. L. and Hagen, L. J. 1977. Reducing wind erosion with barriers, *Transactions of the American Society of Agricultural Engineers*, **20**, 911–915.

Smalley, I. J. 1970. Cohesion of soil particles and the intrinsic resistance of simple soil systems to wind erosion, *Journal of Soil Science*, **21**, 154–161.

Smalley, I. J. 1990. Possible formation mechanisms for the modal coarse-silt quartz particles in loess, *Quaternary International*, **7/8**, 23–27.

Smalley, I. J. and Smalley, V. 1983. Loess material and loess deposits: formation, distribution and consequences, in Brookfield and Ahlbrandt (1983) 51–68.

Smith, B. J., McGreevy, J. P. and Whalley, W. B. 1987. The production of silt-size quartz by experimental salt weathering of a sandstone, *Journal of Arid Environments*, **12**, 199–214.

Smith, B. J., Wright, J. S. and Whalley, W. B. 1991. Simulated aeolian abrasion of Panonian sands and its implications for the origins of Hungarian loess, *Earth Surface Processes and Landforms*, **16**, 745–752.

Smith, D. M., Jarvis, P. G. and Odongo, J. C. W. 1995. Water use by Sahelian windbreak trees, in *Wind erosion in West Africa: the problem and its control*, International Symposium, 5–7 December, Universität Hohenhiem, in press.

Smith, H. T. U. 1945. Giant grooves in northwest Canada (abstract), *Bulletin of the Geological Society of America*, **56**, 1198.

Sneh, A. 1988. Permian dune patterns in northwestern Europe challenged, *Journal of Sedimentary Petrology*, **58**, 44–51.

Sneh, A. and Weissbrod, T. 1983. Size-frequency distribution on longitudinal dune ripple flank sands compared to that of slipface sands of various dune types, *Sedimentology*, **30**, 717–726.

Sorensen, C. J. and Marotz, G. A. 1977. Changes in shelterbelt milage statistics over four decades in Kansas, *Journal of Soil and Water Conservation*, **32**, 276–281.

Spaan, W. P. and van den Abeele, G. D. 1991. Wind-borne particle measurements with acoustic sensors, *Soil Technology*, **4**, 51–63.

Stapff, F. M. 1887. Karte des unteren !Kuisebthales, *Petermanns Mitteilungen*, **33**, 202–214.

Stapor, F. W., May, J. P. and Barwis, J. 1983. Eolian shape-sorting and aerodynamic traction equivalence in the coastal dunes of Hout Bay, Republic of South Africa, in Brookfield and Ahlbrandt (1983) 149–164.

Statham, I. 1977. *Earth surface sediment transport*, Oxford University Press, Oxford.

Stebelski, I. 1985. Agricultural development and soil degradation in the Soviet Union: policies, patterns and trends, in Singleton, F. (ed.) *Environmental problems in the Soviet Union and eastern Europe*, Lynne Rienner, London, 71–96.

Steele, R. P. 1985. Early Permian (Rotliegendes) palaeowinds of the North Sea – comment, *Sedimentary Geology*, **45**, 293–313.

Stipho, A. S. 1992. Aeolian sand hazards and engineering design for desert regions, *Quarterly Journal of Engineering Geology*, **25**, 83–92.

Stockton, P. H. and Gillette, D. A. 1990. Field measurement of sheltering effect of vegetation on erodible land surfaces, *Land Degradation and Rehabilitation*, **2**, 77–85.

Stokes, S. and Breed, C. S. 1993. A chronostratigraphic re-evaluation of the Tusayan Dunes, Moenkopi Plateau and southern Ward Terrace, northeastern Arizona, in Pye (1993a) 75–90.

Stokes, S. and Gaylord, D. R. 1993. Optical dating of Holocene dune sands in the Ferris Dune Field, Wyoming, *Quaternary Research*, **39**, 274–281.

Stokes, W. L. 1968. Multiple parallel-truncation bedding planes – a feature of wind-deposited sandstone formations, *Journal of Sedimentary Petrology*, **38**, 510–515.

Stolt, M. H. and Rabenhorst, M. C. 1987. Carolina Bays on the eastern shore of Maryland, I: soil characterization and classification, *Soil Science Society of America, Journal*, **51**, 394–398.

Storrier, R. R. and Stannard, M. E. (eds) 1980. *Aeolian landscapes in the semi-arid zone of south eastern Australia*, Australian Society of Soil Science, Inc., Riverina Branch.

Story, R. 1982. Notes on parabolic dunes, winds and vegetation in northern Australia, *Technical Paper* 43, Division of Water and Land Resources, Commonwealth Scientific and Industrial Organization.

Sturt, C. 1849. *Expedition into central Australia*, London.

Suslov, S. P. 1961. *Physical geography of Asiatic Russia* (translated by N. D. Gershevsky), Freeman, London.

Suzuki, T. and Takahashi, K. 1981. An experimental study of wind abrasion, *Journal of Geology*, **89**, 23–36.

Swart, D. H. and Reyneke, P. G. 1988. The role of driftsands at Waenhuiskrans, South Africa, *Journal of Coastal Research* (Special Issue), **3**, 97–102.

Sweet, M. L., Nielson, J., Havholm, Karen and Farrelley, J. 1988. Algodones dune field of southern California: case history of a migrating modern dune field, *Sedimentology*, **35**, 939–952.

Swift, D. J. P., Molina, B. F. and Jackson, R. G. III 1978. Intermittent structure of the atmospheric boundary layer made visible by entrained sediment: example from the Copper River Delta, Alaska, *Journal of Sedimentary Petrology*, **48**, 897–900.

Syers, J. K. and Walker, T. W. 1969. Phosphorus transformations in a chronosequence of soils developed on wind-blown sand in New Zealand, I. Total and organic phosphorus, *Journal of Soil Science*, **20**, 57–64.

Szczypek, T. and Wach, J. 1993. Human impact and intensivity of aeolian processes in the Silesian–Cracow upland (southern Poland), *Zeitschrift für Geomorphologie Supplementband*, **90**, 171–178.

Taira, A. and Scholle, P. A. 1979. Origin of bimodal sands in some modern environments, *Journal of Sedimentary Petrology*, **49**, 777–786.

Talbot, M. R. 1980. Environmental responses to climatic change in the West African Sahel over the past 20 000 years, in Williams, M. A. J. and Faure, H. (eds) *The Sahara and the Nile*, Balkema, Rotterdam, 37–62.

Talbot, M. R. 1984. Late Pleistocene rainfall and dune building in the Sahel, *Palaeoecology of Africa*, **16**, Balkema, Rotterdam, 203–214.

Talbot, M. R. and Williams, M. A. J. 1978. Erosion of fixed dunes in the Sahel, Central Niger, *Earth Surface Processes and Landforms*, **3**, 107–113.

Taylor, K. C., Lamorey, G. W., Doyle, G. A., Alley, R. B., Grootes, P. M., Mayewski, P. A., White, J. W. C. and Barllow, L. K. 1993. The 'flickering switch' of Late Pleistocene climatic change, *Nature*, **361**, 432–436.

Tchakerian, V. P. 1994. Palaeoclimatic interpretations from desert dunes and sediments, in Abrahams and Parsons (1994) 631–643.

Tear, F. J. 1925. Sand dune reclamation in Palestine, *Empire Forestry Journal*, **4**, 24–38.

Tetzlaff, G. and Peters, M. 1986. Deep-sea sediments in the eastern equatorial Atlantic off the African coast and meteorological flow patterns over the Sahel, *Geologische Rundschau*, **75**, 71–79.

Tetzlaff, G., Peters, M., Janssen, W. and Adams, L. J. 1989. Aeolian dust transport in West Africa, in Leinen and Sarnthein (1989) 185–201.

Thom, A. S. 1971. Momentum absorption by vegetation, *Quarterly Journal of the Royal Meteorological Society*, **97**, 414–418.

Thomas, D. S. G. 1984. Ancient ergs of the former arid zones of Zimbabwe, Zambia and Angola, *Transactions of the Institute of British Geographers*, **9**, 75–88.

Thomas, D. S. G. 1987. The roundness of aeolian quartz sand grains, *Sedimentary Geology*, **52**, 149–153.

Thomas, D. S. G. 1988. Analysis of linear dune sediment-form relationships in the Kalahari Dune Desert, *Earth Surface Processes and Landforms*, **13**, 545–553.

Thomas, D. S. G. (ed.) 1989a. *Arid zone geomorphology*, Belhaven, London.

Thomas, D. S. G. 1989b. Aeolian sand deposits, in Thomas (1989a) 232–261.

Thomas, D. S. G. and Goudie, A. S. 1984. Ancient ergs of the southern hemisphere, in Vogel (1984) 407–418.

Thomas, D. S. G. and Shaw, P. A. 1991. *The Kalahari environment*, Cambridge University Press, Cambridge.

Thomas, D. S. G. and Shaw, P. A. 1993. The evolution and characteristics of the Kalahari, southern Africa, *Journal of Arid Environments*, **25**, 97–108.

Thomas, D. S. G. and Tsoar, H. 1990. The geomorphological role of vegetation in desert dune systems, in Thornes, J. B. (ed.) *Vegetation and erosion*, John Wiley and Sons, Chichester, 471–489.

Thompson, C. H. and Bowman, G. M. 1984. Subaerial denudation and weathering of vegetated coastal dunes in eastern Australia, in Thom, B. G. (ed.) *Coastal geomorphology in Australia*, Academic Press, London, 263–290.

Thorne, C. E. and Darmody, R. G. 1980. Contemporary eolian sediments in the alpine zone, Colorado Front Range, *Physical Geography*, **1**, 162–171.

Tiessen, H., Hauffe, H. K. and Mement, A. R. 1991. Deposition of Haramttan dust and its influence on base saturation of soils in Ghana, *Geoderma*, **49**, 285–299.

Tiller, K. G., Smith, L. H. and Merry, J. H. 1987. Accessions of atmospheric dust east of Adelaide, South Australia, and the implications for pedogenesis, *Australian Journal of Soil Research*, **25**, 43–45.

Tinker, J. 1977. Sudan challenges the sand dragon, *New Scientist*, **73**, 448–450.

Tinus, R. W. (ed.) 1976. *Shelterbelts on the Great Plains*, Publication 78, Great Plains Agricultural Council.

Tricart, J. 1984. Evidence of Upper Pleistocene dry climates in northern South America, in Douglas, I. and Spencer, T. (eds) *Environmental change and tropical geomorphology*, Allen and Unwin, London, 197–217.

Tseo, G. 1990. Reconnaissance of the dynamic characteristics of an active Strzelecki longitudinal dune, south-central Australia, *Zeitschrift für Geomorphologie NF*, **34**, 19–36.

Tseo, G. 1993. Two types of longitudinal dune fields and possible mechanisms for their development, *Earth Surface Processes and Landforms*, **18**, 627–643.

Tsoar, H. 1974. Desert dunes, morphology and dynamics, El-Arish, northern Sinai, *Zeitschrift für Geomorphologie Supplementband*, **20**, 41–61.

Tsoar, H. 1978. *The dynamics of longitudinal dunes*, Final Technical Report, European Research Office, United States Army, London, DA-ERO 76-G-072.

Tsoar, H. 1982. Internal structure and surface geometry of longitudinal (seif) dunes, *Journal of Sedimentary Petrology*, **52**, 823–832.

Tsoar, H. 1983a. Dynamic processes acting on a longitudinal (seif) dune, *Sedimentology*, **30**, 567–578.

Tsoar, H. 1983b. Wind tunnel modelling of echo and climbing dunes, in Brookfield and Ahlbrandt (1983) 247–260.

Tsoar, H. 1985. Profiles analysis of sand dunes and their steady state signification, *Geografiska Annaler*, **67A**, 47–59.

Tsoar, H. 1986. Two-dimensional analysis of dune profiles and the effect of grain size on sand dune morphology, in El-Baz and Hassan (1986) 94–108.

Tsoar, H. 1989. Linear dunes – forms and formation, *Progress in Physical Geography*, **13**, 507–528.

Tsoar, H. 1990a. Grain-size characteristics of wind ripples on a desert seif dune, *Geography Research Forum*, **10**, 37–50.

Tsoar, H. 1990b. New models for the formation of linear sand dunes – a discussion, *Geography*, **75**, 144–147.

Tsoar, H. 1990c. Trends in the development of sand dunes along the southeastern Mediterranean coast, *Catena Supplement*, **18**, 51–58.

Tsoar, H. and Blumberg, D. 1991. The effect of sea cliffs on inland encroachment of aeolian sand, *Acta Mechanica Supplementum*, **2**, 131–146.

Tsoar, H. and Møller, J.-T. 1986. The role of vegetation in the formation of linear sand dunes, in Nickling (1986) 75–97.

Tsoar, H. and Pye, K. 1987. Dust transport and the question of desert loess formation, *Sedimentology*, **34**, 139–154.

Tsoar, H., Rasmussen, K. R., Sørensen, M. and Willetts, B. B. 1985. Laboratory studies of flow over dunes, in Barndorff-Nielsen *et al.* (1985) 327–350.

Udden, J. A. 1894. Erosion, transportation and sedimentation performed by the atmosphere, *Journal of Geology*, **2**, 318–331.

Udden, J. A. 1914. Mechanical composition of clastic sediments, *Bulletin of the Geological Society of America*, **25**, 655–744.

Udo de Haes, H. A. and Wolters, A. R. 1992. The golden fringe of Europe: ideas for a European coastal conservation strategy and action plan, in Carter *et al.* (1992b) 525–530.

Uematsu, M., Duce, R. A., Prospero, J. M., Chen, L., Merrill, J. T. and McDonald, R. L. 1983. Transport of mineral aerosol from Asia over the North Pacific Ocean, *Journal of Geophysical Research*, **88C**, 5343–5352.

United Nations Conference on Desertification (UNCOD) 1977. *Sahel Green Belt Transnational Project, Item 5 of the provisional agenda, Action Plan to Combat Desertification. Feasibility study*, UNCOD, Nairobi, 29 August–September 1977.

United Nations Environment Programme (UNEP) 1992. *World atlas of desertification*, Edward Arnold, London.

Valentin, C. 1991. Surface crusting in two alluvial soils of northern Niger, *Geoderma*, **48**, 201–222.

Van Burkalow, A. 1945. Angle of repose and angle of sliding friction: an experimental study, *Bulletin of the Geological Society of America*, **56**, 669–707.

Vandenberghe, J. 1993. Changing conditions of aeolian sand deposition during the last deglaciation period, *Zeitschrift für Geomorphologie Supplementband*, **90**, 193–207.

Van der Maarel, E. 1981. Environmental management of coastal dunes in the Netherlands, in Jefferies, R. L. and Davy, A. J. (eds) *Ecological processes in coastal environments*, Blackwell, Oxford, 543–570.

Van der Meulen, F. 1990. European dunes: consequences of climatic change and sea level rise, in Bakker *et al.* (1990) 209–223.

Van der Meulen, F. and Jungerius, P. D. 1989. Landscape development in Dutch coastal dunes: the breakdown and restoration of geomorphological and geohydrological processes, in Gimmingham *et al.* (1989) 219–229.

Van der Meulen, F., Jungerius, P. D. and Visser, J. (eds) 1989. *Perspectives in coastal dune management*, SPB Scientific, The Hague.

Van der Meulen, F., Witter, J. V. and Ritchie, W. 1991. Precepts, approaches and strategies, *Landscape Ecology*, **6**, 7–13.

Van Vliet-Lanoë, B., Seppälä, M. and Käyhko, J. 1993. Dune dynamics and cryoturbation features controlled by Holocene water-level change, Hietàtievat, Finnish Lapland, *Geologie en Mijnbouw*, **72**, 211–224.

Van Zoest, J. 1992. Gambling with nature? A new paradigm of nature and its consequences for nature management strategy, in Carter *et al.* (1992b) 503–515.

Veenstra, H. J. and Winkelmolen, A. M. 1971. Directional trends in Dutch coversands, *Geologie en Mijnbouw*, **50**, 547–558.

Verlaque, Ch. 1958. Les dunes d'In Salah, *Traveaux de l'Institut de Recherches sahariennes*, **17**, 12–58.

Verstappen, H. Th. 1968. On the origin of longitudinal (seif) dunes, *Zeitschrift für Geomorphologie NF*, **12**, 200–220.

Verstappen, H. Th. 1970. Aeolian geomorphology of the Thar Desert and palaeo-climates, *Zeitschrift für Geomorphologie Supplementband*, **10**, 104–120.

Vilenskii, D. G. 1963. *Soil science*, Israel Program for Scientific Translations, Jerusalem.

Vincent, P. 1984. Particle size variation over a transverse dune in the Nafd as Sirr, central Saudi Arabia, *Journal of Arid Environments*, **7**, 329–336.

Visher, G. S. 1969. Grain size distributions and depositional processes, *Journal of Sedimentary Petrology*, **39**, 1074–1106.

Vogel, J. C. (ed.) 1984. *Late Cainozoic palaeoclimates of the southern hemisphere*, Balkema, Rotterdam.

Walker, J. D. and Southard, J. B. 1982. Experimental study of wind ripples, *Proceedings of the 11th Congress, International Association of Sedimentologists*, Hamilton, Ontario, 65.

Walker, T. R. 1979. Red color in dune sand, in McKee (1979a) 52–81.

Walmsley, J. L. and Howard, A. D. 1985. Application of a boundary-layer model to flow over an eolian dune, *Journal of Geophysical Research*, **90** (D6), 10 631–10 640.

Walther, J. 1924. *Das Gesetz der Wüstenbildung in Gegenwart und Vorzeit*, Reimer, Berlin.

Wang Yue and Gong Guangrun 1994. Sand sea history of the Taklimakan for the past 30,000 years, *Geografiska Annaler*, **76A**, 131-144.

Ward, A. W. and Greeley, R. 1984. Evolution of yardangs at Rogers Lake, California, *Bulletin of the Geological Society of America*, **95**, 829–837.

Ward, J. D. 1984. *Aspects of the Cenozoic geology in the Kuiseb Valley, central Namib Desert*, Unpublished PhD thesis, University of Natal.

Warren, A. 1968. *The Qoz region of Kordofan*, PhD Dissertation, University of Cambridge.

Warren, A. 1970. Dune trends and their implications in the central Sudan, *Zeitschrift für Geomorphologie Supplementband*, **10**, 154–179.

Warren, A. 1971. Dunes in the Ténéré Desert, *Geographical Journal*, **137**, 458–461.

Warren, A. 1972. Observations on dunes and bi-modal sands in the Ténéré desert, *Sedimentology*, **19**, 37–44.

Warren, A. 1976a. Morphology and sediments of the Nebraska Sand Hills in relation to Pleistocene winds and the development of eolian bedforms, *Journal of Geology*, **84**, 685–700.

Warren, A. 1976b. Dune trend and the Ekman spiral, *Nature*, **259**, 653–654.

Warren, A. 1988a. The dynamics of network dunes in the Wahiba Sands; a progress report, in Dutton (1988) 169–181.

Warren, A. 1988b. A note on vegetation and sand movement in the Wahiba Sands, in Dutton (1988) 251–255.

Warren, A. 1988c. The dunes of the Wahiba Sands, in Dutton (1988) 131–160.

Warren, A. 1995. Changing understandings of African pastoralism and environmental paradigms, *Transactions of the Institute of British Geographers*, **20**, 193–203.

Warren, A. and Knott, P. 1983. Desert dunes: a short review of needs in desert dune research and a recent study of micrometeorological dune-initiation mechanisms, in Brookfield and Ahlbrandt (1983) 343–352.

Wasson, R. J. 1983a. The Cainozoic history of the Strzelecki and Simpson dune fields (Australia), and the origin of the desert dunes, *Zeitschrift für Geomorphologie Supplementband*, **45**, 85–115.

Wasson, R. J. 1983b. Dune sediment types, sand colour, sediment provenance and hydrology in the Strzelecki-Simpson dunefield, Australia, in Brookfield and Ahlbrandt (1983) 165–195.

Wasson, R. J. 1984. Late Quaternary palaeoenvironments in the desert dunefields of Australia, in Vogel (1984) 419–432.

Wasson, R. J. and Hyde, R. 1983. Factors determining desert dune type, *Nature*, **304**, 337–339.

Wasson, R. J. and Nanninga, P. M. 1986. Estimating wind transport of sand on vegetated surfaces, *Earth Surface Processes and Landforms*, **11**, 505–514.

Wasson, R. J., Rajaguru, S. N., Misra, V. N., Agarwal, D. P., Dhir, R. P., Singhvi, A. K. and Kameswara Rao, K. 1983. Geomorphology, Late Quaternary stratigraphy and palaeoclimatology of the Thar dune field, *Zeitschrift für Geomorphologie Supplementband*, **45**, 117–151.

Wasson, R. J., Fitchett, K., Mackey, B. and Hyde, R. 1988. Large-scale patterns of dune type, spacing and orientation in the Australian continental dunefield, *Australian Geographer*, **19**, 89–104.

Watson, A. 1985. The control of wind blown sand and moving dunes: a review of the methods of sand control in deserts, with observations from Saudi Arabia, *Quarterly Journal of Engineering Geology*, **18**, 237–252.

Watson, A. 1986. Grain-size variations on a longitudinal dune and a barchan dune, *Sedimentary Geology*, **46**, 49–66.

Watson, A. 1988. Desert gypsum crusts as palaeoenvironmental indicators: a micropetrographic study of crusts from southern Tunisia and the central Namib Desert, *Journal of Arid Environments*, **15**, 19–42.

Watson, A. 1990. The control of blowing sand and mobile desert dunes, in Goudie (1990) 35–86.

Watts, W. A. and Wright, H. E. Jr. 1966. Late-Wisconsin pollen and seed analysis from the Nebraska Sand Hills, *Ecology*, **47**, 202–210.

Weber, K. J. 1987. Computation of initial well productivities in aeolian sandstone on the basis of a geologic model, Leman gas field, UK, in Wilman, R. W. and Weber, K. J. (eds), *Reservoir sedimentology*, Special Publication 40, Society for Economic Palaeontology and Mineralogy, Tulsa, Oklahoma, 333–354.

Wehmeier, E. 1980. Desertification processes and groundwater utilisation in the northern Nefzaoua, Tunisia, in Meckelein (1980) 125–143.

Wellendorf, W. and Krinsley, D. H. 1980. Wind velocities determined from the surface textures of sand grains, *Nature*, **283**, 372–373.

Wells, S. G., McFadden, L. D. and Dohrenwend, J. C. 1987. Influence of late-Quaternary climatic changes on geomorphic and pedogenic processes on a desert piedmont, eastern Mojave Desert, California, *Quaternary Research*, **27**, 130–146.

Wentworth, C. K. 1922. The shape of beach pebbles, *United States Geological Survey Professional Paper* **131C**.

Werner, B. T. 1990. A steady-state model of wind-blown sand transport, *Journal of Geology*, **98**, 1–17.

Werner, B. T. 1994. Computer simulation of eolian dunes, in Desert Research Institute (1994) 111.

Werner, B. T. and Haff, P. K. 1988. The impact process in aeolian saltation: two-dimensional simulations, *Sedimentology*, **35**, 189–196.

Werner, B. T., Haff, P. K., Livi, R. P. and Anderson, R. S. 1986. The measurement of eolian ripple cross-sectional shapes, *Geology*, **14**, 743–745.

Westoby, M., Walker, B. and Noy-Meir, I. 1989. Range management on the basis of a model which does not seek to establish equilibrium, *Journal of Arid Environments*, **17**, 235–239.

Westphal, D. L., Toon, O. B. and Carlson, T. N. 1987. A two-dimensional numerical investigation of the dynamics and microphysics of Saharan dust storms, *Journal of Geophysical Research*, **92**, 3027–3049.

Whalley, W. B. (ed.) 1978. *Scanning electron microscopy in the study of sediments*, Geobooks, Norwich.

Whalley, W. B., Smith, B. J., McAlister, J. J. and Edwards, A. J. 1987. Aeolian abrasion of quartz particles and the production of silt-size fragments: preliminary results, in Frostick and Reid (1987) 129–138.

Wheaton, E. E. 1992. Prairie dust storms: a neglected hazard, *Natural Hazards*, **5**, 53–63.

White, G. F. 1986. '*The Future of the Great Plains*' revisited, *Great Plains Quarterly*, **6**, 84–93

White, K. H. and Drake, N. A. 1993. Mapping the distribution and abundance of gypsum in south-central Tunisia from Landsat Thematic Mapper data, *Zeitschrift für Geomorphologie NF*, **37**, 309–325.

Whitehead, E. E., Hutchinson, C. F., Timmermann, B. N. and Varady, R. G. (eds) 1988. *Arid lands: today and tomorrow*, *Proceedings of an International Research and Development Conference*, Westview Press, Boulder, CO.

Whitney, M. I. 1978. The role of vorticity in developing lineation by wind erosion, *Bulletin of the Geological Society of America*, **89**, 1–18.

Whitney, M. I. 1983. Eolian features shaped by aerodynamic and vorticity processes, in Brookfield and Ahlbrandt (1983) 223–245.

Whitney, M. I. 1985. Yardangs, *Journal of Geological Education*, **33**, 93–96.

Whitney, M. I. and Brewer, H. B. 1968. Discoveries in aerodynamic erosion with wind tunnel experiments, *Michigan Academy of Science, Arts and Letters*, **53**, 91–104.

Whitney, M. I. and Dietrich, R. V. 1973. Ventifact sculpture by windblown dust, *Bulletin of the Geological Society of America*, **84**, 2561–2581.

Whitney, M. I. and Splettstoesser, J. F. 1982. Ventifacts and their formation: Darwin Mountains, Antarctica, *Catena Supplement*, **1**, 175–194.

Wiggs, G. F. S. 1993. An integrated study of desert dune dynamics, in Pye (1993a) 37–46.

Wiggs, G. F. S., Livingstone, I., Thomas, D. S. G. and Bullard, J. E. 1994. The effect of vegetation removal on airflow structure and dune dynamics in the southwest Kalahari, *Land Degradation and Rehabilitation*, **5**, 13–24.

Wiggs, G. F. S., Livingstone, I. and Warren, A. 1995. The role of streamline curvature in sand dune dynamics: evidence from field and wind-tunnel measurements, *Earth Surface Processes and Landforms, in press*.

Wigner, K. A. and Peterson, R. E. 1987. Synoptic climatology of blowing dust on the Texas South Plains, *Journal of Arid Environments*, **13**, 199–209.

Wilburg, P. L. and Rubin, D. M. 1989. Bed roughness produced by saltating sediments, *Journal of Geophysical Research*, **94**, 5011–5016.

Willetts, B. B. and Phillips, C. J. 1978. Using fences to create and stabilize dunes, *Coastal Engineering*, **2**, 2040–2050.

Willetts, B. B. and Rice, M. A. 1983. Practical representation of characteristic grain shape in sands: a comparison of methods, *Sedimentology*, **30**, 557–565.

Willetts, B. B. and Rice, M. A. 1985a. Wind-tunnel tracer experiments using dyed sand, in Barndorff-Nielsen *et al.* (1985) 225–242.

Willetts, B. B. and Rice, M. A. 1985b, Inter-saltation collisions, in Barndorff-Nielsen *et al.* (1985) 83–100.

Willetts, B. B. and Rice, M. A. 1986. Collision in aeolian transport; the saltation/creep link, in Nickling (1986) 1–19.

Willetts, B. B. and Rice, M. A. 1989. Collisions of quartz grains with a sand bed: the influence of incident angle, *Earth Surface Processes and Landforms*, **14**, 719–730.

Willetts, B. B., Rice, M. A. and Swaine, S. E. 1982. Shape effects in aeolian grain transport, *Sedimentology*, **29**, 409–417.

Williams, G. P. 1964. Some aspects of the eolian saltation load, *Sedimentology*, **3**, 257–287.

Williams, J. J., Butterfield, G. R. and Clark, D. G. 1990. Rates of aerodynamic entrainment in a developing boundary layer, *Sedimentology*, **37**, 1039–1048.

Williams, M. A. J. 1994. Cenozoic climatic changes in deserts: a synthesis, in Abrahams and Parsons (1994) 644–670.

Williams, M. A. J., Abell, P. I. and Sparks, B. W. 1987. Quaternary landforms, depositional environments and gastropod isotope ratios at Adrar Bous, Ténéré Desert of Niger, south-central Sahara, in Frostick and Reid (1987) 105–125.

Willis, A. J. 1989. Coastal sand dunes and biological systems, in Gimmingham *et al.* (1989) 17–36.

Wilson, I. G. 1971. Desert sandflow basins and a model for the development of ergs, *Geographical Journal*, **137**, 180–199.

Wilson, I. G. 1972a. Aeolian bedforms – their development and origins, *Sedimentology*, **19**, 173–210.

Wilson, I. G. 1972b. Universal discontinuities in bedforms produced by the wind, *Journal of Sedimentary Petrology*, **42**, 667–669.

Wilson, I. G. 1973. Ergs, *Sedimentary Geology*, **10**, 77–106.

Wilson, P. 1979. Experimental investigation of etch-pit formation on quartz sand grains, *Geological Magazine*, **116**, 477–482.

Wilson, P. 1988. Recent sand shadow development on Muckish Mountain, County Donegal, *Irish Naturalists' Journal*, **22**, 529–531.

Wilson, P. 1989. Nature, origin and age of Holocene aeolian sand on Muckish Mountain, Co. Donegal, Ireland, *Boreas*, **18**, 159–168.

Wilson, P. 1991. Sediment clasts and ventifacts from the north coast of Northern Ireland, *Irish Naturalists' Journal*, **23**, 442–446.

Wilson, P. 1992. Trends and timescales in soil development and coastal dunes in the north of Ireland, in Carter *et al.* (1992b) 153–162.

Winkelmolen, A. M. 1971. Rollability, a functional shape property of grains, *Journal of Sedimentary Petrology*, **41**, 703–714.

Wintle, A. G. 1993. Luminescence dating of aeolian sands – an overview, in Pye (1993a) 49–58.

Wippermann, F. K. and Gross, G. 1986. The wind-induced shaping and migration of an isolated dune: a numerical experiment, *Boundary-Layer Meteorology*, **36**, 319–334.

Witter, V., Jungerius, P. D. and ten Harkel, M. 1991. Modelling water erosion and the impact of water repellency, *Catena*, **18**, 115–125.

Wolfe, S. A. and Nickling, W. G. 1993. The protective role of sparse vegetation in wind erosion, *Progress in Physical Geography*, **17**, 50–68.

Woodruff, N. P. and Siddoway, F. H. 1965. A wind erosion equation, *Proceedings of the Soil Science Society of America*, **29**, 602–608.

Worster, D. 1979. *Dust Bowl: the southern High Plains in the 1930s*, Oxford University Press, Oxford.

Wyrwoll, K.-H. and Smyth, G. K. 1985. On using the log hyperbolic distribution to describe the textural characteristics of eolian sediments, *Journal of Sedimentary Petrology*, **55**, 471–478.

Wyrwoll, K.-H. and Smyth, G. K. 1988. On using the log normal (?HYPERBOLIC) distribution to describe the textural characteristics of eolian sediments: reply, *Journal of Sedimentary Petrology*, **58**, 161–162.

Yaalon, D. H. and Dan, J. 1974. Accumulation and distribution of loess-derived deposits in the semi-arid desert fringe area of Israel, *Zeitschrift für Geomorphologie Supplementband*, **20**, 91–105.

Yaalon, D. H. and Ganor, E. 1980. Origin and nature of desert dust, in Péwé, T. L. (ed.) *Desert dust: origin, characteristics and effects on man*, Special Paper **186**, Geological Society of America.

Yaalon, D. H. and Laronne, J. 1971. Internal structures in eolianites and paleowinds, Mediterranean coast, Israel, *Journal of Sedimentary Petrology*, **41**, 1059–1064.

Yair, A. 1990. Runoff generation in a sandy area – the Nizzana Sands, western Negev, Israel, *Earth Surface Processes and Landforms*, **15**, 597–609.

Yair, A. 1994. The ambiguous impact of climatic change at a desert fringe: northern Negev, Israel, in Millington and Pye (1994) 199–228.

Yalin, M. S. 1977. *Mechanics of sediment transport*, Pergamon, Oxford.

Young, J. A. and Evans, R. A. 1986. Erosion and deposition of fine sediments from playas, *Journal of Arid Environments*, **10**, 103–115.

Yu, B., Hesse, P. P. and Neil, D. T. 1993. The relationship between antecedent regional rainfall conditions and the occurrence of dust events at Mildura, Australia, *Journal of Arid Environments*, **24**, 109–124.

Zachar, D. 1982. *Soil erosion*, Developments in Soil Science, **10**, Elsevier, Amsterdam.

Zhang Linyan, Dai Xuerong and Shi Zhentao, 1991. The sources of loess material and the formation of the Loess Plateau in China, *Catena Supplement*, **20**, 1–14.

Zhao Songqiao 1988. Human impacts on China's arid lands: desertification or de-desertification, in Whitehead *et al.* (1988) 1127–1138.

Zhirkov, K. F. 1964. Dust storms in the steppes of western Siberia and Khazakhstan, *Soviet Geography*, **5**, 33–41.

Zhu Zhenda, 1984. Aeolian landforms in the Taklimakan Desert, in El-Baz (1984a) 133–143.

Zobeck, T. M. 1989. Fast-vac: a vacuum system to rapidly sample loose granular material, *Transactions of the American Society of Agricultural Engineers*, **32**, 11 316–11 318.

Zobeck, T. M. 1991. Soil properties affecting wind erosion, *Journal of Soil and Water Conservation*, **46**, 112–118.

Zobeck, T. M. and Fryrear, D. W. 1986a. Chemical and physical characteristics of windblown sediment. I. Quantities and physical characteristics, *Transactions of the American Society of Agricultural Engineers*, **29**, 1032–1036.

Zobeck, T. M. and Fryrear, D. W. 1986b. Chemical and physical characteristics of windblown sediment. II. Chemical characteristics and nutrient discharge, *Transactions of the American Society of Agricultural Engineers*, **29**, 1032–1036.

Zobeck, T. M. and Popham, T. W. 1990. Dry aggregate size distribution of sandy soils as influenced by tillage and precipitation, *Soil Science Society of America, Journal*, **54**, 198–204.

Zon, R. 1935. Shelterbelts: futile dream or workable plan? *Science*, **81**, 392.

INDEX